MetroFarm

The Guide to Growing for Big Profit on a Small Parcel of Land

Michael Olson

TS Books

MetroFarm: The Guide to Growing for Big Profit on a Small Parcel of Land
by Michael Olson

Publisher: TS Books, PO Box 1244 Santa Cruz, CA 95061

Library of Congress Catalog No.: 93-61496

ISBN 0-963787-60-8

99   98   97   96   95   94   6   5   4   3   2   1

Printing code: Right most double-digit number is the year of book's printing; right most single-digit number, the number of book's printing

MetroFarm's Michael Olson

Author Michael Olson and daughter Kelsey work a vineyard near Rogue River, Oregon. Photo by Marlene Olson.

AGRICULTURALIST

Olson cultivated his first field at the age of six with what appeared to him, at the time, to be the world's largest tractor. He has since participated in the commercial production of beans, beets, blueberries, cattle, garlic, hay, oats, shallots, strawberries, turf grass and wheat in the states of Montana, Oregon, and California. Olson's most recent project has been to help establish a commercial vineyard of Cabernet wine grapes in Southern Oregon.

BUSINESS PERSON

Olson designed, blended and packaged a fertilizer for container-grown house and garden plants; certified and registered this product as a "specialty fertilizer" with the State of California; sold this product to the national lawn and garden market. As an advertising Account Executive and General Sales Manager, Olson has over 12 years experience helping local, regional and national businesses compete in the Monterey Bay Area metropolitan marketplace.

JOURNALIST

Olson authored, photographed and/or produced feature stories for San Francisco Chronicle's *This World*, San Francisco Examiner's *Image*, NBC Television's *NBC Magazine with David Brinkley*, KQED Public Television's *Express*, as well as various horticulture and general-interest books, calendars and magazines. Olson is currently Executive Producer of *Talk Agriculture*, a Central California radio talk show which presents agriculture-related issues and personalities to a prime time audience.

Credits

PUBLISHER
TS Books

PRODUCT DIRECTOR
Marlene Olson

PHOTO & TEXT EDITOR
Michael Olson

BOOK & COVER DESIGN
Marlene & Michael Olson

COVER PHOTOGRAPHY
Pete Burnight
Michael Olson

COVER COMPUTER COMPOSITION
Tammi Goldstein

COMPUTER CONSULTANT
Louis Goldstein

SOFTWARE CONSULTANT
Tammi Goldstein

TEXT & CHART COMPOSITION
  Kathleen Eastly
  Kristi Hill

ILLUSTRATIONS
  All line drawing illustrations, unless otherwise noted, are by
  Marlene Olson.

PHOTOGRAPHS
  All photographs, unless otherwise noted, are by
  Michael Olson.

PHOTO REPRODUCTION
  Kathleen Eastly
  Kristi Hill
  Paul Miller
  Scott Murray
  Sabra Schneider

COPY READERS
  Bill Nolan
  Dolores Cardinal
  Floyd Cardinal
  Marlene Olson
  Kathleen Eastly
  Tammi Goldstein
  Tawnya Napoli

Acknowledgements

Metrofarmers everywhere, who opened their doors and welcomed our inquiries.

Nancy Mack Tappan, whose complete acceptance and determined advocacy helped nourish *MetroFarm* during lean times.

Floyd Cardinal, who was detained by security agents of the former Soviet Union while photographing Moscow's free markets for these pages.

Students and staff of the University of California, Santa Cruz, who generously gave their time and technologies of agriculture, computer engineering and publishing.

Staff and associates of TS Books, whose long hours and incredible efforts made a what-could-be a what-is.

Kelsey Olson, who laughingly and lovingly tolerated a father who went to work at 4 am and came home late and tired.

Marlene Olson, who said, "Yes! We can do it!" and proved it up by arranging interviews, setting up photo shoots, asking strategic and tactical questions, proofing text, hiring assistance, computerizing production, designing the look, contracting printers and calling sellers.

For my parents Irene Nauman Olson and Robert C. "Pat" Olson

Give me a fish
and I will eat today.
Teach me to fish
and I will eat always.

-Unknown-

MetroFarm at a Glance

Table of Contents

Chapter 8: Establishing Production _____ 205

PART IV: CONVERSATIONS

Charts, Illustrations & Photos

CHAPTER EIGHT

CHAPTER NINE

CHAPTER TEN

GERD SCHNEIDER NURSERY

PLANT CAROUSEL

HEYNS COUNTRY GARDENS

POGONIP FARMS

THE FLOWER LADIES

PREFACE

MetroFarm was born a mere pamphlet, high on a hill overlooking the Rogue Valley of Southern Oregon.

The occasion, ironically enough, was the taking of an establishing shot for an NBC News documentary called "Armed for Armageddon" (*NBC Magazine with David Brinkley*; Vernon Hixson, Producer). The documentary was to tell the story of individuals who had fled big cities to escape what they believed was the imminent collapse of law and order.

Most of the documentary was already "in the can," including camouflaged computer consultants blasting away at pop-up targets in sunlit Oregon woods, a determined explosives expert with plans to blow up every bridge leading into Southern Oregon ("I'll keep those crazy city people out!") and sheep anxiously bleating at a thought-reform farm for affluent suburban delinquents.

We still needed to establish the story's location for viewers, which explained why we were filming from atop the hill.

"Beyond you can see the Rogue Valley of Southern Oregon," intoned Jack Perkins, the NBC announcer, as the camera panned to reveal how the delightfully fertile valley had been carved into small parcels of well-fenced, readily-defended survival retreats. As I looked down from the hilltop, however, the small uncultivated parcels of land suddenly struck me as being— well— ridiculous. I turned to Nancy Mack Tappan, our guide and publisher of Janus Press, and said, "Nancy, let's make a pamphlet on ways to farm the small parcels of land!"

Looking Back

There were many reasons why I would make a proposal of this nature. One reason was an early childhood spent on my grandparents' farm near Belfry, Montana, where I had cultivated my first field at the age of six with what, at the time, seemed to be the world's largest tractor.

Another reason was hot summers of hard work for extra school money on row crop farms near Billings, Montana. Hot summers of hard work, of course, resulted in a strong desire to see the world, which was soon satisfied by the US Navy ("No weeds to hoe on Navy ships, son!").

As a young sailor I was quickly captivated by the space-intensive farms I observed while bicycling through the Japanese countryside. The tiny Japanese farms, which averaged less than three acres in size, supported a population of 800 people per square mile; whereas farms back home in Montana, which averaged over 100 acres, supported a population of less than one person per square mile!

Later, as a college student swept up in America's Cultural Revolution, I periodically escaped back to farming for hard work and common sense. Summers found me working berry farms in Oregon's Willamette Valley, where I learned the economies of chartering a business jet to fly one case of blueberries to the East Coast for an advertising photo session; and a Montana ranch, where long hard days of harvesting wheat, tending cattle and mending machinery were rewarded with home-cooked meals, long horse back rides through lingering prairie twilight and deep sleep.

And later still, while earning a degree in English and Chinese Literature from the University of California, Santa Cruz, I discovered the small parcel of University-owned ridgetop being farmed by Alan Chadwick, the charismatic ex-Shakespearean actor whose students had dropped classes in Calculus, Molecular Biology and History of Consciousness to study Composting 1A. Though not a Chadwick student, I did enjoy watching as that barren ridgetop became one of the most attractive and fecund small farms in the world.

Yes, there were many reasons why the incredibly fertile and surprisingly fallow survivalist retreats would spark my hilltop proposal. But if all were reduced to one, it would be the Montana frontier maxim of "Grow or go!"

Nancy Mack Tappan, whose grandfather had founded the Mack Truck Company, said, "Great idea!" to the hilltop proposal. "Let's get to work on this right away!"

Finding Precedent

By the time we had finished the NBC documentary and returned to our various home bases, the small pamphlet idea had grown into a substantial book project complete with a major Midwest book distributor, an Empire State Building-based publicist and a press date.

My wife and partner, Marlene, and I, together with a small cadre of dedicated assistants and associates, began a systematic search for successful prototypes.

Our initial search turned up many individuals farming small parcels of land. Most, however, invested in their enterprise, made a few crucial— and often silly— mistakes and then moved on to different vocations. Others invested, muddled through somehow and earned very little.

Our prospecting then hit pay dirt with real farmers. Though none conspicuously displayed their success— no designer label clothing, manicured fingernails, imported luxury cars— each survived and prospered by farming a small parcel of land. Indeed, they earned up to eight times the average income in the United States on as little as an acre of land. We called them "metrofarmers" because each was tightly focused on producing for a metropolitan marketplace.

Given the bigger-is-better direction of modern times, you might see a successful metrofarmer as an exception rather than as a rule. Look more closely. Today *micro*processor-powered laptop computers displace room-sized mainframes. Steel *mini*mills, five percent the size of their giant competitors, prosper by selling steel for one-third the price. Owner-operated *micro*breweries sell high quality beer at twice the price paid for mass-produced labels. Neighborhood *mini*marts flourish around the corner from giant discount warehouses.

Today, competition for the consumer dollars is between the very big and the very small. The middle ground, like the one occupied by my grandparents' farm, is being squeezed into oblivion.

Nor is the small-is-better technology new. Look back, for one example, 300 years to the *troque* farmers of King Louis XIV's Paris. (The French word *troque* means "trade" or "barter.") The high-density production technologies developed by *troque*— or *truck*— farmers gave life to a thriving trade in fresh farm products between major capitals of Europe.

Furthermore, evidence suggests big money was made on small farms even during the height of the big-is-better movement. The 1982 Census of Agriculture shows that the most productive farms in the United States, in terms of dollar value of crops per acre, were located in New York City's Borough of the Bronx. The second most productive farms, which generated an average of $76,421 per acre, were located in San Francisco. And so on.

Going to Work

What is the secret? How do metrofarmers grow for big profit on a small parcel of land? Do they have a special crop? A perfect location? A good climate? Government subsidies? Grants from Grandma? If research could identify the secret of the metrofarmers' technology, the book project could be completed in a twinkling of an eye!

Our research, however, revealed only diversity. Successful metrofarmers are young and old, male and female, married and single, from backgrounds rich and poor. They own land outright and lease it year to year. They farm organically and conventionally. They grow small fruits on prairie bench lands, house plants in coastal valleys, flowers on steep wooded hillsides, vegetables in city greenbelts and ornamentals in neighborhoods of million dollar homes. They sell to neighbors at the farmers market and to exclusive department stores on the other side of the country.

Finally, one took me aside and said, "Look, Michael, we do not have any secret. We just have to do everything right around here and we have to do it all at the same time! It's... it's... it's concurrence that counts around here!" I then realized all metrofarmers were trying to say the same thing: The secret is concurrence!

I also realized this was a story of far greater significance than one of merely earning a few dollars. This was the story of surviving and prospering in the chaos and confusion of open markets. There, before our eyes, were individuals competing for consumer dollars with some of the largest and best financed corporations in the world. And winning! I was about to set a world record for missing a book deadline!

It has been 12 years since I made the hilltop proposal. The Empire State Building-based publicist has since moved on to make her millions on other, less problematic books. The giant Midwest book distributor was bought out by an international business publication, folded, merged and finally downsized into a neighborhood bookstore. Nancy Mack Tappan formed a partnership with Vernon Hixson which now produces some of the finest Cabernet sauvignon grapes in all of Oregon on her former survivalist retreat. Nancy and Vernon became the godparents of our daughter, Kelsey.

The years are an expression of belief: You can grow for big profit on a small parcel of land. You can survive and prosper. *MetroFarm* is your guide. It is presented in four parts and each part has a specific mission:

Part I, Strategy, will help you understand the principles of agriculture and business. It will help you develop a strategy to successfully compete against the Goliaths of the marketplace.

Part II, Production, will help you survey the market, evaluate and control land, select crops, organize a business and establish production.

Part III, Marketing, will help you prepare for and sell into the Goliath's-potato-against-your-potato no-holds-barred world of the free market.

Part IV, Conversations, will show the personal investment others have made to overcome obstacles and win the competition for consumer dollars.

MetroFarm provides information. With information you can grow for big profit on a small parcel of land. No book, however, can give you the initiative to start seeds, the ingenuity to integrate resources nor the determination to see things through. If you have the initiative, ingenuity and determination, turn the page!

PART I

STRATEGY

INTRODUCTION

Y ou can earn a substantial income by farming a small parcel of land. You can be old or young, rich or poor, married or single. You can own land outright or lease it year to year. You can succeed with small fruits on prairie bench lands, house plants in coastal valleys, flowers on steep wooded hillsides, vegetables in city greenbelts and ornamentals in neighborhoods of million dollar homes. There is only one catch: You must compete with other farmers for consumer dollars.

Competition among farmers is the foundation upon which most human activity is conducted. People somehow lost track of this elementary fact. Farming became the object of big city humor and farmers the butt of many jokes. The attitude is even reflected in our language, as illustrated by *Webster's Third*: "farmer 4a: an ignorant rustic: YOKEL, BUMPKIN b: clumsy stupid fellow: DOLT c(slang): a green hand inexperienced or incompetent at the trade at which he is working."

This way of looking at the competition is decidedly unrealistic. Today's farmer must be a skilled technician, proficient in the technologies of agriculture and business. Lacking skills, the farmer will fall victim to costly error. For example, an associate was recently asked to evaluate forty acres of solar hydroponic greenhouses. The limited partners in this enterprise, a group of wealthy professionals, invested

$20 million to build an infrastructure of greenhouses, solar ponds and plumbing fixtures. The general partners, a lawyer and a salesman, managed the operation until $500,000 worth of tomatoes froze during a cold spell.

Today, "yokels" and "bumpkins" are culled from the field with the weeds. Jack Neuman, an Illinois farmer, put it most succinctly in a conversation with *Time* magazine: "It used to be that if you had a child who wasn't too bright, you'd say, 'Son, you're going to be a farmer.' Nowadays, if that dumb kid comes along volunteering to farm, you've got to say, 'Oh, a man in hell would love to drink a glass of ice water, but it just can't be done.'"

Though farming is now dominated by big business, skilled technicians and vast amounts of capital, you can successfully compete within this industry. Part I of *MetroFarm* will help you get started. Chapters 1 and 2 will help you develop a good understanding of the fundamental principles of agriculture and agribusiness. Chapter 3 will help you develop a strategy— a metrofarm strategy— which will enable you to win the competition for consumer dollars by farming a small parcel of land.

CHAPTER **1**

Understanding Agriculture

[Agriculture] is a spectable science of the very first order. It counts among its handmaids the most respectable of sciences, such as Chemistry, Natural Philosophy, Mechanics, Mathematics generally, Natural History, Botany. In every College and University, a professorship of agriculture, and the class of students, might be honored as first.

Thomas Jefferson
in a letter to David Williams,
November 14, 1803

There is a recipe for rabbit stew which begins with the admonition: "First, catch the rabbit." Good advice: no rabbit, no rabbit stew. Here begins the recipe for metrofarming: First, develop a good understanding of the science and art of growing crops. No crops, no metrofarm.

Though agriculture is a 10,000 year-old technology, many of its innovations have been made possible by scientific breakthroughs of the last century. The innovations can be seen on any real farm. Visit a California tomato farm and see six-ton machines harvesting genetic hybrid tomatoes specially bred to ripen on the same day. Visit an Iowa corn farm and find the farmer hard at work cultivating numbers at a computer terminal. Visit an organic vegetable farm and see tiny laboratory-bred wasps attacking insect pests.

Modern high-technology agriculture can seem imposing, if not downright frightening, to an aspiring farmer. But before you throw your hands up in confusion, consider a fact: Agriculture has, as its irreducible essence, the growth and development of green plants.

Green plants provide energy for all organic forms of life on earth. Even highly specialized farm operations, like automated cattle feedlots, merely convert the growth of green plants into animal crops. Plants therefore provide the key for unlocking the complexities of modern agriculture. By developing a good understanding of how plants grow and develop in nature, you can easily understand how they are grown and developed by using the various technologies of modern agriculture.

HOW PLANTS GROW AND DEVELOP

Plants are classified as autotrophic organisms because they are self (auto) nourishing (trophic). They take simple elements and inorganic compounds from the environment and convert them into the more complex organic compounds of life.

Heterotrophs such as people and animals, by contrast, require complex organic compounds produced by autotrophs to complete their metabolic synthesis. This distinction is made because people often claim to have grown plants. "I grew these tomatoes from seed," they will say. Actually, the converse is closer to the truth: Plants provide for the growth of people and animals.

The autotrophic metabolism of green plants is the foundation upon which the technologies of agriculture are developed. Greatly simplified, the process works as follows:

Our physical world is composed of approximately 105 inorganic elements. Though most of the elements have been identified in various examinations of plant tissue, we know, at this writing, of only 16 which are essential for plant growth. The 16 elements are made available to plants through the metabolic processes of photosynthesis and respiration.

Photosynthesis is the process by which plants convert radiant energy from the sun and nutrients from the earth into stored chemical energy. (Photo means "light;" synthesis means "put together.") Energy conversion process occurs in microscopic bodies called chloroplasts. These bodies contain chlorophyll, the substance which gives plants their characteristic green color. Chlorophyll absorbs certain wavelengths of radiant energy. Energy is then used to split water and carbon dioxide molecules and then to recombine them into carbohydrate compounds, oxygen and water.

Following are two formulae which may help you to better understand photosynthesis:

Carbohydrate compounds— or sugars— produced in photosynthesis are

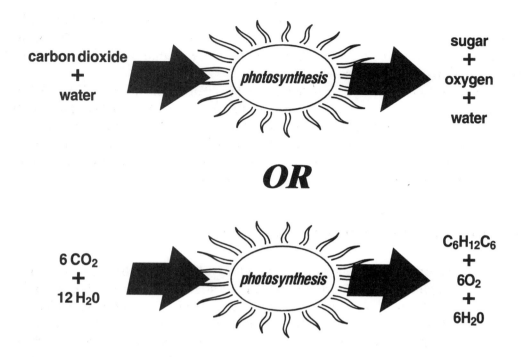

FIG 1.1 Photosynthesis is the process by which plants convert radiant energy from the sun and nutrients from the earth into stored energy.

forms of water soluble chemical energy which plants use to synthesize organic plant tissue. The carbohydrate compounds are the fuel for the fire, so to speak.

Photosynthesis yields three of the 16 essential nutrients: carbon, hydrogen and oxygen. The remaining 13 essential nutrients come from the soil and consequently are often called "soil nutrients." Soil nutrients are classified according to the amounts used in plant growth and development. There are three primary soil nutrients: nitrogen, phosphorus and potassium; three secondary soil nutrients: calcium, magnesium and sulfur; and seven trace soil nutrients: zinc, iron, manganese, copper, boron, molybdenum and chlorine.

Soil nutrients are made available to plants through the physical and chemical processes of respiration. Respiration is governed by plants' environment. If you take a moment to examine a living plant, you will see it has no heart, lungs nor any organ to provide for an independent circulatory system. So, unlike animals and people, plants must rely on elements of their environment to provide necessary impetus for circulating nutrients. In other words, growth and development are fixed by elements of the plants' environment.

Both photosynthesis and respiration occur at the cellular level. As the processes are difficult to see, they are difficult to understand. A plant's circulatory system, however, can be readily perceived by following the flow of the transpiration stream. The transpiration stream is the cyclical flow of water from the atmosphere into soil, from soil into plant and from plant back into the atmosphere.

In figure 1.2, the transpiration stream is represented by the clockwise circulation of arrows. The circulation of water through a plant is like blood circulating within our bodies. By following the flow of the transpiration stream, you can "see" how plants synthesize 16 inorganic elements into organic plant tissue.

First, water enters the soil from the atmosphere. Soil, if healthy, is populated by various organisms called decomposers. Decomposers vary in size from microscopic bacteria to larger— and hence more familiar—organisms like earthworms. Through metabolic activity of decomposers, organic compounds are converted back into inorganic soil nutrients.

Soil nutrients are held in place by humus. Humus is the composted organic matter in soil which has two colloidal properties: It is a negatively-charged mass which attracts and binds positively-charged nutrients, and it

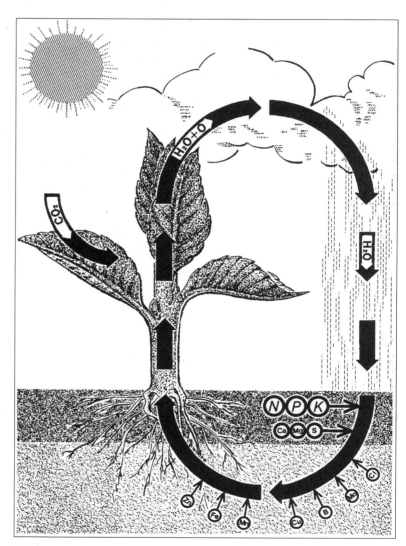

FIG 1.2 The transpiration stream is the cyclical flow of water from atmosphere into soil, from soil into plant and from plant back into atmosphere. Illustration by Marlene Olson.

is a sticky substance which traps and holds partially decomposed nutrient-rich matter within its mass. Soil nutrients are then made available through further decomposition of the humus mass. Clay is the other soil colloid and it has about one-third the colloidal power of humus. Combined colloidal power of humus and clay defines the soil's capacity to hold nutrients and moisture.

Second, plants purchase nutrients from soil. This transaction is conducted in much the same way we purchase food from the grocery store. Roots travel through soil to find proper nutrients. Each root is surrounded by a mass of hydrogen ions earned through respiration. When the proper nutrient is found, it is purchased through an exchange of ions: One plant ion is exchanged for one nutrient ion.

Nutrients and water are then absorbed into roots and drawn up into the plant via the xylem. The solution is then absorbed into the plant's protoplasm, where, combined with carbohydrate compounds and oxygen produced in photosynthesis, it is made into compounds like amino acids and proteins, which are the building blocks of living cells.

Finally, excess water, oxygen and extraneous matter are transpired back into the atmosphere through stomata, which are tiny openings which open and close in response to the plant's environment. Environmental factors which regulate the flow of the transpiration stream include air and soil temperature, relative humidity and wind velocity. When conditions are favorable, such as warm air and soil temperatures with a dry wind, the transpiration stream will flow rapidly. When conditions are unfavorable, such as cold air and soil temperatures with wet, stagnant air, the transpiration stream will flow slowly.

The transformation of a living cell into a mature plant involves both growth and development. While growth is simply an increase in the number of cells, development consists of differentiation, specialization and organization of cells. Each cell within an individual plant begins life with a set of identical genetic information stored in its deoxyribonucleic acid (DNA). Cells develop and become unique through regulation of stored genetic information.

Genetic information is switched on and off— like electricity in a microprocessor— in response to instructions contained within the cell itself, instructions from other cells or instructions from the environment. Hormones such as auxins, cytokinins and gibberellins are chemicals which control movement of genetic information within a plant.

Through photosynthesis, respiration and the switching on and off of genetic information, a plant grows and develops from seed to seedling, from seedling to mature plant and from mature plant to seed. By developing a good understanding of natural processes, you will have earned the ability to understand how crops are grown and developed with technologies of modern agriculture.

HOW PLANTS ARE GROWN AND DEVELOPED

In nature, the autotrophic metabolism of plants generally provides enough growth and development to enable reproduction. Natural growth, however, is insufficient to support the needs of a population of hungry heterotrophs. People, therefore, intervene in the natural process by improving factors which regulate autotrophic metabolism. In doing so, people earn the right to say, "I grew this plant from seed."

Three agricultural technologies have evolved during the past 10,000 years to improve crop growth and productivity. They consist of controlling environment, directing growth and development, and programming genetic intelligence.

Controlling The Environment

Plants grow and develop in direct response to their environment. Environment includes radiant energy from the sun, which provides light and heat; soil, which provides water, nutrients and support; air, which provides carbon dioxide and oxygen; and competition, which affects the availability or quality of other elements. It stands to reason growth rate and productivity can be improved by controlling the environment. Elements of environment most amenable to control are soil, water, temperature, light and competition.

SOIL: A farmer judges soil by its capacity to promote growth. There are three characteristics of good soil: First, it must have a sufficient and available supply of essential nutrients; second, a balanced *pH;* and third, a friable tilth (or loose texture). When soil is deficient in essential nutrients, fertilizers are added or a cover crop of legumes planted. When nutrients are in sufficient supply but locked-up because of an unbalanced

FIG 1.3 Steve Raycraft (left) and the author drill prototype holes for a vineyard of Cabernet Sauvignon in Southern Oregon. Photo by Marlene Olson.

chemistry, *pH* is corrected by adding amendments like dolomitic lime. If soil has poor physical structure, tilth is improved by applying organic matter, tilling deep or improving drainage.

WATER: Growing plants are approximately 90 percent water by weight. Water affects growth and development three ways: First, water is a raw material for photosynthesis, providing essential nutrients of hydrogen and oxygen. Second, water provides turgor— or pressure— which maintains cell rigidity. Third, water picks up nutrients from the soil, delivers them to growing cells and then removes oxygen and by-products of metabolic synthesis by carrying them back into the atmosphere. There can be too little water in the environment, in which case irrigation systems are established; there can also be too much water, in which case drainage is improved.

TEMPERATURE: Temperature affects nearly every physical and chemical process of plant growth and development. Some examples: Air temperature regulates the flow of the transpiration stream; soil temperature regulates the availability of essential nutrients; and air and soil temperatures, together with other factors, regulate the plant's cycle of growth.

There are many techniques for controlling temperature. The most important one is to match a crop with a favorable location and site. To grow a good crop of Red Delicious apples, for example, one could match environmental needs of apple trees with a good location— Washington State— and a good site— the Columbia River Basin. There are also many on-site techniques for controlling temperature. They include mulches, which can be used to raise or lower soil temperatures; greenhouses, which can be used to warm air and soil; and shade-houses, which can be used to cool air and soil.

LIGHT: There are two ways in which light affects plant growth and development. Intensity, together with temperature, regulates photosynthesis and respiration. By controlling light intensity, growth rate can be improved and development directed. For example, fast growing plants like corn are situated in open fields to maximize exposure to light, while slower growing plants like the ones in the fern family are situated in shade-houses to minimize exposure.

Light duration regulates a plants' life cycle. By controlling the period of light, plant development can be directed. For example, if one wishes to induce a greenhouse full of flowers to produce blossoms by a certain date, one can reduce the period of light the flowers receive, thereby tricking the flowers into initiating their flowering cycle.

COMPETITION: Growing crops compete with weeds, insects, animals, diseases and other competitors for available resources. Some competitors reduce productivity by taking the crop's nutrients, water and sunlight. Others, such as the botrytis fungus, attack the crop directly. The most important technique for controlling competitors is to match a crop with a favorable location and site. If one wishes to grow wine grapes, for example, one should select a site on a sunny slope and avoid the ones in damp bottom lands which promote botrytis.

There are three additional strategies for controlling competitors: One strategy is to establish a healthful, well-balanced environment and then rely on this balance to prevent any single competitor from growing out of control. This strategy is a fundamental tenet of organic farming. Another strategy is to dominate an environment with pesticides and cultural practices. This strategy is used by greenhouse horticulture and large-scale farm operations. The third strategy, called integrated pest management (IPM), combines a healthful, well-balanced environment with periodic control measures, like insecticides, if required.

Directing Growth And Development

In nature, growth reflects a plant's struggle to survive and reproduce itself. In agriculture, natural growth is often considered unproductive for one reason or another, so growth is improved and redirected. Redirected growth may result in faster growth rate, more desirable shape and appearance, differentiation in the cycle of growth (e.g.: inducing a plant to flower) and increased yield. There are physical, biological and chemical controls for directing crop growth and development.

FIG 1.4 By grafting scions from standard trees to rootstocks from dwarfed trees, standard-sized fruit can be produced on dwarf-sized trees.

PHYSICAL CONTROLS: Training and pruning are techniques which have been used for thousands of years to redirect the shape, size or orientation of plants. Training consists of bending, twisting or tying a plant to a supporting structure like an espalier. Pruning is the judicious removal of unwanted plant parts to improve growth in more desirable parts. Physical controls are associated with the culture of longer-living crops like trees and vines. Fruit orchards, vineyards and ornamental nurseries

FIG 1.5 Training consists of bending, twisting or tying a plant to a supporting structure to encourage specific growth and development.

are examples of farm operations which use physical controls to redirect growth and development.

BIOLOGICAL CONTROLS: A plant can also be redirected by manipulating its biological processes. By grafting the scion (the above ground portion) of a standard apple tree with a rootstock of a dwarfed variety, normal-sized apples can be produced from a dwarfed tree. Hardening-off is another biological technique. Young cuttings and seedlings propagated in a greenhouse are often acclimated to the outdoors before being subjected to the stress of an open field. Though biological controls require a considerable amount of sensitivity and knowledge, they provide nurseries, orchards and even large-scale field operations like potato farms with one more means to direct crop growth and development.

CHEMICAL CONTROLS: Chemical controls represent a comparatively recent advance in agriculture. In the past 50 years, researchers have identified many organic chemicals which have a broad spectrum of effects when applied to growing plants. Examples include indoleacetic acid for stimulating root growth, gibberellins which induce seed stalk formation and pollination and auxin-derivatives which set fruit in some species and thin the fruit set in others. Though most chemicals are used as pesticides to control environmental pests, many are now used in redirecting growth and development of the crop itself.

Programming Genetic Intelligence

Genotype is the double helix of genes stored in the deoxyribonucleic acid (DNA) molecule of each living plant cell. Genotype is like a computer software program— it provides the intelligence a plant needs to grow and develop in response to stimuli from its environment. The interaction between genotype and environment results in growth characteristics called phenotype. Examples of phenotypic response include a plant's shape, color, bouquet, nutrient content and rate and duration of growth.

Phenotypic responses vary from species to species— the difference between apples and oranges— and from plant to plant— such as the difference between two Red Delicious apple trees growing on the same tree. Different phenotypic responses within a single plant variety represent the varietal's attempt to find a better way to survive and reproduce itself. Each plant, in other words, is an experiment in survival. The experiment yields information which is programmed into the plant's genotype and passed on to the next generation.

In nature, as Darwin asserted, phenotypic responses are selected through survival of the fittest. Responses which improve a plant's ability to survive are passed on, while those which inhibit survival are not. Under cultivation, phenotypic responses are selected through "survival of the hoe." Responses which increase productivity are selected, while those which do not are weeded out. The hoe enables a farmer to program a genotype in much the same way a keyboard enables a computer programmer to write a software program.

Favorable phenotypes are propagated with either sexual or asexual techniques. With sexual techniques, favorable responses are selected from male and female plants and then propagated via the male-female reproductive cycle. With asexual techniques, favorable responses are selected from a single plant and then propagated by clonal reproduction of cuttings or tissue cultures.

Programming genotype can be as simple as selecting the best from this year's

FIG 1.6 Phenotypic responses include shape, color, bouquet, nutrient content and rate and duration of growth, as demonstrated by the color differences in Red and Golden Delicious apples.

crop to generate seeds for next year's crop or as complex as inducing mutations with X-ray radiation, selecting favorable mutations and then propagating the selection into a population of clones.

Recent advances in recombinant DNA technology have increased our ability to program genotypes. The technology enables one to splice genes from two different sources onto a common double helix, thereby creating a new genotype with favorable phenotypic responses.

By selecting favorable phenotypes from wild grasses, our primitive forebears created wheat. Modern technology speeds-up the programming process. Consider, for example, the dwarf peach tree. The dwarf peach cultivar was imported from China around 1915 because its phenotypic response of dwarfed growth presented great possibilities for farmers. The dwarf tree bore fruit in half the normal time, produced twice as much fruit per acre and was easier— and therefore cheaper— to cultivate. However, the dwarf produced a bitter-tasting, off-color fruit. The dwarf's genotype needed to be reprogrammed before farmers could realize its potential.

When 70 years of cultural selection failed to isolate favorable taste and color responses in the dwarf peach, scientists at the University of California, Davis, applied the turbocharger of recombinant DNA technology. They spliced genes which control dwarfing together with genes from another varietal which control taste and color. In a split-second, relatively speaking, they reprogrammed the dwarf genotype so it bears sooner, produces more, is easier to cultivate and yields colorful, sweet-tasting peaches.

There are many plant cultivators and few plant breeders. Nevertheless, genetic software programs are available to every metrofarmer through the medium of seed catalogs. By carefully selecting from among the phenotypic responses presented in the colorful pages, each metrofarmer becomes a programmer of plant genetics.

REVIEW

Agriculture is the science and art of growing crops. Technological innovations of the past 100 years have made possible tremendous increases in agricultural production. The technological innovations, which have as their foundation the natural growth and development of green plants, consist of ways to control environment, direct growth and development, and program genetic intelligence. You need a good understanding of these technologies to succeed as a metrofarmer. No crops, no metrofarm.

CHAPTER ONE EXERCISES

i   Invest in a large loose-leaf binder, paper and section dividers. Title this workbook *(Your Name)'s Metrofarm Project*. Label four of the section dividers with Strategy, Production, Marketing and Information Sources. Place the section dividers and paper in your work book.

ii   Begin collecting information relating to possible directions for your metrofarm project and place the information in your workbook. Cut articles from newspapers and magazines. Take notes on conversations and broadcast news stories. Call a noted authority and ask questions. Fill your empty pages with hard information.

iii   Research the art of bonsai. Write brief answers to the following questions and place the answers in your metrofarm workbook: What is a bonsai? What is the significance of bonsai to people living in congested metropolitan areas? How is a bonsai made? How is a bonsai tended?

iv   Adopt the bonsai of your choice. If one is not available for a reasonable price from a local retailer, use the skills researched in the previous exercise to make your own.

v   Move the bonsai to different locations around your home. Evaluate the flow of the transpiration stream through the bonsai to determine which location is best. Care for this bonsai as if you planned to give it to a great grandchild in fifty years.

Understanding Agribusiness

God almighty first planted a garden, and indeed it is the purest of human pleasures. It is the greatest refreshment to the spirits of man.

Sir Francis Bacon

The farmer is as much a business man as the man who sits upon the board of trade and bets upon the price of grain.

William Jennings Bryan

Though agriculture is practiced by gardeners and farmers alike, the similarity ends there. Farmers must develop a proficiency in an additional technology— one of converting green plants into green money.

Gardening is an avocation. As Sir Francis Bacon said, it provides us with "the greatest refreshment." It also supplements our food supply, adds color to our lives and is an excellent way to spend Sunday afternoon. Because gardening is so refreshing, we tend to project the feeling onto the business of farming. This is not a realistic projection.

Farming is a vocation. People farm so they can purchase food for their table, a roof over their head and clothes for their back. As a vocation, farming has an intensity of character unlike gardening. If the need is there, farmers work Sunday morning before breakfast; and if the need is still there, they work Sunday evening long after the dinner dishes have been removed from the kitchen table.

You may have the technological proficiency to be a rose gardener. Your roses may be so perfectly shaped, brilliantly hued and delicately scented that people will travel for miles to see them on display with blue ribbons at the county fair. But this ability alone is not sufficient to enable you to prosper as a rose farmer. You must also compete with other rose farmers for consumer dollars.

The farm is an enterprise built upon a foundation of simple business principles. Develop a good understanding of the principles. Your metrofarm business can then grow naturally, like green plants under the sun.

HOW A BUSINESS GROWS AND DEVELOPS

Agribusiness is a big word. In recent years the word has become synonymous with large corporations. For example, consider the word's use in this paragraph culled from *The Wall Street Journal*:

> Throughout the agribusiness community, the watchword these days is bigness. During much of the 1970's, booming farm exports helped make agribusiness a growth industry and encouraged new competitors. Now, the farm-belt recession is pushing many smaller contenders out of business, and the biggest merchants and processors are taking advantage of bargain-basement prices to grow through acquisition. Only four months after Cargill bought Seaboard's mills, two other agribusiness giants, ConAgra Inc. and Peavey Co., proposed a merger.

In the same newspaper, financial institutions vie to serve agribusiness giants by advertising business services like collateral management, joint-venture captive financing, tax-leasing, trade receivable passthrough certificates, credit indemnification, off-balance sheet financing, mortgage passthroughs and non-tangible asset financing.

As a consequence of this kind of media reporting, agribusiness may seem a complex and imposing prospect to an aspiring farmer with a small parcel of land. Let's take advantage of some literary license and a pilot's license to reduce this complex and confusing image to something manageable.

From high above California's Central Valley, the agribusiness capital of the world, you can see farms of every conceivable shape and size. Also visible are features of the agribusiness infrastructure— the barns, silos, greenhouses, roads, canals and power transmission lines. Dip a wing and see vehicles creeping along section roads, servicing farms with equipment and supplies and transporting farm products to processors, distributors and consumers. Look into the distance and see a large factory receiving rice from farmers, processing it into breakfast cereal, packaging it in fancy boxes and shipping it to distant cities.

This flight reveals a significant fact: Agribusiness is an industry built upon the productivity of individual farms. Other enterprises within this industry, including the huge equipment and fertilizer manufacturers, bankers, truckers, processors and brokers, merely support— and live off surplus from— working farms.

Farmers invest time, money and know-how. These investments are readily defined: Time is the investment of being there when needed, as in, "Farmers spend time in the field." Money is the stock of accumulated assets used to pay for expenses, as in, "Farmers spend money on seeds, equipment and land payments." Know-how is the investment of technological skill, as in, "Farmers know how to convert energy from the sun and nutrients from the earth into marketable products."

Return is the difference between what it costs to grow a crop and what is received when the crop is sold. Generating a positive return is the photosynthesis of farming. Without a positive return, farms— and agribusiness— would wither and die like green plants without sun.

Business, like agriculture, is an ancient technology. Our primitive forebears must have wondered about the economic benefits of sowing wild grass seed on the slopes of the Fertile Crescent. Their question: "Will sowing these wild grass seeds yield a return, or should we continue to invest our time hunting and gathering?" Had this enterprise failed, our forebears might have starved. But it was a success and with their newly acquired capital of surplus food, our forebears specialized and developed society.

The technology of business has improved dramatically over the years. Entrepreneurs can now use financial tools like "trade receivable passthrough certificates" and "off-balance sheet financing" to get the job done. Yet the fundamental question remains essentially unchanged: "Will sowing this crop yield a profitable return, or should I invest in something else?"

Though this question may seem difficult, remember that business would not have survived as a disciplined technology by being difficult. On the contrary, business survives because it offers the simplest and most readily perceived method of achieving a desired objective.

Business is competition. Consumers compete for the supply of a product and producers compete to supply the product. John Stuart Mill, in *Political Economy,* expressed the relationship between supply and demand as follows: "The Law is, that the demand for a commodity varies with its value and that value adjusts itself so that the demand shall be equal to the supply."

When supply and demand achieve an equilibrium, competition is governed by survival-of-the-most-efficient. Efficient farmers supply more products, of higher quality, at less expense than do inefficient ones, who must either borrow additional inputs from outside sources or go out of business.

The true measure of efficiency is the margin between the cost per unit of production— be the units bushels or barrels, tons or ounces— and the unit's market value. Following is an equation which expresses this measure of efficiency:

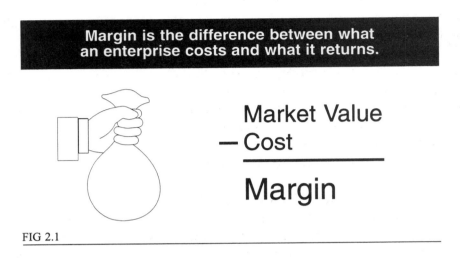

Margin is the difference between what an enterprise costs and what it returns.

$$\frac{\text{Market Value} - \text{Cost}}{\text{Margin}}$$

FIG 2.1

Consider an example: Farmer Alpha produces a flat of strawberries for $1.00 and sells the flat to a neighborhood fruit stand for $2.00. Alpha's margin is $1.00, for a 100 percent return on his investment. It costs Farmer Bravo $3.00 to produce a flat of strawberries for the same fruitstand. Bravo's margin is a minus $1.00, for a 100 percent loss on his investment. The margins tell us Alpha is the more efficient producer.

The margin between costs and returns is a measure of efficiency which translates well into any language. Sit down behind the loan desk of the neighborhood bank for a moment and listen to Alpha and Bravo describe their respective aspirations and needs. Which strawberry farmer do you understand?

Agriculture has a measure. When sound cultural practices are applied to a living crop, the crop will grow, develop and yield more than if it were left to struggle on its own in the competitive environment of nature. The business of agriculture also has a measure. When sound business practices are applied to an operating farm, the farm will grow, develop and return more profit than if it were left to struggle on its own in the competitive environment of the marketplace.

HOW TO PROJECT BUSINESS GROWTH AND DEVELOPMENT

Jack Thompson owns and operates a $1.7 million vegetable and cotton farm near Bakersfield, California. When asked for the secret to his success by *Farm*

Futures magazine, Jack replied, "Our secret, if that's what you want to call it, is that we've had the right crystal ball, and knew enough not to plant losers."

Mr. Thompson's crystal ball is not a magical instrument; it is a practiced skill which enables him to make educated estimates as to the results of his business decisions. Business persons need to see into the future. Without this ability, investments placed at risk today may not return a profit tomorrow.

Mr. Thompson used financial projections to make the all-important strategic decision of which crops to plant. But projections are also used to make ordinary tactical decisions which fill a work day. By way of illustration, consider the often-used image of a farmer contemplating a handful of topsoil. Advertising agencies worldwide use this image to imply a deep spiritual revelation. If we could listen to the farmer's innermost thoughts, however, we would likely hear, "What investments must I make to maintain the fertility of this soil?" And, "Will my investments return a profit or a loss?"

The prerequisite for making accurate projections is the ability to gather and process information. Current trends must be accounted for by keeping accurate records for costs of production and values of crops. Future trends must be accounted for by keeping track of forces which can alter prices and values, such as the ever-changing tastes of consumers, weather patterns, energy prices, value of currency, government policy and foreign relations.

Project the Profit

The objective of a projection is to accurately assess the feasibility of a business decision. By way of illustration, consider the notion of developing the one-acre vacant lot down your street:

The owner of this property, a doctor in a distant city, is willing to lease the lot for the cost of his property tax. Form a partnership to develop the lot into a part-time metrofarm. To enhance your speculation, plan to grow a sweet-tasting crop with nice round figures— like watermelon. Call your business the "Melon Company."

Before investing a lot of time and money in the Melon Company, evaluate whether the opportunity is real or, like fool's gold, shiny but worthless. Answer the question here, on paper, so you can save precious resources and neighborly relations. Call this look into the future your "profit projection."

Base your projection on industry-wide figures obtained by calling a cooperative county extension agent in the large melon-producing district of California's Kern County. Use my phone and ask for three figures: per acre costs for production ($750), operations ($650) and marketing ($350); units produced per acre (50,000 pounds of watermelons); and market value per unit (6 cents per pound on sales to melon brokers). Now, use the figures to look into the watermelon industry:

ADD FOR TOTAL COSTS: First, determine the total per-acre cost of doing business by adding production, operating and marketing costs. Production costs include inputs required to grow melons, such as seeds, soil, water and fertilizer. Operating costs include inputs required to operate the business, such as rent, labor, equipment and insurance. Marketing costs include the storage, transportation and

Add to find the total cost of an enterprise.	
THE FORMULA	THE MELON COMPANY
Cost of Production	$ 750
Cost of Operating	$ 650
+ Cost of Marketing	+ $ 350
Total Cost	$1750

FIG 2.2

advertising inputs required to sell melons. Simple addition reveals the melon industry's per-acre cost averages $1750.

DIVIDE FOR COST PER UNIT: Next, determine the cost per unit of production by dividing the per-acre cost by the production per acre. Simple division says the industry average cost is 3.5 cents per pound.

Divide to find the cost per unit.	
THE FORMULA	THE MELON COMPANY
$\dfrac{\text{Per-Acre Cost}}{\text{Production Per Acre}} = \text{Cost Per Unit}$	$\dfrac{\$1750}{50,000 \text{ Pounds}} = 3.5¢ \text{ Per Pound}$

FIG 2.3

SUBTRACT FOR MARGIN: Finally, determine the watermelon industry's average margin by subtracting the cost per unit of production from the unit's market value. Simple subtraction reveals the industry's average margin is a profit of 2.5 cents for every 3.5 cents invested.

Subtract to find the margin per unit.	
THE FORMULA	THE MELON COMPANY
Market Value of Product — Cost of Product ――――――― Profit or Loss	6.0¢ Per Pound — 3.5¢ Per Pound ――――――― 2.5¢ Per Pound

FIG 2.4

Your look into the melon industry has provided valuable information regarding potential for the Melon Company. You learned how much it costs the competition to produce and how much they receive for their melons. You even managed a peek at their profit margin. These figures are very important because they define competitive pressures under which you must operate.

If your projection revealed a loss or a thin margin, you could simply forget the idea of a Melon Company before investing valuable time and money. However, the look revealed the possibility of generating a 71 percent net return on investments, which is reason enough for convening another meeting of the Melon Company.

Project the Prosperity

A business person survives by operating efficiently. But it is not enough to merely survive; one must also prosper.

Prosperity, Webster tells us, is "the condition of being successful"; and success, as we all know, is relative to personal ambition. Success for one might be having enough to live a peaceful, independent lifestyle; for another it may mean enough to buy a new split-level home with two luxury cars in the garage. Though success is relative to personal ambition, successful business persons share one common attribute: Each has a goal.

A goal provides a business person with two very important benefits: First, it compels one to convert a vague personal ambition into a concrete objective. For example, by converting "I want to earn a lot of money" into "I want to earn $500,000 a year," a wistful desire becomes a concrete objective. Second, it provides incentive. The farmer is the single most important element in the business of agriculture. It is the farmer who places resources of time, money and know-how at risk. Without incentive, the farmer will withhold the resources and sleep late in the morning. With a good incentive, the farmer will wake-up early, invest and compete.

To understand these benefits, convene another meeting of the Melon Company. In a profit projection, you calculated the potential to earn 2.5 cents for every 3.5 cents of invested inputs. Though the opportunity sounds survivable, you do not want to invest $1750 of precious resources to merely survive. You want to prosper. Begin with a prosperity goal.

Since your full-time job does not provide enough surplus income for the trip to the South Seas you have been discussing with the neighbor over the backyard fence, use this daydream as your goal. Resolved: You and the neighbor will agree to exchange a few months of part-time work at the Melon Company for four tickets to Tahiti. (Spouses may go if they agree to help tend the melons!) The tickets for two weeks in Tahiti will cost $5,000, room and board included. Goal quantified, incentive established.

Can the Melon Company enable you to achieve this goal? Use the industry-wide figures to make a prosperity projection.

MULTIPLY FOR TOTAL MARGIN: First, determine the potential net profit for an acre's worth of watermelons by multiplying the average per unit profit by the average number of units produced on an acre. Simple multiplication reveals a potential per-acre net profit of $1,250.

Multiply to find the total margin.	
THE FORMULA	THE MELON COMPANY
Number of Units Produced X Profit Per Unit _____ Total Profit	50,000 Pounds X 2.5¢ Per Pound _____ $1,250 Total Profit

FIG 2.5

DIVIDE FOR REQUIRED ADJUSTMENT: Finally, determine what you must do to achieve your prosperity objective by dividing the $5,000 goal by the average per acre profit of $1,250. Simple division indicates you must increase productivity by 400 percent to achieve the Melon Company's goal.

Divide for the Required Adustment	
THE FORMULA	THE MELON COMPANY
$\dfrac{\text{Profit Goal}}{\text{Profit Projected}} = \text{Required Adjustment}$	$\dfrac{\$5000}{\$1250} = +400\%$

FIG 2.6

A 400 percent increase in productivity is a formidable objective. Our projections, after all, are based on averages achieved by some of the largest melon growers in the world. There are several courses of action to take: First, you can forget the idea of a Melon Company and invest elsewhere. Second, you can try to duplicate the big growers' efforts and, if successful, satisfy yourselves with a week at the lake. Third, you can develop a strategy to achieve the 400 percent increase in productivity needed to achieve the objective.

REVIEW

The farm is a business which grows and develops in the competitive environment of the marketplace. Profit supplies the energy for growth and development. The farmer, as business person, must always stand up to the question: "Will the farm return a profit or should I invest my resources elsewhere?" The farmer uses financial projections to answer the question. Financial projections are the implements of business. Use the implements to project the consequence of your actions.

CHAPTER TWO EXERCISES

i   In very general terms, explain why you want to metrofarm. Say, for example, "I want a second job so I can earn additional income." Or, "I want to quit my full-time job at the office and become self-sufficient." Write this objective at the top of a new page and place it in the "Strategy" section of your metrofarm workbook.

ii   In very specific terms, assign your goal a dollar value. If, for example, your goal is to become self-sufficient, say, "My goal of self-sufficiency will require an income of $_____ per year." Write the quantified objective down immediately below the general one completed in exercise *i*.

iii   On a clean piece of paper, write a one-sentence definition for the true measure of business efficiency. Place the definition in the strategy section of your metrofarm workbook.

iv   Find three business proposals for the same kind of product. Any product will suffice. (You can find racks of proposals in the office of a financial services broker.) Read each proposal carefully and research words you do not understand.

v   Make a profit projection for the proposals examined in the previous exercise and determine which proposal would be the most efficient one in which to invest your capital. Include the calculations in the "Strategy" section of your metrofarm workbook.

Developing a Metrofarm Strategy

Your position that a small farm well-worked and well-manned will produce more than a larger one ill-tended, is undoubtedly true in a certain degree. There are extremes in this as in all other cases. The true medium may really be considered and stated as a mathematical problem: Given the quantum of labor within our command, and land ad libitum offering its spontaneous contributions: required the proportion in which these two elements should be employed to produce the maximum.

Thomas Jefferson to Charles Peale
April 17, 1813

A difference of opinion exists as to which is the best way to compete for consumer dollars: Some maintain big is best. Advocates include the former U.S. Secretary of Agriculture Earl Butz, who campaigned throughout the land with the cry, "Get Big or Get Out!" *MetroFarm*, on the other hand, maintains you can prosper by farming a small parcel of land.

Skeptical? After all, the competition looks like Goliath, "whose height," the Bible says, "was six cubits and a span. He had a helmet of bronze on his head, and he was armed with a coat of mail, and the weight of the coat was five thousand shekels of bronze. And he had greaves of bronze upon his legs, and a javelin of bronze slung between his shoulders."

You can beat Goliath Farms Inc. in the competition for consumer dollars if, like David, you enter the competition armed with a well-conceived and coordinated strategy.

This chapter will help arm you. "Economies Of Scale" describes the field of competition and the economic factors which give rise to the Goliaths of farming. "The Grand Strategy" tells how to compete against those who have more land, better equipment and more money. "Business Strategies" describes ten specific courses of action you can take to achieve your objective.

CONSIDER THE ECONOMIES OF SCALE

At the age of six, I soloed my first working tractor on my grandparents' farm near Belfry, Montana. This 320-acre farm was a diversified business: Sugar beets, beans, peas, hay and grain grew in the fields; cattle, sheep and horses grazed in the pastures; hogs wallowed in their pen; ducks, geese and chickens pecked away in their coops; apple, plum and cherry trees grew in the orchard; and a variety of table produce grew in a garden about half the size of a football field. This type of diversified farm is rapidly becoming the dinosaur of agribusiness. The reason for its impending demise may be gleaned from an economic theory called "economies of scale."

How Farmers Use Money

American farmers, like my grandparents, achieved a considerable degree of success on their labor-intensive farms. Though few became cash-rich, many owned their land, home, equipment, livestock and automobiles. Their kitchen tables were filled with good food. It was a stable return on the investments of time and know-how.

This stable return, however, was insufficient to meet the needs of living in a modern, city-dominated economy. To buy new tractors or televisions or to send children to college, American farmers needed to increase their cash income. The Green Revolution provided a means to that end.

"Green Revolution" is a popular term for a 40-year period of scientific discovery and technological innovation. Said *National Geographic* magazine, "In a single lifetime, U.S. agriculture has advanced more than in all the preceding millenniums of man's labor on the land."

Products of the Green Revolution included new pest and disease resistant crops; fertilizers which improved plant growth and yield; pesticides which protected crops from a competitive environment; and new machinery which reduced labor-intensive chores such as cultivating soil and harvesting crops. The mechanical tomato harvester, for example, could harvest 25 tons of tomatoes in one hour.

Though Green Revolution technology provided a way to increase production, it was not magic. It did not eliminate needs of growing plants nor needs of farmers to generate a return on their investments. The new technology merely changed the nature of how farmers invested their resources.

The essential business of a farmer, as Jefferson's letter to Charles Peale states, is to examine "...the relative value of land and labor," so as to know "... the proportion in which these two elements should be employed to produce the maximum." This calculation of value is an important step in projecting a farm's ability to generate profit.

The United States has been, with the exception of a few short periods of time, a growing country with an expanding economy. There were wars to be fought, cities to be built and a rapidly expanding manufacturing sector to be managed. Because of this prolonged growth, competition for skilled labor has been intense and the value of skilled labor, relatively high.

Money, on the other hand, was readily available and inexpensive due to inflating land values and low interest rates. Farmers could borrow on equity in their land and let inflation pay the mortgage. These conditions made an investment of time— or labor— relatively more expensive than an investment of money.

Products of the Green Revolution provided farmers with means to substitute inexpensive money for expensive time. How did farmers make this new way of investing pay at the bottom line?

How The Use of Money Affects Farm Size

Green Revolution technologies require farmers to adopt a systems approach to production and marketing. For example, if a farmer buys a new tomato harvester, he also must buy the trailers, bulk bins, forklift tractors and maintenance equipment which make the harvester an effective tool. If he buys the new tomato hybrids, he must also buy fertilizers, pesticides and water to grow the hybrid according to its accompanying specifications. Cash costs of this tomato production system are consequently very high.

To generate a return on this investment, the tomato farmer has to increase the size of his farm. The more tomatoes he produces, the less each tomato costs to produce. By growing larger the farmer can realize this economy of scale.

At what size does a farm achieve its optimum economies of scale? The Cooperative Extension of the University of California has published economies of scale surveys over the years for various crops. By placing an imaginary farmer— Alpha by name— into the surveys, we can see how the products of the Green

Revolution have affected farm size over the years.

When farmer Alpha returned from World War II, he could make a good living, pay his debts and send his children to college by farming tomatoes on 100 acres in California's Central Valley. By 1960, Alpha needed 800 acres; by

FIG 3.1 A difference of opinion exists as to which is the best way to compete for consumer dollars. Some say, "Get big or get out!"

ed 800 acres; by 1970, he needed 2,640 acres to maintain that same good standard of living.

Alpha dreamed of having more wealth and prestige and Cooperative Extension advisers told him how to achieve the dream. "The research to date," Extension publications claimed, "has indicated that the firms operating in today's agriculture do not increase their cost (per unit of production) as they become larger." So Alpha's farm grew beyond its optimum economies of scale size and became Goliath Farms Inc. (I recently interviewed a tomato farmer operating on *10,000 acres* in California's Central Valley.)

The Green Revolution and personal ambition have changed the size and scope of the American farm. The medium-sized, diversified farm is rapidly being replaced by larger and larger farms growing fewer and fewer crops. Numbers tell the story: In 1963, there were more than 4,000 farmers growing processing tomatoes in California; ten years later there were 597.

If you grow tomatoes, you will compete with giant tomato farms in California, Texas, Florida and Mexico. If you grow cut-flowers, you will contend with giant, subsidized greenhouses in Columbia. If you grow apples, you will vie with giant orchards of Red Delicious in the Pacific Northwest. How can a person farming a small parcel of land overcome low-margin, high-volume, money-intensive Goliaths and win consumer dollars?

DEVELOP A GRAND STRATEGY

Metrofarming is a business for intelligent people, and intelligent people do not charge into competition with a Goliath Farms Inc. without a way to win. There is a strategy to win. It was first described some 2,000 years ago by the Chinese military strategist Sun Tzu and has been used throughout history by those who were forced to compete against larger, better-equipped adversaries.

Though Sun Tzu described competition in terms of warfare, his strategy, with due apologies, can be paraphrased and used for metrofarming:

> *Now a metrofarmer may be likened to water, for just as flowing water avoids the heights and hastens to the lowlands, so a metrofarmer avoids strength and competes with weakness.*
>
> *And as water shapes its flow in accordance with the ground, so a metrofarmer manages his success in accordance with the situation of his competition.*
>
> *And as water has no constant form, there are in metrofarming no constant conditions.*
>
> *Thus, a metrofarmer who competes by modifying his tactics in accordance with the competition's strength may survive and prosper.*

Mao Tse Tung also borrowed from Sun Tzu during his successful rebellion against the more-powerful Kuomindang. Said Mao, "The enemy advances, we retreat. The enemy camps, we harass. The enemy tires, we attack. The enemy retreats, we pursue."

Metrofarming's grand strategy, therefore, is to win the competition for consumers' dollars by avoiding strength and attacking weakness. The following observations of strengths and weaknesses will help you evaluate and know the Goliaths which dominate your field of competition.

Recognize Strengths

Being big has its advantages, some real and others illusory. Separate the real from the imagined by identifying factors which enable competitors to produce more efficiently. Large-scale competitors typically have six economies of scale strengths to protect against. They are:

SPECIALIZATION OF RESOURCES: As a farm grows in size, its opportunities for specialization— by man and machine— provide economic advantage.

When my grandfather retired from his farm he took a part-time job on a much larger farm down the road. This operation was large enough to allow him to become the irrigation specialist. As a result of this specialization, he was able to develop a more efficient irrigation system for the farm, thereby reducing its irrigation costs.

Bill Richards' 8,500 acre Ohio farm was made possible by the decision to specialize in corn and soybeans. By concentrating on two crops, instead of many, Richards was able to quickly identify and adopt the latest techniques, such as "no till" cultivation, which allow him to approach his goal of having one hired-hand per 1,000 acres.

When labor and equipment are focused on one chore instead of many, they can be used more effectively, thereby reducing the farm's cost per unit of production.

UTILIZATION OF EQUIPMENT: Big operators can make efficient use of equipment by spreading costs over a sufficiently large volume of output. The latest tomato harvester, for example, costs approximately eight times the average income in the United States. In addition to the harvester, one must also purchase necessary ancillary equipment— bins, trucks and packing facilities— and new tomato hybrids. Though the final tab on this equipment is very high, costs can be readily justified when spread over many acres of tomatoes.

COST OF INPUTS: Volume discounts are available to individuals who buy in bulk. In addition, for certain types of purchases, like tractors, the price per unit of capacity, such as horsepower, is often less when a larger unit is purchased. Price discounts apply to almost every aspect of a farm operation, from borrowing money to buying seed, from purchasing fuel to supplying paper clips for the office. The bigger the purchase, the bigger the discount. Consider the cost of money: Capital markets favor selling large because the cost of making loans does not increase in proportion with the size of loans (*i.e.*, it does not cost a thousand times more to loan five million dollars than five thousand). Consequently, the big borrower is awarded prime interest rates while the little borrower is charged retail.

GOVERNMENT PROGRAMS: The federal government currently spends about $40 billion a year on various farm programs. While the programs are ostensibly designed to benefit all farmers, large-scale farmers typically benefit the most. There are two reasons for this advantage.

First, government programs are

FIG 3.2 Economies of scale enable large-scale farmers to carry Very Important People to 37,000 feet, where fertile new business grounds may be cultivated.

often set up on a per-acre or per-unit basis. The more acres a farmer has, the more return he gets from the government. The 1983 Payment-in-Kind (PIK) surplus-reduction program is one good example of this largess: For every acre a farmer took out of production, the government returned an acre's worth of surplus crop. Whereas the average American farm received a PIK subsidy of $3,922, the J.G. Boswell Company, a large California farm, received a PIK subsidy of $3.7 million.

Second, large-scale farmers can manage politicians and bureaucrats more effectively by spreading influence costs over more units of output. Records from California's Secretary of State, for example, show a number of large-scale farmers flew a recent governor around the state during his election campaign. Though the costs of a business jet exceed the ability of most farmers, the business jet can become a cost-effective farm implement when its costs are spread over many thousands of acres.

Returns generated from a well-tended relationship with government can be extraordinary. While the J.G. Boswell Company farm was receiving its $3.7 million PIK payment for not growing crops in 1983, the Army Corps of Engineers pumped water out of its flooded fields for an estimated additional savings of $10 to $12 million.

Government assistance, like a double-edged sword, cuts both ways: It takes money out of your pocket and puts it into the pockets of your competitors.

PREFERENTIAL TAX TREATMENT: Large-scale farmers often benefit through provisions in the tax code. Periodically, for example, allowances for investment tax credits and accelerated depreciation are programmed into the tax code to encourage farmers to purchase equipment and construct buildings. The more expensive the purchase, the greater the tax credits and depreciation write-off. Though tax benefits are also available to small-scale farmers, most cannot take full advantage because they do not have the resources to buy, nor the capacity to use, the equipment and buildings.

MARKET ACCESS: The wholesale market for farm commodities requires a large volume with a uniform quality. By using specialized equipment and large fields, large-scale farmers can readily satisfy those requirements. Larry Galper, who grows strawberries on 290 acres in California's Pajaro Valley, put it most succinctly: "You can't grow higher quality food than we grow here — but I'm talking about visual quality, because that's all people buy." Mr. Galper can produce large quantities of strawberries— each berry looking very much like the next— on 290 acres of land. His crop can easily be sold through a wholesale and retail distribution chain which extends across the country and around the world. Small-scale farmers often do not have the capacity to satisfy the prescription buying habits of mass-markets and so may be effectively locked out of the markets.

Recognize Weaknesses

If the mere glimpse of a Goliath Farms Inc. is enough to keep you from planting the first seed, there is good news in the laws of physics, which demand symmetry: If there is a north pole, there must also be a south pole; if there is a positive charge, there must also be a negative charge; if there are strengths, there must also be weaknesses. The following observations will help you find the weaknesses in your competition's position.

OVERBURDENED MANAGEMENT: Plants and animals are living organisms. The single most important element in determining productivity of their growth and development is the amount of attention given them by the farmer. As a farm grows in size, the farmer becomes increasingly removed from his crops, thereby placing a burden on his ability to manage.

Consider the progression: As Farmer Alpha's tomato farm grows in size, he grows from the person standing in the field, to the manager driving around in a pickup, to the number cruncher sitting behind a desk, to the executive flying off to government in a business jet. What of Farmer Alpha's tomatoes? They are left in the care of field workers, who do not make policy, but merely carry it out.

Bureaucratic management typical of large-scale businesses creates many vulnerabilities. At a recent Pacific Horticultural Show, the largest and most lavish display was sponsored by a multinational corporate nursery. Though this giant enjoyed all the advantages of money and size and could readily afford lavish displays, it was owned by people very much removed from the growth and development of nursery stock. "They may be big, but they sure produce some poor quality plants!" was a comment heard many times on the showroom floor. As if to lend credence to the observation, the large corporation was forced to sell its nursery operations shortly thereafter— at a considerable loss.

Independent farmers, agricultural economists and corporate managers agree: The most efficient production unit is the farm run by its owner.

Before retiring, Russel Giffen, who operated a 100,000-acre farm in California, told the *San Francisco Examiner*: "I believe the very most efficient farm is if you can get the combination I am going to bring up now. Suppose a young couple wants to go into farming. If the wife has gone to high school and can do a little bit of accounting, she will keep the books, records, and do the shopping. The man is free to climb on a tractor and do all the heavy work. If you have that combination, both workers, and both bright, you can't beat that combination; I can't beat it, Tenneco can't beat it."

Tenneco Incorporated apparently agrees. In its annual report of November 1976, Tenneco said, "From the standpoint of efficiency, there is no effective substitute for the small-to-medium sized independent grower who lives on or near his land."

Bureaucratic management produces crops which, while uniform in size and appearance, tend to lack the aspects of quality known as character. Many consumers enjoy and demand products with character. Guy Clark expressed this longing in a popular song called *Home Grown Tomatoes*: "There's only two things that money can't buy— that's true love and homegrown tomatoes."

RELIANCE ON MONEY-INTENSIVE INPUTS: Farmers get big by substituting money for time. When money is readily available, as in times of inflation, money-intensive inputs, like huge tomato harvesting machines, can be an effective means of generating a big return. But when times swing into deflation, the bill for the same money-intensive inputs can leave one dangerously exposed.

Look back to the inflationary 1970's: Between 1975 and 1980, the value of farmland nationwide increased by 90 percent, while farmland in Iowa alone increased 138 percent. This inflation created a reservoir of new money from which the farmer could draw. Said Senator Tom Harkin of Iowa: "We had bankers going up and down the road like Fuller Brush salesmen during the '70's. They couldn't get farmers to borrow enough." And borrow they did. Through the 1970's, farm debt increased from less than $50 billion to around $200 billion.

An Iowa banker, in a conversation with *Barron's* magazine, described growth of this debt frenzy as follows: "In 1950, debt service was 10 percent of farm income, today (1979) it is 70 percent."

A small change can create big problems in a business environment dominated by debt. According to a study by Iowa State University, a bushel of corn selling for $2.50 will support farmland values of $1100 an acre. If the bushel drops 25 cents to $2.25, however, it will support farmland values of only $675 an acre.

When commodity prices fell in the early 1980's, many overextended farmers found themselves in a position best described by Louis XIV, who said, "Credit supports agriculture like a rope supports a hanged man."

RELIANCE ON DEFERRED PAYMENT: Large-scale producers have been able to gain significant advantage by deferring many true costs of production. When costs come due, large producers' competitive position will erode. To fully understand this vulnerability, consider the following observations on costs of fertilizers, topsoil, chemicals and water.

Fertilizers: In 1840, the German chemist Justus von Liebig proved plants could be nourished rapidly by applying mineral salts in solution directly to plants' roots. This breakthrough allowed for tremendous increases in crop production. By 1954, mineral salts accounted for more than 97 percent of fertilizers used in the United States.

Synthetically-derived fertilizers, like anhydrous ammonia (NH_3), have been relatively inexpensive when compared to organic fertilizers because fossil fuels and money required to make them have been relatively inexpensive. But there is another cost to synthetic fertilizers and it has been deferred.

Nitrogen, in certain forms, is a toxic substance. Nitrates (NO_2), for example, can impair circulation of oxygen in blood ("blue-baby syndrome") and cause vitamin deficiencies. The National Science Foundation estimates plants use only 30 percent of a farmer's application of nitrogen fertilizer; the remaining 70 percent escapes into soil, rivers, aquifers and air. To date, nobody has paid to clean up the mess. This deferred cost, however, may come due.

During a recent conference sponsored by the California Fertilizer Association, James Helmer, Unit Chief of California's Department of Food and Agriculture, talked about various fertilizer regulations which relate to manufacturers. One of his messages was, "Fertilizer contamination is not a problem of today; it is a problem of tomorrow or the day after tomorrow. You (the manufacturers) are responsible and the burden shall fall on your shoulders. You better start looking for a solution today."

If the bill for cleaning up fertilizer contaminations goes to manufacturers, they will raise prices to farmers, who will in turn raise prices to consumers. Consumers will then either pay to clean up the mess or shop for an alternative, such as the one offered by a small-scale organic grower.

Topsoil: John Deere invented the moldboard plow in 1833. This plow turned weeds back into the soil where they would decompose and nourish growing crops. Today plows are grouped together into huge gangs and pulled across land with giant tractors, creating deep furrows which are the hallmark of modern cultivation technique.

The cost of this technique has been determined by adding up manufacturing, sales, interest, insurance, fuel and maintenance costs. Another cost, however, has been omitted.

The plow exposes topsoil to erosive forces of wind and rain. By some estimates, the plow has contributed to the loss of up to 50 percent of the nation's topsoil. In Iowa, where 10 tons of topsoil per acre are lost every year, it costs two bushels of topsoil to produce one bushel of corn. So much topsoil washes into the Mississippi River, said William C. Moldenhauer of Purdue University in an interview with *Newsweek*, that "40 acres of good cropland goes past Memphis every hour."

The United States, which has so much, could always afford to lose topsoil. However, Lester Brown, head of the Worldwatch Institute, recently claimed that productivity of 34 percent of American cropland is declining because of excessive erosion. Though this loss of productivity can be masked by increasing use of fertilizer, this method compounds erosion by salinizing the soil.

Farmers who defer the true costs of large-scale cultivation practices will eventually pay through reduced productivity, increased fertilizer costs and loss of land to wind and rain. They will pass new costs on to consumers, who will either pay higher prices or shop for an alternative.

FIG 3.3 The buyers' fear of contaminated food is continually reinforced by accounts of toxic leaks, spills and disasters.

Pesticides: There are ways to reduce soil erosion. One is called no-till. Instead of turning weeds under with a plow, farmers kill them with herbicides and then drill seeds into unplowed ground.

Every year, farmers in the United States use an estimated one billion pounds of some 11,000 registered chemical poisons such as herbicides. The price of pesticides has been determined by adding manufacturing, capital, middleman and transportation costs. Other costs, however, have been deferred.

Because of the amount and frequency of their use, pesticides become concentrated in the environment. Some examples: Pesticides have been found in the water table of 23 of California's 58 counties (2500 wells). Ethylene dibromide

(EDB) has been detected by environmental researchers in 300 of Florida's drinking-water wells. And 44 percent of fresh fruit and vegetables in a recent test by the Natural Resource Defense Council contained residues of 19 different pesticides (produce purchased from San Francisco Bay Area markets).

Concentrations create controversy, with some authorities maintaining pesticides are safe and effective, and others saying pesticides are dangerous and a threat to our well-being. In fact, many people are afraid of pesticides, and their fear is reinforced by news accounts of chemical spills and disasters. One stark example is the 1984 leak of methyl isocyanate gas from Union Carbide's pesticide plant in Bhopal, India. This leak killed 2,500 people and injured another 150,000. Accounts from the scene were graphic. Said *Time* magazine, "At the factory, dead bodies were still on the ground, being picked up and loaded aboard a waiting truck. Everywhere one turned, people were retching, bent over horribly, racked by violent coughing that brought a red froth to their lips."

Fearful people have a way of becoming active people and active people demand regulations, restrictions and cleaning-up of their environment. In a recent poll of California residents, the Field Institute found that 77 percent of the respondents think manufacturers should be made to pay for costs of cleaning up their toxic chemicals.

If manufacturers pay to clean pesticides from the environment, they will pass costs on to farmers, who will in turn pass costs on to consumers. Consumers will then either pay the true costs of using pesticides or shop for an alternative.

Water: Three percent of the water on earth is usable, and agriculture uses approximately 85 percent of the usable supply. Said Mark Twain, "Whiskey is for drinking; water is for fighting."

Large scale production systems use irrigation techniques which, according to the U.S. General Accounting Office, deliver only 50 percent of the water to crops. But in addition to inefficient irrigation practices, precious water is also used to "wash" soluble salts and chemicals away from the crop root zone. As a result of irrigation and soil washing practices, it now costs approximately 1,400 gallons of water to produce a quarter-pound hamburger, bun, fries and soft drink. How is it possible to exchange pocket change for something which costs 1,400 gallons of the world's most valuable resource? The answer is simple: The true costs of water have been deferred.

Consider the Ogallala Aquifer, the vast underground reservoir extending from South Dakota to Texas, which provides water for one-fifth of the nation's croplands. The Ogallala's water table has been falling at a rate of three feet per year and the aquifer is now— optimistically speaking— about one-half full. Various estimates claim crop production based on Ogallala water will be reduced by 40 percent within several decades. California's aquifers are being depleted at the rate of two million acre-feet per year. The ground-level in some areas of the

FIG 3.4 Said Mark Twain: "Whiskey is for drinking; water is for fighting!"

San Joaquin Valley has fallen by 25 feet. The figures, if even remotely correct, say we are cashing in a water savings account to replenish a depleted water checking account.

The value of water grows as its supply is diminished. This value can be judged by the extent to which people will go to get water. Weigh this scheme put forward in *Fusion* magazine: "Starting in southern Alaska and extending into western Montana, surplus runoff water would be collected from the Rocky Mountains and delivered for use in southern California, Arizona, Texas, and Mexico. It would also be directed eastward through channels leading to the Great Lakes and Mississippi River. Using appropriate nuclear-explosive construction methods, the project would cost about $130 billion and take 10 to 15 years to complete."

True cost of water development schemes has been paid by taxpayers. But as the projects tend to benefit Goliath Farms Inc., taxpayers are becoming increasingly more reluctant to pay the bill. Said a recent editorial in the *San Jose Mercury News*, "It's time to stop the gravy train. Taxpayer subsidies for water projects should be phased out, and governments should institute a policy of requiring all users to pay the full marginal cost of new water supplies— that is, the actual cost of developing and delivering additional water."

If Goliath Farms Inc. must pay true costs for its water, it will pass the costs on to consumers. If consumers have to pay true costs for this water, they will likely shop for a less expensive alternative.

RELIANCE ON GOVERNMENT PROTECTION: Governments are traditionally very kind to large-scale businesses. They protect against uncertainties of the market with taxpayer sponsored programs, provide preferential tax treatments and allow for deferment of true costs of production inputs. Going into business with government, however, can prove to be a very fickle affair.

When the United States needed a "food weapon" to counter OPEC's "energy weapon," government urged U.S. farmers to expand production. Earl Butz, then secretary of agriculture, spread the word: If you plant fencepost to fencepost, the government will finance your expansion and loan money to other nations so they can buy your crops. Butz's admonition came in no uncertain terms: "Get big or get out."

Farmers listened to Butz and became big operators specializing in strategic crops like corn, wheat, cotton and soybeans. Between 1972 and 1982, production increased from 289 million acres to 413 million acres. For a time, the government's strategy seemed to pay. Said Butz, in a 1977 debate with agricultural writer Wendell Berry in *CoEvolution Quarterly*, "So we made a net plus contribution last year in American agriculture to our balance of payments of $12 billion. Believe me, that's rather important in this year when our overall balance of payments is running about a negative $25 billion."

Times change. In the 1980's, the United States found itself with an annual budget deficit of $200 billion and an annual trade deficit around $100 billion. When farmers went to government for help with debt burdens and crop surpluses, they found times had really changed. One hundred years ago, 45 percent of the people in the United States lived on farms; today, only four percent live on farms. Politicians govern by consent of the governed, and the governed have moved into town. City-dwellers simply do not want their incomes redistributed to a Goliath Farms Inc. in the countryside.

Budget Director David Stockman delineated the government's new strategy: "For the life of me I cannot figure out why the taxpayers of this country have the responsibility to go in and refinance bad debt that was willingly incurred by consenting adults who went out and bought farmland when the price was going up and thought that they could get rich, or who went out and bought machinery and production assets because they made a business decision that they could make money."

In all likelihood, government will remain best of friends to the giants of agribusiness. However, one day tax payers will become reluctant to pay the bill for this friendship. When the day comes, the giants' competitive position will disintegrate.

The larger and more complex a farm becomes, the more vulnerable it becomes, regardless of how omnipotent its superior size might make it appear. As Sun Tzu said, "For if he prepares to the front, his rear will be weak, and if to the rear, his front will be fragile. If he prepares to the left, his right will be vulnerable and if to the right, there will be few on his left. And when he prepares everywhere he will be weak everywhere."

DEVELOP BUSINESS STRATEGIES

David avoided a hand-to-hand fight with Goliath to strike the giant's unprotected forehead with a rock fired from his slingshot. How are you going to overcome your competition for consumer dollars?

Business strategies provide courses of action you can take to win. Each business strategy is developed, as Sun Tzu would say, "to manage success in accordance with the situation of the competition."

Following are examples of ten metrofarm business strategies. Each provides a way to attack Goliath Farms Inc. by having lower costs, higher prices or both lower costs and higher prices. Each business strategy is expressed first as a general principle and then exemplified by using it to demonstrate how to achieve goals set for the Melon Company (see Chapter Two, Business Projections).

In the last meeting of the Melon Company, you established a goal of earning two weeks in Tahiti by growing melons part time on the one-acre vacant lot across the street. To achieve this goal, you need to be 400 percent more productive than the giant watermelon growers in California's Central Valley. How are you going to succeed?

Increase the Yield

The giant melon growers of California's Central Valley produce an average of 50,000 pounds of watermelons per acre. Your Melon Company's objective is to grow 150,000 pounds per acre.

It will not be possible to achieve this 300 percent increase in yield by duplicating the competition's production technologies. There is not enough equipment, land or money to attempt their kind of watermelon production. Instead, adopt and modify low-cost, high-density production technologies developed over centuries by small-scale growers throughout the world.

This 300 percent increase in yield will not come without cost. You will have to increase the investments of time and know-how to design and build a high-density production system. But economies of scale will enable you to spread the additional costs over 150,000 pounds of melons, thereby allowing for a significant net gain in productivity as well as yield.

Potential customers will lean on the fence and say, "My goodness, this is the most amazing watermelon patch I have ever seen. How much are your melons?"

Improve the Selection

Large-scale growers produce only the few melon varieties which are suitable for mass-market distribution. The Melon Company, on the other hand, will produce varieties which are suitable for satisfying local tastes.

Grow giant melons for family picnics, regular melons for individuals who mistrust anything new and midget melons which can fit into the refrigerator of a widower. Grow green melons, green and yellow-striped melons and golden melons. Grow red-fleshed melons with seeds and golden-fleshed melons without seeds.

Though seeds for the Melon Company's exciting new varieties will cost an additional penny or so per pound, they will enable you to raise prices three or four pennies per pound, for a significant net gain in productivity.

Though potential customers may buy melons for less money at the supermarket, many will find your selection an attractive value. "Have you tasted the Melon Company's golden seedless yet?" they will ask each other. "You just have to try one today!"

Raise the Quality

The competition's large fields, big equipment, hired-hands, vast amounts of chemical inputs and mass-market distribution tend to produce melons of bureaucratic uniformity. The Melon Company will attack this vulnerability by producing melons with character and taste.

This objective can be achieved by conscientiously giving each melon plant your personal attention. Though you will have to invest more time, each plant in your one-acre field will grow and develop to its maximum potential. This strategy, if successfully implemented, will yield a more interesting— and hence more valuable— product. You will then be able to increase prices and reduce marketing costs, thereby realizing a significant increase in net productivity.

When customers slice into their first Company melon, they will find a firm, sweet flesh which actually smells and tastes like watermelon. Even the neighborhood skeptic will say, "I haven't tasted watermelon like that for 30 years!"

Time the Sale

Watermelon prices plunge 400 percent and more when big operators hit the market in early summer. The Melon Company will, to the extent possible, avoid this glut by producing for early and late season prices.

This objective can be achieved by using season-extending devices like starter greenhouses with heated seed beds, frost caps, shade houses and day-neutral varieties. You will have to invest more time, know-how and perhaps some extra money to make this strategy work. But off-peak season melon prices will readily allow for a huge increase in net productivity.

Prospective customers on their way home from work on a warm spring day will say, "I'll buy a melon from the Melon Company and surprise the family."

Capitalize on Location

The competition grows melons in California's Central Valley, Mexico and other favorable climates and then ships them into your market. The Melon Company will take advantage of this distance and grow locally.

This strategy will increase productivity in several ways: First, it will allow you to avoid costs of shipping 75 tons of melons thousands of miles to market. Second, your fresh, vine-ripened melons will taste much better than melons two weeks off the vine and will consequently have more value. And third, you will gain the pennies-per-pound advantage of community support.

Community support of local businesses— which circulates money back into the community instead of away into distant hands— provides many financial advantages: Neighbors will lend a helping hand or barter for the use of their tractor; bureaucrats will bend a rule or two in your direction; and, if you run a clean, photogenic operation, the local TV and newspaper will run stories about you and the Melon Company. You may have to invest more know-how to make up for what may be a less-than-ideal growing climate, but the reduced costs and increased prices will bring huge increases in productivity.

Days before harvest the headlines of the local newspaper will read, "The Melon Company: Local Business Set To Harvest Green by Growing Delicious Melons."

Become Self Reliant

The competition employs professional managers, chemical application specialists, equipment operators, field hands, truckers, brokers and others to grow and distribute its melons. Avoid this huge payroll and do the Melon Company's work yourself.

For a true indication of how self-employment will affect productivity, a reasonable hourly wage should be calculated for your time. Instead, take some TV time, gossip time and even hammock time and give it— gratis— to the business. If you need additional help along the way, hire local youth on a part-time basis. By avoiding huge costs associated with "official" labor,the Melon Company will generate a substantial increase in productivity.

Prospective customers will watch your dedicated work and say, "The owners of that Melon Company are hard-working, conscientious people. I think I'll try one of their melons."

Conserve Cash

The competition relies on money-intensive inputs like four-wheel-drive tractors, crop-dusting airplanes and large tractor-trailer rigs to grow melons and ship them to market. Avoid money-intensive inputs and rely instead on your old pickup truck and ingenuity.

To make this strategy work, resolutely avoid all forms of "new-paint disease." Instead of buying a new truck, for example, wash the old truck and then have your friend the sign-maker paint a new Melon Company emblem on the side. By deriving status from the quality of your product and from your ingenuity in "making-do," rather than from the size of your debt, you will reduce production costs and achieve significant gains in productivity.

Potential customers will see you driving down the street and say, "Did you see how the Melon Company people rigged their beautiful old truck to haul watermelons? How ingenious!"

Employ Neighborly Technologies

The competition's melon plants— situated one after the other, in row after row, on acre upon acre of saline soil— must be nurtured with mineral salt fertilizers and protected with powerful chemical poisons. Attack this vulnerability by growing with more neighborly technologies.

The Melon Company is situated on one small, readily-managed parcel of land. Nurture the melon plants with soil built with legume cover-crops, manures, composts and other nutrient-rich materials collected in your spare time. Protect the melons by keeping your plot clean and by pouncing on pests with the appropriate chemical-free technologies.

FIG 3.5 Large-scale production requires the extensive use of mineral-salt fertilizers, powerful pesticides and abundant supplies of water. Attack this weakness and win the consumer dollars.

Neighborly technologies increase productivity in several ways: First, they reduce or eliminate the cash costs associated with mineral salt fertilizers and pesticides. Second, they increase the perceived value and consequent price of your melons. And finally, they help maintain good relations with the neighborhood by eliminating the need to send periodic clouds of toxic chemicals wafting down the street.

Potential customers will see your organic label and say, "No poisons! I'll take three!"

Fight for Position

Large-scale competition relies on government to defer the true costs of many production and marketing inputs. Attack this weakness by aggressively demanding government stop protecting competition with money it collects from you in taxes.

This strategy, if successfully implemented, will help in two ways: First, when government taxes you and gives to your competition, it naturally detracts from your ability to compete and improves your competition's ability. Second, when government protection is removed and competition is made to pay the true costs of its production inputs, the cost of its melons will jump beyond belief. Consider what the cost of the competition's melons would be if it were made to pay true costs for water? For cleaning fertilizers and pesticides from the environment?

You will have to invest more time and know-how to successfully implement this strategy. You will have to aggressively lobby government, community groups and neighbors. If you succeed, however, your efforts will yield a greater increase in Melon Company productivity than all other strategies combined.

When buyers discover the true price of the competition's melons they will say, "Let's buy a Melon Company melon. They have the best price in town!"

Control Distribution

Large-scale competition sells melons to brokers, who sell to packers, who sell to shippers, who sell to distributors, who sell to wholesalers, who sell to retailers, who sell to consumers. The Melon Company will avoid this army of expensive intermediaries and sell direct.

The Company will sell melons to people who enjoy picking their own, to people driving up to its little roadside stand, to retail stores around town, to the college cafeteria, to restaurants, to consumers at farmers markets and so on. You will have to invest more time, money and know-how to develop the markets.

Nevertheless, selling direct will allow you to earn 90 cents of the consumers' dollar instead of 30 cents.

The customer will pick up a Melon Company melon, look you in the eye and say, "I'll take this one!"

Each of the ten business strategies considered can increase the Melon Company's productivity. If all 10 are combined into one well-conceived and coordinated attack, and more developed and implemented as you proceed, your Company can easily increase productivity by 400 percent over per-acre productivity achieved by giant watermelon growers in California's Central Valley.

To be certain, many things can happen. Neighborhood kids may sneak into the patch for a midnight feast! Yet your Company can succeed, because when it comes to using land, no operation is more efficient than the operation of a small-scale farmer. In a series of studies on farm productivity in developing nations, the World Bank concluded small owner-operated farms, using few products of modern technology, were three to 14 times more productive per acre than were large high-technology farms of the Green Revolution.

If you develop production and marketing skills to achieve a 1400 percent superiority, you can quit your full-time job, lease another acre or two of land and become a big operator on a watermelon metrofarm.

REVIEW

Conventional farmers have adopted technological innovations of modern agriculture as a means to increase their cash income. To benefit from technologies, farmers spread their costs over more units of production. As a consequence of their need to realize economies of scale, fewer and fewer farmers grow fewer and fewer crops on larger and larger parcels of land.

Competition for consumer dollars is now dominated by giant producers. Though the giants appear omnipotent, they are, in fact, extremely vulnerable to metrofarmers who operate with a well-conceived and coordinated strategy to avoid strength and attack weakness. This strategy provides means to win the consumer dollars by being the most efficient producer. In the final analysis you, like other small-scale farmers, can be 1400 percent more productive than giants of high-technology agriculture.

CHAPTER THREE EXERCISES

i   Shop your market for an interesting product. Any product is fine, as long
       as it interests you. Write the name of this product on the top of a new page
       of paper and insert the page into the "Strategy" section of your metrofarm
       workbook.

ii   Identify the dominant supplier of this product in your market. Ask a couple
       of retailers and wholesalers, in a friendly sort of way, and they will likely
       provide the supplier's name and address. Write the supplier's name and
       address on the page used for exercise *i*. (*e.g.,* WIDGETS, Goliath Farms,
       Inc., Anywhere, USA).

iii   Immediately below the product and supplier headline developed for the
       preceding exercises, divide the page into two vertical columns. Label the
       left column "Strengths." To the best of your ability, identify the reasons
       why the supplier has become dominant and list each reason in the "Strengths"
       column.

iv   Label the right column "Weaknesses." To the best of your ability, identify
       where and when the dominant supplier is unable to satisfy the market's
       demand. (For inspiration, read "The Plant Carrousel" in Part IV, Conver-
       sations.) List each deficiency in the "Weaknesses" column.

v   On a second workbook page, list five ways to avoid the dominant supplier's
       strengths and five ways to attack its weaknesses. Write a brief description
       of each strategy.

PART II

PRODUCTION

INTRODUCTION

S trategy tells us what to do; tactics tell how to do it. Tactical planning consists of determining how to employ the resources of markets, land, crops, business organization and production systems to generate maximum revenue from minimum cost.

Since no two metrofarms operate with the same resources, metrofarm production systems are, like fingerprints, unique. Nevertheless, metrofarmers share the same tactical problem. To wit: "How can I produce enough products on this small parcel of land to make the enterprise worthwhile?"

To answer the problem, successful metrofarmers emulate a great technology of modern times— processing of electronic information. Consider: Farms are factories which convert radiant energy from the sun and nutrients from the earth into stored chemical energy of crops. Similarly, computers are factories which convert electrical current into binary digits ("bits") of information.

In the pioneering days of computing, bits were managed by vacuum tubes. The fragile glass tubes enabled engineers to build ENIAC, the world's first electric computer. ENIAC was a remarkable factory. It contained 18,000 vacuum tubes, filled a thirty by fifty foot room and weighed 60,000 pounds. Though ENIAC thrilled the world with its capacity to solve problems, it consumed too much electricity (the equivalent of 100 lighthouses), gave off too much heat, broke down frequently and was too expensive to build and maintain.

Scientists struggled to find a more efficient way to manage electricity. This struggle resulted in the 1947 invention of the transistor. The transistor was far more efficient than the vacuum tube and the newly found efficiency sparked an explosion in computer technology.

Today, millions of transistors are etched into microprocessors smaller than your fingernail. The "computers-on-a-chip" interpret and process instructions, perform arithmetic calculations and remember data, just like the giant ENIAC. Unlike ENIAC, however, microprocessors use very little electricity, give off little heat, are inexpensive to build and maintain and perform millions of calculations per second.

Large-scale farms tend to be the ENIACs of agriculture: They consume too much energy, give off vast amounts of waste (some of which is toxic), break down frequently with credit crunches and price squeezes and are extremely expensive to build and maintain.

Metrofarms are the microprocessors of agriculture. Chapters 4 through 8 of *MetroFarm* will help you develop the "farm-on-a-chip" production tactics needed to produce more crops, of higher quality, at less cost than can the ENIAC's of vacuum-tube agriculture.

Surveying the Market

*If I don't have a market for a crop,
I don't grow it.*

Charles Heyn
Billings, Montana

The first step toward efficient production is taken long before earth is turned or seeds planted. The first step is a survey of the market.

The market is where supply meets demand. This interaction yields information upon which production is based. A survey is the process by which information is gathered and evaluated. Since the market continually changes with the ebb and flow of supply and demand, surveying is a day-to-day, week-to-week, month-to-month and year-to-year process.

Farmers are vulnerable without good information, regardless of how efficient their production might be. This elemental fact was demonstrated one cold January day at a seminar on alternative crops sponsored by the County Extension Agency of Yellowstone County, Montana. The seminar was directed to the area's sugar beet farmers, who were among the world's best producers. Their problem was the market. Consumer concerns about excess calories, development of corn and artificial sweeteners and a glut of cheap imported sugar had led to a collapse of demand. Farmers were left out in the cold without a contract and extension agents were trying to help.

Charles Heyn (Part IV, Conversations) was scheduled as the seminar's featured speaker. What kind of advice would Heyn give to the assembled big operators?

Farmers filled the room early, forcing the extension agents to move the meeting to another room twice as large. Farmers filled that room to standing-room only and then spilled out into the hall. An extension agent opened the seminar with a talk on successful farming. "Successful farming," he said, "is having the right product, in the right form, in the right place, at the right time, in the right quality, in the right quantity, at the right price." The assembled farmers, all of whom appeared to have the wrong of everything, shifted uncomfortably at the wisdom.

It was then announced Charles Heyn would not be able to attend, but instead had sent along a bit of advice. The extension agent unfolded a slip of paper and read, "If I don't have a market for a crop, I don't grow it." Tough words for the assembled farmers, but a very logical thing for Heyn to say. Successful metrofarmers like Charles Heyn survey their market very closely. Surveys enable farmers to detect changes in supply and demand and make required adjustments in time to profit.

Your ability to succeed as a metrofarmer depends upon the quality of your information. This chapter will help you find the right information. The first section surveys kinds of information the market can yield and sources from which they can be gathered. The remaining four sections survey market sectors available to metrofarmers and fundamental characteristics of operating within each sector.

ORGANIZE A SURVEY

The survey is a critical examination of the market to obtain the kind of exact, unbiased information it takes to make objective decisions. Decisions to be made in production of crops can be divided into two categories, cultural and business.

Cultural decisions are those required to facilitate the growth of crops. Information required to make good cultural decisions is often easy to obtain: If plants are wilting, they need water; if plants are turning yellow, they need fertilizer.

Business decisions are those required to generate a return on investments. The information required to make good business decisions is often difficult to obtain. To paraphrase the military strategist Karl von Clausewitz, "A great part of the information obtained from the market is contradictory, a still greater part is false and by far the greatest part is of doubtful character."

This section will help you infiltrate the market and return with the kind of exact, unbiased information it takes to win the competition for consumer dollars.

Select Information

The Montana extension agent put it most succinctly: "Successful farming is having the right product, in the right form, in the right place, at the right time, in the right quality, in the right quantity, at the right price." The agent was not succinct in telling what was right, however. Only the market can provide the right information. Following are examples of how the market can tell you what is right and what is wrong. (The farmers discussed are the ones featured in Part IV, Conversations.)

THE RIGHT PRODUCT: The market tells which product is right and which wrong by revealing level of consumer demand, potential for future demand and competition's strengths and weaknesses in satisfying the demand.

Had sugar beet farmers discussed in the introduction to this chapter known about weakness in demand for sugar, foreign competition's strengths in producing it and an ever-increasing number of artificial sweeteners, they might have been able to find a profitable alternative before their market collapsed. Charles Heyn, on the other hand, sold directly to consumers who told him exactly what they wanted. Heyn was consequently able to stay home on the cold January day and take reservations for the next season's crops instead of seeking help from county extension agents.

THE RIGHT FORM: Form is the shape and structure of a product, as distinguished from the material of which it is composed. The market tells which form is right by revealing the level of consumer demand for each form.

The foliage (houseplant) trade in Medford, Oregon, was dominated by large nursery operations based in Central California. Jan LaJoie surveyed retail stores which stocked this product and discovered an "Achilles Heel." Two and one-fourth inch potted plants could not take the long trip up from California without a significant deterioration in quality. Though consumers loved the two and one-fourth inch package, few would pay for damaged goods. LaJoie attacked this vulnerability by producing two and one-fourth inch potted plants in Medford. The small, delicate and undamaged little houseplants enabled her to capture a share of the market.

THE RIGHT PLACE: Place is the market sector in which a product is sold. The right place is the market sector which returns most profit for least cost. Many variables must be evaluated to determine which sector is right, and only the market can provide the kind of information it takes to make an evaluation.

When Charles and Lois Heyn began metrofarming, they sold through grocery stores in nearby Billings, Montana. Though the Heyns were able to sell all of their products through retailers, it cost too much time and money to drive into town and service each outlet. They surveyed the market and determined that they could generate more income by having customers drive out to their farm. The

Heyns then executed an orderly retreat from the retail sector and established a strong and profitable position in the direct sector.

THE RIGHT TIME: The right time to sell is when demand is high and supply low. The market tells which time is right by telling when its need for a product is greatest.

The Heyns' principal competition in the vegetable and small fruit trade was located in distant growing regions like California's Central and Imperial Valleys. The Heyns learned when produce arrived in the market and planned accordingly. To achieve their objectives, the Heyns used frost caps and hot beds which allowed for an early start on the season. When the competitor's product finally arrived, after a thousand mile truck ride, the Heyns were waiting with fresh-from-the-garden products already for sale.

THE RIGHT QUALITY: Quality refers to essential characteristics like taste, appearance, nutrient content, fragrance and uniformity. Quality standards for many farm products are determined by logistics of mass-marketing and mass-marketing typically requires uniformity. The right quality, however, is eventually determined by consumers in the market.

Weddings provide a good market for the cut-flower industry. To facilitate mass-market distribution, flower growers typically produce a large volume of a few standard varieties like carnations, roses and baby's breath. The Flower Ladies surveyed brides and detected a demand for a different kind of quality. To take advantage of the demand, they grew more than 200 varieties of old-fashioned flowers. Their diversity enabled them to construct floral arrangements with 20 shades of blue flowers or ones which had the fragrance of frosting on a wedding cake.

THE RIGHT QUANTITY: Quantity refers to the amount of products sent to market. Quantity standards for many farm products are governed by logistics of mass-distribution. These standards do not favor small-scale producers. Still, the market can reveal where the right quantity may, for example, be less than a 10,000 case minimum.

Much trade in the ornamental nursery industry consists of high-volume transactions among producers, wholesalers and retailers. Volume sales give large producers an economies of scale advantage over small producers. Gerd Schneider surveyed this market and detected a need for limited numbers of unique ornamentals and propagation services. By offering unique products and services to various nurseries, landscape firms and government agencies, Schneider developed a satisfied clientele. This clientele, in turn, helped Schneider develop a trade in mainstream nursery products like ornamental juniper.

THE RIGHT PRICE: Price refers to the value assigned a product. The right price is determined by interaction of supply and demand. No marketplace displays this interaction more clearly than a farmers market. Farmers ask a price and customers either pay or walk away. Jeff Larkey sells through several farmers

Market Information

 THE RIGHT PRODUCT — The right product is the crop which can be grown with a substantial margin of profit.

 THE RIGHT FORM — The right form is the shape and structure of a product which will allow that product to be readily sold.

 THE RIGHT PLACE — The right place is the market sector that will allow for the greatest return on investments of time, money and know-how.

 THE RIGHT TIME — The right time to sell is when demand for a product is high and supply is low.

 THE RIGHT QUALITY — The right quality consists of elements of taste, appearance, nutrient content, fragrance, uniformity and cultural practices demanded by consumers and by logistical and legal considerations of the market.

 THE RIGHT QUANTITY — The right quantity is the amount demanded by consumers and by logistical considerations of the market.

 THE RIGHT PRICE — The right price is the highest one allowed by the interaction of consumer demand, market supply and how quickly the seller wants to make a sale.

 THE RIGHT STANDARDS & PROGRAMS — The right standards and programs are those legal restrictions placed on production and distribution of a crop by government and/or industry.

FIG 4.1 "Successful farming is having the right product, in the right form, in the right place, at the right time, in the right quality, in the right quantity and at the right price." From a County Extension agent in Yellowstone County, Montana.

markets. By carefully surveying the level of consumer demand each market day, Larkey obtains the best price for his varied line of products. Were he to sell all of his products through a produce broker, Larkey would have to accept whatever price the broker offered.

THE RIGHT STANDARDS AND PROGRAMS: The County Extension agent with a list of "rights" failed to mention "the right standards and programs," which are the legal restrictions established by government and industry. Legal restrictions can influence one's ability to compete within a market and only a careful survey of the market can reveal where restrictions are in effect. There are state and federal standards for grading, packing, labeling and inspecting many kinds of farm products. Articles 1380.18 and 1380.19 of California's Administrative Code, for instance, define what kind of containers can be used to ship apples to market.

Marketing orders, councils and commissions can also affect flow of products into the market. The programs are legally sanctioned entities in which growers, processors and handlers work with government to facilitate and regulate the market for specific crops.

There are, at this writing, some 53 state and federal marketing programs in California alone. These programs have authority to expand product demand through advertising, promotion and consumer education; conduct research and development relating to production, processing and distribution; regulate product quality, condition, grade and size; restrict the volume of products flowing into the market; eliminate unfair trade practices; and control or eradicate insects, predators, diseases or parasites.

All producers of a crop must comply with standards voted in by the majority of producers. During a recent drought, California farmers were ordered to destroy an estimated 500 million pounds of peaches and nectarines because they were slightly smaller than federal standards permit. Complained Dan Gerawan, of Gerawan Farming, the largest peach and nectarine grower in the world, "Even my trained pickers can't tell the difference between acceptable fruit and the fruit that is too small!"

Standards were established to facilitate the kind of mass-distribution in which a buyer on one side of the country can purchase from a seller on the other side and be assured of quality, honesty and fair practices. However, standards can also govern a farmer with five boxes of apples to sell a corner grocery.

A careful survey of standards and regulations can reveal how to cut through expensive red tape. By establishing a "certified" farmers market, Jeff Larkey and his fellow metrofarmers were able to obtain a legal exemption from government grading, labeling and packaging standards. This exemption enabled him and his fellow farmers to sell apples and many other products without having to comply with expensive standards and programs.

Select Information Sources

There are many sources of information. Some provide valuable insights into what is happening in the market; others mislead and confuse. Only by listening to a number of sources can one obtain the kind of exact, unbiased information it takes to win the competition for consumer dollars.

BUYERS AND SELLERS: The most important information sources are buyers and sellers. Buyers reveal the level of demand, when demand is greatest, need for a specific forms or packages and prevailing standards for quality. Sellers can reveal the competition's strengths and weaknesses, their pricing policies and, through sales histories, when the need for a product is greatest.

When Jan LaJoie started the Plant Carrousel, her objective was to establish a plant rental service for restaurants and doctors' offices in Medford, Oregon. When she discussed the possibility with another local grower, he suggested she grow foliage starts for the retail trade instead. LaJoie then surveyed retailers in the Medford area to confirm the opportunity. She asked how many foliage plants each retailer sold and how much they paid for each form and type of plant. Each retailer provided LaJoie with the information she requested, thereby enabling her to make the kind of accurate financial projections it takes to succeed.

TRADE ASSOCIATIONS: Farmers, like most business people, establish associations to protect and promote their industry. While general interest associations like the Farm Bureau serve all farmers, special interest associations like the California Certified Organic Farmers serve farmers who share a common interest. Associations can provide a wealth of information.

Hanging on a wall at the California Association of Nurserymen's headquarters is a fascinating photograph of Gerd Schneider and several other individuals dressed in large Mexican sombreros at an Association function. Why would Schneider, who is all business, dress for fun? Because he was placing himself in the center of the information flow and, judging from his expression, having some fun, too!

NEWSLETTER SERVICES: Newsletter services provide specialized information to individuals in need. There are government publications like *Market News Service* and private sector publications like *Organic Market Report,* published by the California Certified Organic Farmers.

Newsletters collect information on market prices, sales volume and other market conditions, publish the information in either print or electronic formats and then distribute it to interested parties. Since the information is collected from a number of sources, it tends to reflect broad market prices and trends. Specific prices and trends vary from market to market, city to city and state to state.

NEWSPAPERS, MAGAZINES, BOOKS: Reading is a precursor of success because it provides a wealth of information.

Newspapers, for example, offer "Food" sections which detail new trends in demand, such as baby carrots of "nouvelle cuisine." General interest articles also point the way to market trends. For many years, articles discussed the link between high-fat foods and heart disease. Then, quite suddenly and dramatically, consumers began reducing their consumption of fatty beef.

Magazines also provide general and specific information. Rodale's *New Farm*, for instance, offers specific cultural information, "The End of Corn Rootworm: It's As Near As Your Next Crop," as well as insights into general market trends, such as "They're Looking Forward: New Production And Marketing Methods Give These Farmers Reason To Be Optimistic."

Trade associations publish magazines and newsletters which provide specific information on production and marketing of respective crops. Trade publications can be extremely specific. For example, the American Peanut Research and Education Society publishes *Peanut Science*, which provides information on production, storage, processing and marketing of peanuts.

Books have the capacity to provide the most detailed kinds of information on market trends. Books have, for example, the structural patience to examine the 50-year trend in marketing apples in upper New York State. Though unable to track daily trends in markets, books can tell what has or has not worked in the past, and what has worked in the past may well work again in the future.

The Flower Ladies, when fledgling entrepreneurs, were dedicated readers. Newspapers, periodicals and books generated many new ideas for building their wedding business. Their success soon reversed the information flow and they became the success story featured in many local and regional periodicals.

GOVERNMENT AGENCIES: Federal, state and county governments provide many good sources of market-related information.

The United States Department of Agriculture (USDA) aids in the formation of marketing cooperatives. The California State Department of Food and Agriculture's Direct Market Program helps develop farmers markets, harvest trails, roadside stands, U-Picks and other farmer-to-consumer forms of marketing. Cooperative Extensions extend the research of land grant colleges and the USDA to farmers through workshops, seminars, conferences, field demonstrations, tours and other educational programs. And County Agricultural Commissioners keep track of state and federal standards for grading, packaging and inspection.

Jeff Larkey tapped into the governmental flow of information while developing area farmers markets. This information helped build several good farmer-to-consumer markets and helped local farmers gain a legitimate exemption from industry-mandated packaging standards.

LIBRARY: Libraries may well be the most under-utilized of information resources. A good library contains a number of valuable information sources, including newspapers from different cities, periodicals from today and yesterday, books, state and federal publications and librarians. Librarians are trained in the art

Information Sources

BUYERS
- Timing of product demand cycle.
- Desire for specific form, shapes or color.
- Prevailing standards of quality.

SELLERS:
- Competitive strengths and weaknesses of other producers.
- Prevailing pricing policies.
- Timing of product marketing cycles.
- Recent trends in demand.

TRADE GROUPS:
- Recent technical innovations.
- Practical marketing insights.
- Access to other information sources.
- How to protect against unnecessary government restrictions.
- How to access applicable support programs.

ONLINE NEWS:
- Recent trends in market demand.
- Current market prices.
- Recent technical innovations.
- Recent government restrictions and/or support programs.
- Access to other information sources.

NEWSPAPERS, ETC.:
- Recent trends in demand.
- Recent pricing trends.
- Specific cultural information.
- Broad historical trends in consumer demand.

GOVERNMENT:
- Access to in-depth research on crop production technologies.
- Aid in marketing, including help in forming co-ops, farmers' markets, etc.
- Help with specific production and marketing problems.
- Grading, packaging and inspection standards.

PUBLIC LIBRARIES:
- Names and addresses of relevant persons and organizations.
- Demand and supply trends in neighboring markets.
- Historical trends in supply and demand.

FIG 4.2 Only by surveying a number of sources can one obtain the kind of exact, unbiased information it takes to win the competition for consumer dollars.

Three Steps to an Effective Market Survey

DEFINE THE OBJECTIVE

State your objective in one clearly written sentence. Use this objective to keep your survey efforts focused.

WRITE THE QUESTIONS

A. Limit the number of questions to the minimum required to achieve your objective.

B. Develop a logical beginning, middle and end sequence to the list of questions.

C. Ask questions that will elicit specific answers. (e.g., "How many pounds per week do you sell?")

D. Provide the respondent with a sample for questions that call for subjective answers.

	poor				excellent
Is the quality good?	1	2	3	4	5

E. Avoid questions that suggest an answer. (e.g., "Don't you think that this would be better than that?")

F. Be sensitive when asking about emotionally-charged issues such as age and income.

TEST AND REVISE THE SURVEY

A. Test your survey on several people who are likely to have information.

B. Evaluate how much useful information each question produced.

C. Revise or eliminate questions that did not produce useful information.

D. Retest the survey.

FIG 4.3 Surveying for information is the essential chore of every business person.

and science of tracking down information. If you want to know how many varieties of apples were grown commercially in 1931, ask a good research librarian and he or she will find the answer.

DUE DILIGENCE: Information received from others may be biased, and biased information can be worse than no information. All information should be cross-checked by personal observation. This cross-checking is called "due diligence."

Suppose a salesperson of kiwi (*Actinidia Chinensis*) starts offers a financial projection for the future of a fresh kiwi market in your neighborhood. Since his commissions are based on your purchase, chances are good his projection will be optimistic. Before purchasing starts, conduct a due diligence survey. Ask local retailers how well kiwi sells. Ask consumers why they buy kiwi, how kiwi are used and how many kiwi they buy each month. Due diligence is the first hand checking of the validity of information received from others.

SURVEY THE WHOLESALE SECTOR

Gathering information is the first step toward becoming a successful metrofarmer. The next step is surveying the four market sectors in which you may compete. Our first look is at the wholesale sector.

The wholesale sector consists of intermediaries who buy large units of a product for resale in smaller units to other wholesalers or retailers. Brokers and agents are wholesalers who arrange for others to buy and sell, but who do not take physical possession of the product.

Categories of Wholesalers: The wholesale sector is divided by function into shipping-points, terminal-markets and jobbers.

Shipping-points buy large quantities from farmers in major growing areas and then resell smaller quantities to terminal-markets in metropolitan areas. Shipping-points are also called country assemblers.

Terminal-markets buy quantities from shipping-points or farmers and then resell smaller quantities to jobbers and retailers within the metropolitan markets. There are integrated and nonintegrated terminal-markets.

Integrated terminal-markets, like a supermarket chain's warehouse, are one element in a large business. Wholesalers generally work through "prescription buying," in which case the larger company's retail needs are defined and then orders to satisfy the needs are issued to a country assembler or, in some isolated instances, a farmer. Integrated terminal-markets require a consistent supply, a large quantity and a uniform quality. The wholesaler prescribes needs to sellers when making a purchase. (*e.g.,* "Send me 1700 cartons of AAA mediums Tuesday.") When its needs cannot be satisfied through a prescription buy, an integrated wholesaler might purchase from a nonintegrated one.

A nonintegrated wholesaler (or terminal-market) purchases from country assemblers or farmers and, in turn, sells smaller lots to jobbers and retailers.

Jobbers buy in terminal-markets and sell to unaffiliated retail outlets like "Mom and Pop" grocery stores, restaurants, hotels and public institutions. Influence of independent jobbers in the wholesale market has been eroded by integrated supermarket chains and convenience stores. As a consequence, many terminal-market firms now do their own jobbing.

Categories of Wholesale Transactions: Two kinds of transactions are con-ducted in wholesale markets, complete and incomplete. A complete transaction con-sists of exchanging all rights to a product for a cash payment. Example: Alpha sells Bravo one bushel of apples; Bravo gives Alpha $10 in cash; transaction complete.

An incomplete transaction consists of exchanging partial rights to a product for a deferred payment. Example: Alpha sells Bravo one bushel of apples; Bravo gives Alpha five dollars and a promissory note for five more; Alpha retains partial rights to the bushel of apples; transaction incomplete.

There are three kinds of incomplete transactions common in wholesale markets: consignment, account handling and pooling.

In a consignment, the intermediary arranges a sale for a percentage of the price. Example: Bravo talks Charlie into buying 100 bushels of Alpha's apples for $500. Bravo keeps 10 percent of the sale price— or $50— for his commission.

In account handling, the intermediary sells the product for a fixed charge. Example: Alpha sells apples to retailers who shop at Bravo's warehouse. Bravo charges Alpha 20 percent of the selling price for each carton of apples sold in the warehouse.

In pooling, the intermediary combines crops of different farmers into one unit and remits payment to each, based on an average selling price. Example: Bravo combines apples of Alpha, Charlie and Delta; puts all apples into "Bravo's Best" cartons; and then sells the cartons, for different prices, to Foxtrot, Golf and Hotel. Bravo calculates the average selling price for all apples and remits payments based on the average price to Alpha, Charlie and Delta.

Generally speaking, the wholesale sector evolved to service large farms and mass markets. Metrofarmers typically do not have a strong position in selling to large integrated wholesalers because most lack the capacity to satisfy the terms of a prescription buy. Nevertheless, the wholesale sector does provide a way for small town farmers to reach big city markets.

Paul Bowers, an *Allium* grower in Grants Pass, Oregon, sells garlic through a large shipping-point firm in the San Francisco Bay area. This wholesaler picks up the crop and pays cash, leaving Bowers with plenty of time for fishing and travel. If Bowers had to sell his entire crop in the small town of Grants Pass, in competition with the rest of the area's garlic growers, the supply would quickly depress prices and reduce the profitability of his farm.

Bruce Dau, a cole and leaf-crop farmer in Santa Cruz County, California, sells his organically-grown lettuce to consumers as far away as Texas with the help of a shipping-point wholesaler who specializes in organic produce.

Benefits of Selling to Wholesalers

COSTS: The principal benefit of selling to wholesalers is reduction in one's immediate marketing expenses. One or two good wholesalers, for example, may be able to buy one's entire crop, thereby freeing time and money for other endeavors. This is a significant benefit to individuals who have other more important jobs or interests.

RETURNS: Though most wholesalers pay on a deferred basis, sales to a good wholesaler may still provide a relatively quick and reliable way to raise needed cash.

HANDLING: Some wholesalers will contract to harvest, grade, package and store a crop. Given wholesalers' ability to generate economies of scale price savings, their services can result in a significant reduction in one's handling costs.

CREDIT: Some wholesalers provide production financing through special grower-shipper partnerships. However, large integrated wholesalers typically do not enter into partnerships with small-scale farmers unless the crop is judged to be a very special one.

MARKETING ASSISTANCE: Some wholesalers will provide promotional services like point-of-purchase displays which can help sell products and build name recognition.

Limitations of Selling to Wholesalers

RETURNS: The average farmer is said to receive only 30 cents out of every consumer dollar, with the remaining 70 cents going to intermediaries. The middleman's take, in other words, can reduce a 100-acre farm to a 30-acre farm, a 10-acre farm to a 3-acre farm and so on.

CONTROL: Small-scale sellers say wholesalers often pay less than is due, use products as loss leaders or delay payment until prices drop and invoices can be written at new, lower prices. The bottom line is small-scale sellers in the wholesale sector are price takers, instead of price givers.

COMPETITION: It costs more to sell a small volume than a large one. For example, a wholesaler's long-distance telephone expenses would likely be the

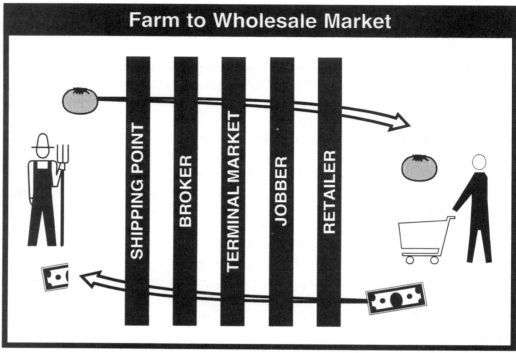

FIG 4.4 The wholesale market consists of intermediaries who buy large units of a product for resale in smaller units to other wholesalers or retailers.

same for selling 10 bushels of apples as it would be for selling 1,000 bushels. When the long-distance phone bill is amortized over total volume, however, it becomes significantly less expensive to sell 1,000 bushels than 10 bushels. Since marketing costs, like telephone bills, are greater for small volumes than large ones, many wholesalers avoid purchasing from small-scale sellers in favor of large-scale ones.

STANDARDS: The prescription buying practices of large wholesalers, like supermarket warehouses, dictate quality and quantity standards. Small-scale sellers who cannot comply with standards of the mass market often receive the lowest price— if any— for their products from wholesale buyers.

INFORMATION: It is difficult for a small-scale producer to know exactly what happens to his crop once it enters the wholesale sector. Wholesalers likely do not have the time it takes to keep everybody informed and so the smallest accounts go without information. A lack of information can create unprofitable and disruptive misunderstandings between a farmer and a wholesaler. Misunderstandings have left many small-scale producers with a healthy distrust of the wholesale sector.

Enter the Wholesale Sector

HAVE THE RIGHT PRODUCT: Successful metrofarmers gain entry into wholesale markets by offering products in which the economies of scale benefits are relatively small or in which the market demand is not sufficient to attract large producers. This is not as difficult as it might seem. Consider, for example, the ubiquitous tomato:

Wholesalers buy large quantities of mass-produced tomatoes in the growing regions of California, Mexico and Florida and then ship the tomatoes into the terminal-markets of metropolitan areas. Tomatoes were hybridized to facilitate mass-production and distribution. As a consequence, tomatoes are now picked green, gassed into redness and then bounced into the supermarket with the texture and taste of tennis balls.

Solar Greens Inc., of Redding, California, entered the market with a vine-ripened tomato produced year-round in hydroponic greenhouses. Consumers tasted the vine-ripened tomatoes and deemed the taste superior to the gassed varieties from Mexico. A wholesaler saw this activity and picked up Solar Greens' tomatoes for distribution to the region's retail trade. Solar Green's working relationship with the wholesaler was made possible by the leverage of its vine-ripened quality.

BE RELIABLE: Profitable farmer-wholesaler relationships require a high degree of reliability. Farmers rely on wholesalers to provide a steady market; wholesalers rely on farmers to provide a steady supply. Many small-scale farmers, especially those growing as a hobby, drift in and out of the market. Many wholesalers consequently avoid doing business with all small-scale farmers. To win over the wholesaler, demonstrate you can be counted on to deliver a steady supply with a uniform quality.

BE INFORMED: The structural constraints small producers face when entering a mass market are often compounded by poor decision-making. The principal cause of poor decision-making is a lack of information. Become informed. Learn the wholesaler's policies for pricing, grading and paying. Find some way of cross-checking the validity of information with other sources in the wholesale sector.

Review the Wholesale Sector

Though the wholesale market tends to be dominated by the prescription buying practices of large integrated marketers, two kinds of opportunities still exist for metrofarmers. The first is convenience: One or two good wholesalers can buy an entire crop, leaving you time for other pursuits. As Paul Bowers, the aforementioned

Allium grower in Grants Pass, Oregon, discovered, wholesalers mean more time for fishing and less time working. The second is market share: A list of good wholesalers can enable you to capture a larger share of the market than would otherwise be possible. Though operating on a comparatively small-scale, Gerd Schneider (Part IV, Conversations) was able to become a significant player in California's horticultural industry by selling through a well-tended list of wholesalers.

SURVEY THE RETAIL SECTOR

The retail sector consists of intermediaries who buy large units from wholesalers and resell individual or family-sized units directly to consumers.

The many retail markets open to metrofarmers include independent supermarkets and grocery stores, restaurants, produce stands, landscape services, nurseries, government institutions and co-op buying clubs. Large integrated retailers, such as supermarket chains, rarely purchase directly from small-scale farmers. (For an example of an exception to the rule, read Jan LaJoie in Part IV, Conversations.)

Transactions conducted between farmers and retailers are similar to the ones conducted between farmers and wholesalers. Short term transactions, like a one time sale, are most often cash-on-delivery. Long term transactions, like the weekly servicing of an account, are often paid on terms, such as net ten days.

Though retail trade is becoming dominated by large supermarket chains and convenience stores, many consumers will always shop for locally-grown products at locally-owned stores. One retail store in this neighborhood, for example, specializes in products of local farmers. During its first years in business, the store's parking lot was filled with old vans decorated with fluorescent paint. Today its parking lot is filled with shiny new luxury cars as well.

Benefits of Selling to Retailers

RETURNS: Small-scale producers can generate more income by eliminating wholesalers and selling direct to retailers. However, this is not free money. One must first assume transportation and sales responsibilities of the wholesaler before benefiting from additional returns.

CONTROL: Farmers gain more control over the marketing process by eliminating wholesalers. In addition to the obvious financial benefits, increased control also yields substantial emotional benefits. As the sugar beet farmers discussed in the introduction to this chapter learned, few conditions are as frustrating as being at the mercy of an unfriendly intermediary.

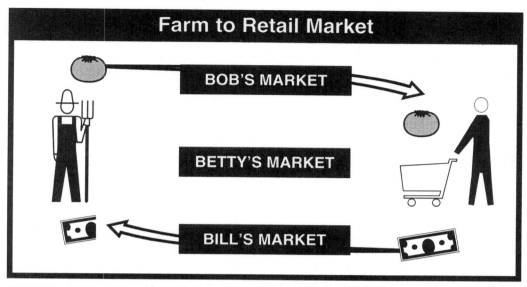

FIG 4.5 The retail market consists of intermediaries who buy large units from wholesalers or farmers for resale in family or individual-sized units to consumers.

MARKET STANDARDS: By selling direct to retailers, small-scale producers can avoid many of the market's legal standards. Standards, as mentioned, are typically directed toward developing a market in which a steady supply, a large volume and a uniform quality are the norm. One locally-owned retail store in the neighborhood buys "free-range" chicken eggs from a nearby farm. Since no two of the eggs are alike, an integrated wholesaler would be reluctant to accept the eggs. The local market, on the other hand, does a brisk trade in the eggs because they taste good and are free of chemical growth stimulants.

RETAILER RELATIONS: Retailers receive several economic benefits by purchasing direct from farmers: They reduce costs for fresher, higher-quality products by reducing or eliminating transportation and packaging costs and enhancing goodwill by supporting local business.

CONSUMER RELATIONS: Consumers realize economic and psychological benefits by purchasing locally-grown products from retailers. They pay lower prices for fresher products, enjoy emotional rewards of consuming locally-grown products, and participate in financial benefits of circulating money through the community instead of through a distant financial institution.

Limitations of Selling to Retailers

SALES VOLUME: Small-scale farmers must contend with storage constraints when trading with local retailers. Retailers trade in individual or family-sized units.

Since most farmers must service a list of retail accounts, they likely do not have capacity to store large surpluses until they can be placed on the sales floor. Storage constraints limit the amount of products which can be sold to each retailer. Since most farmers cannot sell all of their crop to one retailer, they must service a list of retail accounts on a regular basis.

TRANSPORTATION COSTS: The structural constraint mentioned above means individual sellers must spend more time on the road. Somebody must pay the costs and, as Charles and Lois Heyn discovered, they can be prohibitive. The Heyns, who lived 30 miles away from several accounts, switched to selling direct to consumers when transportation costs, in both time and money, proved too much.

SEASONAL CONSTRAINTS: Whereas a local farmer supplies products during the growing season, farmers in distant growing regions supply year-round through wholesalers. Retailers who are open for business all year rely on this availability. If a transaction forces a retailer to choose between a year-round wholesaler and a seasonal farmer, the business might well go to the steady supplier.

RETAILER COSTS: Retailers working directly with farmers may experi-ence the increased costs of locating enough product to supply the demand, of coordinating the timing of purchases and deliveries and of making the required communications and transactions. Increased costs may reduce or eliminate the benefits retailers might otherwise gain in working with uncooperative local farmers.

MASS-MARKET CONSTRAINTS: Local farmers may be subject to state and federal standards when shipping to retailers. The California Department of Agriculture, for example, governs grading, packaging, labeling and inspection of many farm products. Federal and state marketing orders, councils and commissions also restrict quantity and quality of products shipped into the retail sector. Though standards were enacted to facilitate mass-market distribution, they may also regulate a local farmer with six boxes of apples to sell to the corner grocery store.

Enter the Retail Sector

OPEN ENOUGH MARKETS: The retailers' lack of storage facilities means sellers must service a number of accounts with relatively small sales. Consequently, many sellers also sell into other market sectors. Jeff Larkey (Part IV, Conversations) complements his retail sales efforts with sales direct to consumers at local farmers markets. By developing a realistic expectation of how much volume can be moved through each retailer, you can open enough markets to sell your crop.

EVALUATE LOCATION: It costs time and money to service an established list of retail accounts. Costs must be minimized by making certain each retailer is reasonably situated with respect to other retailers on the list. Evaluate how logistics of servicing each retailer will complement handling of the overall list.

BE RELIABLE: Retailers need a steady supply and a consistent quality. If a customer buys a particularly attractive bouquet of flowers at a store one day and returns later to find the flowers are no longer available, she will likely shop elsewhere. Jan LaJoie (Part IV, Conversations) demonstrated her reliability by servicing each retail account personally. "Customers know who I am when I call. I don't have to say, 'This is Jan from the Plant Carrousel.' Sometimes I don't even have to say, 'This is Jan.'"

BUILD A PARTNERSHIP: A farmer/retailer business relationship is a partnership. Only with effort and good intentions do partnerships work smoothly at the outset. Invest time and effort to ensure that all of the partnership's terms are clearly understood at the outset and that a procedure exists for resolving differences.

Review the Retail Sector

The retail sector offers many opportunities for farmers growing in or near a population center. Three of the five farmers featured in Part IV, Conversations sell to the retail market: Schneider sells to nurseries and government institutions, LaJoie to local supermarkets (including integrated supermarket chains) and Larkey to local independent grocery stores.

SURVEY THE DIRECT SECTOR

Direct marketing is the sale of individual or family-sized units directly to consumers.

Though selling direct has not been in the forefront of mainstream commerce during recent years, indications are it may play a very important role in the future. Indications include the ever-increasing cost of highly processed mass-market food, its doubtful taste and nutritional qualities, the possibility of its being contaminated with agricultural chemicals and the increasing frequency of product tampering. Consumers hear and read about the conditions and then consider ways to buy directly from farmers operating in or near a population center.

This section begins with a look at general benefits and limitations of selling direct. It then examines seven popular direct markets and discusses logistics of operating them.

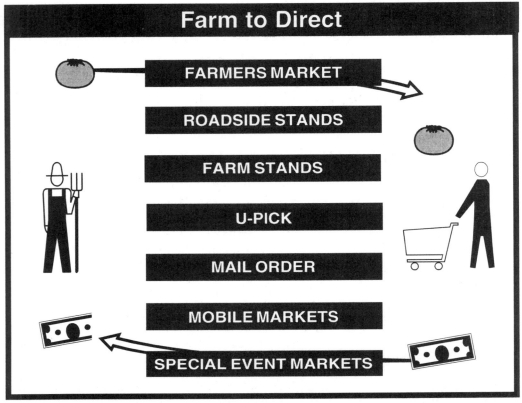

FIG 4.6   The direct market consists of farmers who sell family or individual-sized units directly to consumers.

Benefits of Selling Direct

COSTS AND RETURNS: Sellers reduce costs and increase returns by assuming the marketing functions of wholesalers and retailers which stand between producer and consumer. Eliminating middlemen is always high on the list of ways to generate more income from less production.

CONTROL: Sellers gain more control by selling directly to consumers. Selling direct allows for sale of off-grade products and odd-sized lots which simply could not be sold through the prescription buying practices of the mass market. Control allows farmers to be price setters, instead of price takers.

COMPETITION: Small-scale sellers avoid many of the competition's economies of scale advantages by selling direct. A farmer with ten thousand acres of tomatoes, for example, would not enjoy significant advantages over a farmer with ten acres when selling direct at a farmers' market.

INFORMATION: Selling direct gives farmers many opportunities to stay informed. They can judge consumer reaction to product quality, discover demand for other products and tell consumers how the products were grown and ways in which they can be used.

OPPORTUNITIES: Selling direct often allows one to develop supplemental marketing efforts, such as sales to retail stores or wholesalers. For example, Jeff Larkey (Part IV, Conversations) sells direct at local farmers markets two days per week and then to local retailers and wholesalers the remainder of the time.

SELF RESPECT: Selling direct allows one to benefit from the non-pecuniary rewards of being self-employed. Chief among benefits is the dignity derived when selling one's high-quality product directly to an appreciative consumer. Judging from smiles and attitudes one sees at farmers markets there must be a significant amount of value in self-respect.

Limitations of Selling Direct

RESPONSIBILITIES: Eliminating middlemen means assuming responsibilities of transportation, advertising and publicity, meeting the public, and making the sale. Individuals with other high-paying jobs may feel these additional responsibilities may cancel the financial benefits of eliminating middlemen.

DIPLOMACY: To be a consistently successful direct seller, one must first sell one's self, then the product. The diplomacy required to achieve the two objectives includes being friendly when feeling out of sorts, taking time to inform and convince, and smiling politely when a potential customer says, "Not this time." Some individuals simply do not have the ability to be consistently diplomatic.

INTERRUPTIONS: Some forms of direct marketing, such as farm stands and U-Pick, require the farm be opened to the public. As marketing efforts become more and more successful, they bring more and more people to the farm. Tending to the public is an inconvenience some farmers would rather avoid.

Farmers Markets

A farmers market is a gathering of two or more farmers who sell directly to consumers. (It is also called open-air, curbside, community or green market.) Farmers markets are as old as commerce and continue to provide many small-scale farmers with their primary means of selling a crop.

It appears that farmers markets are becoming increasingly more important, even in the highly concentrated markets of major metropolitan areas. Farmers markets give consumers the opportunity to buy better products at less cost. They give small-scale farmers the chance to sell at retail prices, thereby generating more income from less production. And they provide a good promotional vehicle for

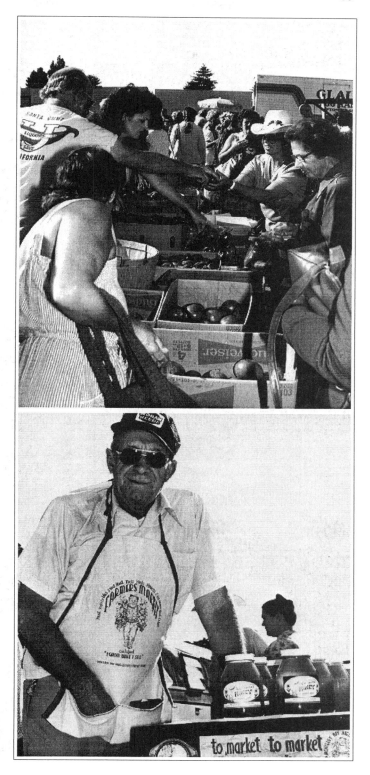

FIG 4.7 "To market, to market,
to the farmer's market."

inner-city neighborhoods and shopping malls.

ORGANIZATION: There are two kinds of farmers markets: peddler and certified. Peddler markets are for-profit businesses which include the sale of nonfarm products. Included in this category are community flea-markets and swap-meets. The world famous Los Angeles Farmers Market is an example of a peddlers market because its merchants sell many products not produced on farms.

Certified farmers markets are nonprofit businesses organized by a group of farmers or community groups such as the Chamber of Commerce. Products sold by participating farmers are certified as being produced by the participating farmers. The resulting nonprofit status allows the market to take advantage of public facilities, the free publicity available through local media and legal exemptions from certain government standards. And finally, it enables the market itself to avoid being taxed.

Certified markets are governed by a board of directors which establishes guidelines and hires a manager. The manager recruits enough farmers to ensure a suitable product line, recruits customers through advertising and publicity and supervises the day-to-day operation of the market itself. The success of a market is, to a great extent, determined by the capabilities of its board of directors and manager.

Both kinds of farmers markets are frequently used by community service groups, like the Chamber of Commerce, to draw people into target neighborhoods or shopping districts.

LOCATION: A market's location is perhaps even more important than the abilities of its management in determining its chances for success. It should be close to population centers and readily accessible. It should have plenty of parking and a clean, enjoyable environment. A centrally located collage parking lot is a popular location for certified markets.

LEGAL RESTRICTIONS: The legal restrictions which govern farmers markets vary from community to community and from state to state. Common restrictions include the need for state certification, zoning variances, business licenses, certified scales and waivers for state-mandated container standards.

FACILITIES: The facilities used in operating farmers markets vary from large enclosed structures with individual stalls to parking lots in which farmers park and set up portable tables. Sanitary facilities required by law are provided by most markets.

CUSTOMERS: People shop at farmers markets for many reasons. Some are looking for lower prices while others are looking for fresher, better-tasting food. Some are out to enjoy the market's carnival-like atmosphere while others are seeking to learn more about the food they eat.

PRODUCT LINES: Many kinds of farm products are put up for sale at farmers markets. Some farmers present a line consisting of a number of popular items like lettuce, tomatoes, peppers, onions, small fruits and nuts. Others present only one or two popular staples like eggs or melons. Also popular are honey, jam, flowers and ornamental plants. If a market is not a certified farmers market, the items produced by farmers are combined with just about anything to turn a sale.

MARKETING FACTORS: Farmers selling at a farmers market compete with each other and with other local retailers for the consumer's dollar. Successful sellers take this competition seriously and present their products in a clean and artistic manner. They are friendly and willing to provide their customers with information and a recipe or two. And, of course, they price their products competitively.

COSTS OF SELLING AT FARMERS MARKETS: The farmers' costs include transportation to and from the market, fees and licenses, shelter, labor and selling materials such as tables, scales, signs, cash box and packaging.

RETURNS OF SELLING AT FARMERS MARKETS: Selling through a farmers market allows one to obtain the best price on top quality goods, to avoid market standards which prevent the sale of off-grade or marginal products and to supplement sales to retailers and wholesalers. Farmers markets are an excellent place to raise cash money and, since they force one to deal directly with consumers, provide a way to discover new demand.

Roadside Stands

Roadside stands evolved with the automobile as entrepreneurial farmers set up small retail oases aside the road to provide motorists with a few moments of relaxation and the opportunity to make a purchase or two.

ORGANIZATION: Roadside stands are for-profit businesses operated by the farmer or by a professional manager and staff. They range in complexity from a table set up under a shade tree to a multifaceted corporation headquartered in some air-conditioned office complex. California's Nut Tree Restaurant, located on Interstate 80 between Sacramento and San Francisco, is an example of a small roadside stand which grew into a multimillion dollar business complete with its own airport.

LEGAL RESTRICTIONS: City and county governments may govern roadside stands with restrictions on zoning, sign size and placement, health permits and business licenses. If the stand is small and features only the products of a farm, restrictions will likely be few; if the stand is large and features a varied product line, restrictions will likely be numerous and more complex.

PRODUCT LINES: Products featured at roadside stands must be sufficiently attractive to lure motorists off the road and out of their automobiles. Good sellers include sweet corn, pumpkins, tomatoes, melons, peaches, nectarines and apples.

Smaller roadside stands may feature one or two products, like apples and melons, produced on the farm; larger stands may feature products of the farm,

FIG 4.8 Roadside stands provide motorists with a relaxing diversion and the opportunity to purchase items of unique interest, like locally grown farm products.

products from neighboring farms, products purchased from the wholesale sector and nonfarm products like craft goods and souvenirs.

Product lines for roadside stands should be coordinated closely with sales. Sweet corn, for example, should be planted on a succession basis so fresh corn will be available on a daily basis.

CUSTOMERS: Motorists stop at roadside stands for a variety of reasons. Some are looking for cheaper prices on fresher, better-tasting goods; some stop for a momentary distraction from a long drive; some stop because the stand is convenient; and some stop because they are curious about what kind of farm products are being produced locally.

LOCATION: A roadside stand's location is the most important factor in evaluating its potential to succeed. The most successful stands, like the Nut Tree, are situated on most-traveled roads. However, being on a well-traveled road is only one aspect of a good location. Access to the road is also important. Motorists must be able to pull into and out of a stand's parking lot without creating traffic obstructions.

MARKETING FACTORS: Successful roadside stands lure customers off the road and into the parking lot. There are two tactics for achieving this objective, pricing and presentation.

Pricing at roadside stands is usually aggressive. Many successful stand operators set prices about 20 percent below prices found in area supermarkets. Another common tactic is the "loss-leader," which consists of selling a product at a loss to increase sales for higher-priced goods. Roadside stands near Castroville, California, the "Artichoke Capital of the World," often advertise artichokes for the extremely low price of "20 for a dollar." When customers see how small the chokes are, they often purchase the larger, more expensive chokes instead.

A good presentation to lure motorists off the road includes an attractively designed stand, graphically pleasing signs, cleanliness and the artistic display of products. Some operators increase the attractiveness of their stands by constructing auxiliary facilities like picnic areas and children's play areas. The Nut Tree, as mentioned, has added an airport, a restaurant, children's play areas, a railroad and many other attractions.

COSTS OF SELLING AT ROADSIDE STANDS: Since roadside stands are for-profit retail outlets with a paid staff, costs tend to be higher than costs incurred in other forms of direct marketing. The costs include facility costs, facility depreciation, full or part-time labor, business licenses and insurance.

RETURNS OF SELLING AT ROADSIDE STANDS: Metrofarmers realize several benefits by selling through roadside stands, including a higher return for their crops, more control of their market and reduced packaging and transportation costs. The stand also allows for the sale of off-grade products at favorable prices. Finally, a roadside stand gives one the opportunity to build additional equity, and equity is a marketable asset. The Nut Tree, for example, is now worth far more than the nut farm from which it grew.

Farm Stands

Farm stands are retail markets established at a farm site. They often resemble roadside stands discussed in the previous section. In farm stand marketing, however, the farm itself becomes the market.

ORGANIZATION: Farm stands often operate on a seasonal basis, with stands opening with the harvest and closing when the harvest has been completed. Farm stand operations are often combined with U-Pick: Some crops are sold at the stand while others are picked by customers in the field. Farm stands range in scope from small family operated businesses— Heyns Country Gardens— to large multi-acre tourist centered complexes— Maui Plantation on the island of Maui.

LOCATION: Since it is a destination market, a farm stand's location is not as critical as the location of a roadside stand. However, it should be close enough to a population center to afford potential customers easy access. Many operators, like Charles Heyn, succeed in drawing customers from 30 to 50 miles away.

PRODUCT LINES: Some farm stands feature crops produced on the farm; others feature products of nearby farms and nonfarm products as well. A common strategy is to feature one or two popular items, such as sweet corn, which draw customers in, and a line of less popular items. One very successful farm stand, the Chino family's Vegetable Shop in Rancho Sante Fe, California, attracts customers with its fresh sweet corn. In addition to sweet corn, the Chinos offer an attractive variety of exotic peppers, tomatoes, melons and beans, which gives customers the chance to stock-up for the week.

CUSTOMERS: Customers must make a special trip to patronize a farm stand. The ones who make the trip are likely in the market for good prices on fresh, high-quality products. Another motive is curiosity. Many people want to know more about the food they eat and about people who grow the food.

FACILITIES: Many farm stands are established with little more than a table or two set up in a garage or processing shed. Then, as customer traffic grows, the business is moved into more elaborate facilities. The Chino family's Vegetable Shop consists of a small open shed under some shade trees next to the parking lot. The Gizdich Ranch, an apple and berry operation in Watsonville, California, located its sales facility in a refurbished red barn. Attention should be given to facilities needed to control customer traffic, including an adequate parking lot and signs.

MARKETING FACTORS: It takes a special incentive to lure customers out of their way to a farm stand. Popular incentives include low prices and volume

FIG 4.9 The farm itself becomes the market when a farm stand is opened for retail sales.

discounts. However, if a farm and its products become popular and subject of much talk about town, pricing may easily go the other way. For example, the Vegetable Shop's sweet corn was recently selling at a 350% premium over the price being asked at a nearby supermarket. (Many customers were standing in line to pay the price!)

Advertising is also an important tool for successful farm stand marketing. Larger stands often use local newspapers, radio and television. Smaller stands simply post signs along well-traveled roads. Good service and a helpful attitude always helps spread a stand's reputation via word-of-mouth (as do bad service and a poor attitude).

Finally, a clean, attractive environment provides the best means for coaxing customers out of the congestion of their cities. The Gizdiches, for example, refurbished their red barn and installed a pie kitchen with a colorful picnic area. Customers at the Gizdich Ranch can now sit down for a slice of deep dish apple pie while enjoying the clean, wholesome environment of a working family farm.

GOVERNMENT RESTRICTIONS: The restrictions a local government might place on a farm stand vary with the stand's size and scope. Small stands which feature only the crops produced on the farm will likely have few restrictions. Large stands which feature added attractions, like the Gizdich's pie shop, may have to secure a business license, a health permit and a zoning variance.

COSTS OF SELLING AT FARM STANDS: Most farmers begin farm stand marketing with a minimum investment in facilities and build as the volume of business warrants. Basic costs include facilities and facility depreciation, labor, which is often supplied by the farmer and family, and customer traffic control. Insurance is becoming a major cost in farm stand marketing, as horror stories abound about the customer who slipped on a pebble and sued for the farm.

RETURNS OF SELLING AT FARM STANDS: In addition to the obvious benefits of eliminating middlemen, selling at a farm stand reduces transportation and packaging costs. It allows for sale of off-grade products at higher prices than would be paid in the wholesale and retail sectors. It enables one to increase the efficiency of labor by employing it on farm chores during slack sales periods. And finally, it gives one opportunity to build additional equity into the farm business.

U-Pick

U-Pick is the most direct form of direct marketing. Customers drive to the farm, walk out into the field, harvest the crop, package it, carry it out of the field, pay for it and then drive it home.

ORGANIZATION: U-Pick marketing is often combined with farm stands and roadside stands. The Gizdich Ranch, mentioned earlier, is a good example. Nita and

Vincent Gizdich sell apples and berries to U-Pick customers. In a refurbished red barn they sell apple and berry pies, juices and frozen berries. At a small shop next door to the barn, they do a brisk trade in early farm antiques. During autumn the Gizdich Ranch is a festival of city people enjoying a day at the farm.

There are many U-Pick themes. One is the "Rent-a-Tree" agreement, in which a farmer leases an entire tree to a customer, tends it through the growing season and then surrenders the fruits of the tree to the customer at harvest time, usually with a guaranteed minimum amount of produce. Variations include "rent-a-cow," "rent-a-row of..." And so on.

Since U-Pick customers harvest the crop themselves, production and sales must be carefully coordinated. Christmas trees, for example, are planted on a succession basis to ensure enough product will be ready each Christmas tree cutting season.

LOCATION: U-Pick farms are generally located close to population centers or along well-traveled roads. However, given customers' need to prepare with appropriate gloves and work clothes, U-Picks generally do not thrive on the casual motorist, as does the roadside stand. Customers may travel a considerable distance to do their picking. U-Picks, like farm stands, are destination markets.

The Gizdich Ranch mentioned above draws customers from Santa Clara Valley and beyond, which is an easy two hour drive. Charles Heyn draws customers 30 to 50 miles and more out of Billings, Montana.

One important element of a successful location is the quality of environment. It is much easier to attract U-Pick customers to a clean, attractive environment than to a dirty, industrial one.

PRODUCT LINES: There are three factors to consider when planning a U-Pick product line.

First is the crop's appeal. A U-Pick crop must have sufficient appeal to induce customers to harvest it themselves. Some crops, like Brussels sprouts, simply do not have the appeal. Favorite U-Pick crops like apples, berries and tomatoes are often picked in quantities large enough for home canning or to provide gifts for friends and neighbors. Other crops, like Christmas trees and pumpkins, are harvested as part of a celebration and are, therefore, somewhat immune to economic considerations which govern everyday purchases.

Second is the logistics of harvest. City people, especially those with children, can be very hard on crops. On a recent visit to the Gizdich Ranch, for example, a family of six harvested its way into an apple tree, taking branches along with apples and nearly destroying the entire tree in their excitement. Some crops are simply too difficult for inexperienced customers to harvest. If a farmer were to turn customers loose on a field of sweet corn, they would peel every ear to see if the corn was ripe, leaving a large number of damaged ears to rot in the sun.

Third is the cropping schedule. Since all U-Pick customers do not arrive on

FIG 4.10 U-Pick customers drive to the farm, walk out into the field, harvest the crop, package it, carry it out of the field, pay for it and then carry it home. U-Pick is the most direct form of direct marketing.

the same day, the cropping schedule must be planned to ensure that a steady supply will be available throughout the season. This can be accomplished with succession planting and use of varieties which ripen at different times. The Gizdiches designed their product line for two seasons of U-Pick per year. Berry crops are featured in summer months; then, after a short breather, apples go into season.

Favorite U-Pick crops include tree and small fruit, Christmas trees, pumpkins and vegetables like beans, peas, tomatoes and cucumbers.

CUSTOMERS: Some U-Pick customers, especially ones who preserve food for the winter, are looking for inexpensive prices for high-quality goods. Others are looking for a pleasant diversion from the high-pressure ways of city living. Perhaps the most important customer base for U-Picks is the city family. The reason, one suspects, is U-Picks provide parents the opportunity to teach children about food while allowing family members to enjoy a pleasant outing together.

Given the preparation and travel time required, U-Pick customers tend to spend more per transaction than they do in other direct markets. However, there are a limited number of people willing to pick and this limitation tends to restrict the number of U-Pick operations in an area.

GOVERNMENT RESTRICTIONS: There are generally few legal restrictions for establishing a U-Pick operation. The local government may require a zoning variance, health permit and a business license.

FACILITIES: The facilities required to operate a U-Pick operation include signs to direct customers off the road, into a parking lot and out to the field, as well as signs which tell what, where and how to harvest. Also required are harvest tools, packaging materials and a checkout stand.

MARKETING FACTORS: Since U-Pick crops must be harvested when ready, advertising is required to get people out to the farm in a timely fashion.

A well-conceived and coordinated campaign of paid advertisements in the local newspaper, radio and/or television will move customers into the fields when needed. Many successful U-Pick operators also use a mailing list compiled from previous customers. The Heyns' U-Pick marketing efforts allowed them to take telephone reservations in January for crops which would not be ready for harvest until June.

One of the most important marketing tools for U-Picks is its environment. A successful environment includes clean, well-groomed fields with amenities, such as play areas for children. The U-Pick farm is an adventure for many city people, and everything a farmer can do to enhance the adventure will help generate more sales.

COSTS OF SELLING THROUGH U-PICKS: There are three costs specific to U-Picks: advertising, because of the need for timely customer traffic; insurance, which can be great because of the propensity of some people to sue at the slightest opportunity; and customer traffic control, because many U-Pickers simply do not understand how to harvest.

RETURNS OF SELLING THROUGH U-PICKS: Having customers harvest the crop allows one to realize a significant reduction in the costs of harvesting, processing, packaging, transporting and selling. The savings mean more income from less production.

In addition, there is the benefit of having people visit and enjoy one's farm. For example, each Christmas season the Fowler Tree Farm, which is a small Christmas tree operation about three miles out of Santa Cruz, California, opens itself to two weekends of extremely hectic U-Pick activity. The entire Fowler clan takes part in supervising traffic (which is backed up for blocks), dispensing saws, collecting money and tying trees onto cars. Though the Fowlers' annual harvest is a serious business, it seems more like a giant family reunion in which the entire community comes to pay its respects.

Mail Order

Mail order marketing has blossomed with computer-generated mailing lists, credit cards, toll-free long distance lines and small parcel delivery services. Sales of food products through private and public mails is now a $500 million a year business in the United States alone.

ORGANIZATION: Mail order businesses are often organized as a simple sole proprietorship or partnership with a small number of popular products. If the products prove successful, the business may be expanded to include a more complex organization with a more comprehensive list of products.

In 1936, a couple of young entrepreneurs in Oregon began selling a beautifully packaged selection of fresh fruit through the mail. Each month Harry and David sent a different fruit to subscribers. They called their enterprise "The Fruit-of-the-Month Club." The Club's annual sales recently topped seventy million dollars, making it an attractive property. The Fruit-of-the-Month Club is now owned by a large corporation.

CUSTOMERS: People patronize mail order for convenience, safety and exclusivity. Convenience is a way to avoid the traffic jams, long lines and the push and shove of supermarket shopping. Safety, especially for the elderly and infirm, is a way to avoid the threat of accidents and muggings. Exclusivity is the quality and originality not readily available on the shelves of mass-market retailers.

PRODUCT LINES: There are obvious limitations to selling through the mails: You can't ship a perishable product like lettuce and expect it to arrive in an edible form. Nor can you ship a low-value product like the radish and expect customers to pay the postage. A mail order product line therefore consists of relatively nonperishable items with a high value. Popular mail order products include fruits, nuts, cheeses, preserved meats, jams, jellies, honey, culinary and medicinal herbs and ornamental and edible plants.

Taylor's Herb Farm is an example of one successful mail order business. The Taylors grow over 100 varieties of culinary, aromatic, ornamental and medicinal herbs on twenty-five acres in Vista, California. Their bulbs, seeds and living plants are shipped via mail and common carriers to customers all over the United States.

GOVERNMENT RESTRICTIONS: Mail order shipments of farm products may be restricted by laws which protect against the spread of harmful insect pests. For example, shipments of fresh fruits from Hawaii to California are prohibited unless they are first cleared through an agricultural inspection. Other possible restrictions include government standards for the grading, packaging and labeling of consumer "gift packages."

LOCATION: As Harry and David demonstrated, the location of a mail order business is not critical. With a relatively efficient post office or common-carrier, even operators in small cities like Medford, Oregon, can establish successful mail order operations which target customers in distant metropolitan centers.

FACILITIES: Since customer traffic is not a factor, there is no reason for the operator of a mail order business to expend a lot of resources on facilities. Many establish their business on the kitchen table and expand into more generous space as the volume of business warrants. The Taylor's Herb Farm mentioned above has grown to the point where it now encourages customers to visit the farm. Quite naturally, the Taylors invested some effort in sprucing up their facilities to create a favorable impression.

MARKETING FACTORS: The two marketing factors to be considered when establishing a mail order business are incentive and advertising.

Incentive means providing a product which is, in some way, special enough so people will send money for it. Package and label designs are also very important. Most mail order products are packaged to give

FIG 4.11 Mail order marketing has blossomed with the technologies of modern times.

customers the impression they have indeed received something special. Harry and David's success was inspired by Fruit-of-the-Month selections which were consistently top quality. And furthermore, the selections were packaged with such taste that Fruit-of-the-Month customers were sold before their first bite.

People will not send money for a product they do not know. Advertising is therefore a critical factor in generating mail order sales. The three elements of a successful advertising campaign are audience, frequency and message. One must reach the right audience enough times with the right message to generate the sales. Popular advertising mediums for small-scale direct mail include the classified section of newspapers and regional magazines. The best medium, of course, is a mailing list of established clients.

COSTS OF SELLING THROUGH THE MAILS: The principal costs associated with mail order include advertising, packaging, shipping materials and labor. Shipping costs are usually paid F.O.B. ("free-on-board" at shipper's warehouse) by the customer. It should be noted that, unless the product is very special, it will take time to build a successful mail order customer base and a profitable return.

RETURNS OF SELLING THROUGH THE MAILS: In addition to the savings generated by eliminating the middleman, mail order affords one an opportunity to sell for the maximum price. (Mail order products are almost always sold for premium prices.) It also allows for sale of other farmers' goods and nonfarm products. Finally, a mail order business affords the opportunity to build additional equity into the farm. Equity can be a valuable asset. R.J. Reynolds, Inc., for example, recently gave Harry and David many millions of dollars for the equity in their Fruit-of-the-Month Club.

Mobile Markets

Mobile markets are retail outlets which move from location to location. Where old world farmers used horse-drawn carriages to slowly sell along city streets; modern farmers use trucks parked at busy intersections. Modern vehicles and busy roads allow for the development of many new marketing opportunities.

Two mobile markets frequent this neighborhood. One sells oranges which are slightly off-grade or, perhaps, are being sold outside of a marketing order. The oranges are excellent juicers and very inexpensive. The other mobile market sells fruits, vegetables and flowers. It is owned and operated by a farm family some 30 miles down the road. The family's market consists of a refrigerated truck, an awning and some tables. Their setup looks good from the road, which helps diminish the fly-by-night image that mobile markets tend to acquire.

ORGANIZATION: Mobile markets are operated by a farmer, a farmer's agent or a peddler. Most are simple sole proprietorships, and some are organized so

FIG 4.12 Mobile marketing is easy to enter in times of surplus and easy to exit in times of shortage. Bottom photo of Moscow tablegrape merchants by Floyd Cardinal.

loosely they do not possess a business license.

CUSTOMERS: People patronize mobile markets for reasons of price, quality, convenience and impulse. Even mobile marketers which do not have fresh, high-quality products attract quality-conscious customers. Recently, while shopping a mobile market parked at a busy intersection, an excited customer was heard to say, "Oh, these peppers look so fresh! I bet they were picked this morning."

PRODUCT LINES: Mobile markets, by nature, are not conducive to featuring a large number of products. Products which are featured should have enough appeal to induce customers to make unplanned purchases. Popular mobile market products include fresh fruits and vegetables, watermelon in summer, pumpkins in the autumn and flowers anytime.

LOCATION: Mobile markets are the essence of niche marketing. Traffic flow and accessibility are two elements of a good niche. Prospective customers should be able to get to the market and back into traffic without causing an accident.

MARKETING FACTORS: As mentioned, mobile markets tend to acquire a fly-by-night image. To lessen the effects of this image, successful sellers use clean decent-looking vehicles, sharp easy-to-read signs and friendly service.

GOVERNMENT RESTRICTIONS: Because they are mobile, many mobile marketers do not purchase a business license or conform to existing zoning ordinances. However, once they have established a good location, they become "sitting ducks" for legal restrictions placed upon normal retail activities. The restrictions may include a business license and sanitation facilities.

COSTS OF SELLING THROUGH MOBILE MARKETS: The mobile market is perhaps the least expensive direct outlet for metrofarmers to develop. Costs, initially at least, are limited to a clean, presentable vehicle, some graphically-pleasing signs and labor.

RETURNS OF SELLING THROUGH MOBILE MARKETS: In addition to the obvious financial benefits of eliminating the middleman, mobile markets can increase one's income by providing an inexpensive sales outlet for production surpluses and off-grade products which otherwise would be difficult to sell. Mobile marketing is easy to enter in times of surplus and easy to exit in times of shortage.

Special Events

All communities celebrate one thing or another with a special activity or event. Examples include county fairs, harvest festivals, religious celebrations, fund-raisers, company picnics and weddings.

Special events provide metrofarmers with many opportunities to sell direct. One common example is Halloween, which affords the opportunity to generate income by setting up a colorful pumpkin patch. The Flower Ladies (Part IV, Conversations) fine-tuned special event marketing into a nine-month per year floral business.

Metrofarmers are in a good position to capitalize on special events because of their proximity to the market and their ability to provide a special crop during a relatively short period of time.

ORGANIZATION: Special event marketing often consists of a partnership between a farmer and an event manager. A civic club, for example, might contract a local farmer to provide sweet corn for the club's booth at the county fair. Brides, for another example, contract The Flower Ladies to grow flowers, construct floral arrangements and deliver the arrangements to their weddings. This partnership always requires management. Both The Flower Ladies and corn growers invest time and effort to coordinate business with customers. Their management of the partnership is a very important element in the success of The Flower Ladies' business.

CUSTOMERS: Special event customers want to enhance their enjoyment of an event through purchase of a product. Consequently, they do not tend to be as cost-conscious as when purchasing products from other markets. Parents and children, for example, increase enjoyment of Halloween by carving pumpkins into jack-o-lanterns. Since Halloween comes but once a year, parents tend to

FIG 4.13 Communities stay viable by celebrating important times with special events, like the Gilroy Garlic Festival in Gilroy, California.

spend more freely when buying their children a pumpkin. Weddings are intended to be a once-in-a-lifetime ceremony. Brides and grooms go to great lengths to enhance enjoyment of their wedding experience by decorating with beautiful flowers. The Flower Ladies' brides often spend more on flowers than on wedding dresses.

PRODUCT LINES: Farm products destined for special events should be engineered to enhance the appreciation of an event. Pumpkins cultivated for Halloween, for example, should be large and round for making good jack-o-lanterns. Sweet corn destined for a booth at the county fair should be picked fresh to ensure its sweetness. Evergreen trees destined for the Christmas U-Pick market should be pruned and shaped to enhance appearance.

LOCATION: Some special events, like Halloween, require production to be located close to concentrations of prospective customers. Other events, like fairs and weddings, do not require close proximity.

GOVERNMENT RESTRICTIONS: Special event markets tend to have few restrictions because of their short duration. Nevertheless, local governments may require a business license and compliance with common-sense requirements like sanitation facilities where applicable.

FACILITIES: Facilities for special event markets depend upon the nature of the event and by logistics of transporting product to consumers. The Flower Ladies' facilities, for example, include a processing shed, a cold storage room and automobiles to transport products to weddings. Half Moon Bay, California, which bills itself as "The Pumpkin Capital of the World," dresses itself up in pumpkin palaces, parades, parties and contests.

MARKETING FACTORS: Ambience and service are two factors which should be considered when selling to special event markets. Products must be engineered to enhance the ambience— or feeling— of the special event. Halloween's pumpkin patches, for example, are often dressed up with caricatures of witches and black cats to enhance the customers' buying experience.

Service is critical because of the limited duration of special events. To ensure the success of their marketing efforts, The Flower Ladies deliver floral arrangements to weddings, place arrangements in proper settings and evaluate the bridal party's reaction to those arrangements. Because timing is so critical, customers, like The Flower Ladies' brides, often contract production ahead of time to ensure delivery.

COSTS OF SELLING TO SPECIAL EVENT MARKETS: Since timing is so critical in special event marketing, the costs of production, processing and transportation may be more extensive than costs incurred when selling to other markets.

RETURNS OF SELLING TO SPECIAL EVENT MARKETS: The principal benefit of selling to a special event is the typically very short marketing season which culminates with a very high return for the crop.

Review the Direct Sector

The successful direct marketer generates more income, gains more control and builds additional equity by assuming functions of wholesale and retail middlemen.

Of the five metrofarmers featured in Part IV, Conversations, only one does not sell direct. Jan LaJoie, who sells principally to the wholesaler and retail sectors, organized her daughters and started a U-Pick strawberry business; Charles and Lois Heyn sell through a farm stand and a U-Pick; Jeff Larkey sells through farmers markets, as well as local retail and wholesale outlets; and The Flower Ladies, as mentioned, have developed a very special special event market.

SURVEY THE COOPERATIVE SECTOR

Small-scale farmers who operate in small towns and villages often lack the ability to reach into lucrative markets of big cities. This limitation may be overcome by forming a marketing cooperative. A cooperative— or co-op— is a nonprofit organization owned and democratically controlled by a group of farmers who share a common interest.

Co-ops have been in existence for many hundreds of years. During the reign of Louis XIV, for example, market gardeners in Paris organized a powerful co-op which protected their "French-Intensive" technology from competitors and gained more favorable prices for their "truck" produce. ("Truck" comes from the French "Troque," which means "trade or barter.") Today, co-ops advance the interests of many farmers in many locations.

Co-ops have the potential to make small-scale producers strong enough to compete in big city markets. Consider, by way of illustration, the Garlic Growers of Southern Oregon— or GGSO. At this writing, GGSO is being organized to serve small-scale garlic farmers of Southern Oregon's Rogue Valley. The Rogue Valley has excellent growing conditions and inexpensive land when compared to San Francisco or Portland. It is an ideal location to grow garlic, onions and shallots.

Many residents of the Valley established small garlic farms to take advantage of location. When their garlic was harvested, however, supply quickly outpaced demand in small town markets and prices plummeted. Consequently, individual farmers looked to lucrative markets of distant big cities. The individuals encountered three obstacles when they arrived in the big city: First, their volume was so small few wholesalers would sit down and talk business. Second, each garlic crop was so unique two crops could not easily be combined into one marketable-sized unit. And third, wholesalers who would sit down and negotiate would not pay a fair price. The individuals did not have much luck selling garlic in the big city.

By organizing the GGSO, the individual farmers of the Rogue Valley provided themselves with the potential to overcome obstacles and win the competition for consumer dollars.

Benefits of Selling Through a Cooperative

By pooling resources, small-scale farmers can obtain economies of scale benefits enjoyed by large competitors. How many benefits can be realized depends upon how many farmers participate, how much each contributes and how well the co-op is managed. A co-op of two farmers may realize a few equipment and labor benefits; a co-op of many farmers may realize a full spectrum of production and marketing benefits.

COSTS: When individuals buy production inputs like seeds and fertilizer in small units, they pay the highest prices. When many combine their needs and purchase large units, they pay the lowest prices.

Garlic requires a substantial amount of phosphorus during formative stages of its growth cycle. If individuals in the Rogue Valley were to buy phosphorus, each would pay full retail. If the individuals combine their orders through a co-op, each would pay only about 60 to 80% of retail. A co-op could save money on a full spectrum of production inputs, thereby reducing costs of farming garlic in the Rogue Valley.

MARKET SHARE: The highest prices are paid in the biggest cities. To sell to wholesale and retail sectors of big city markets, one must provide a large volume and a uniform quality. Individuals in Rogue Valley do not have the ability to satisfy those requirements. Through a co-op, however, each could capture a share of the big city market.

First, a co-op could combine each member's garlic into one large unit. The big city buyer would then exclaim, "You have how many tons? Now we're talking business."

Second, a co-op could standardize production, which would yield a uniform grade of garlic. Standardization would mean each farmer would use the same variety of seeds, soil-building practices, fertilizers, pest controls and so on. A co-op could even regulate the planting schedule of each member, thereby assuring a steady supply of garlic with uniform quality. The big city buyer would then exclaim, "It's all jumbo AAA's? I'm listening."

Finally, by supplying a large volume of garlic with a uniform character and by hiring a part-time sales person, the co-op could eliminate many of the market's intermediaries and sell direct to large integrated supermarket chains. The big city buyer would then say, "Okay, Okay... When can you deliver?"

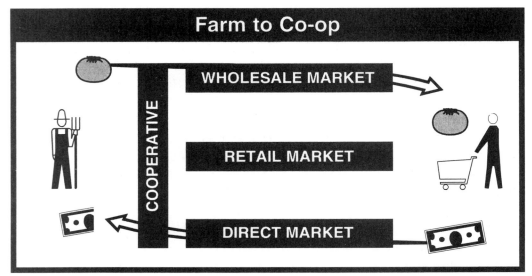

FIG 4.14 Cooperatives are non-profit organizations owned and democratically controlled by individual farmers who share a common interest.

MARKETING: Individual farmers often cannot afford to market products. Consequently, the products of individual farmers may go unsold because buyers simply do not know the products are available. There are four ways in which a co-op's economies of scale can help individual farmers market goods.

First, a co-op can combine advertising needs of its members and buy in volume. If an individual in the Rogue Valley was to buy a newspaper advertisement to promote his or her garlic, it might cost $100. If a co-op was to combine the needs of ten individuals and buy ten advertisements, it might obtain a volume discount of 25 percent. The total cost for the schedule would then be only $750 or $75 per ad.

Second, a co-op's organization enables individuals to share a promotion, thereby reducing costs to a mere fraction. If a co-op was to buy ten advertisements for ten members at $75 each and each advertisement was used to feature the garlic of all ten farmers, then the individual farmer's cost of each ad would be only $7.50.

Third, a co-op can develop and promote a brand name. Building a brand name may not be economically feasible for individuals because of high per-unit costs associated with small volumes. By combining crops of its members, a co-op could establish the "Rogue Valley" brand of garlic. It could then take "Rogue Valley" into the big city and, through its economies of scale, promote it to prospective consumers. After hearing and seeing "Rogue Valley" a number of times, consumers would be more apt to accept it as a name brand garlic.

Fourth, a co-op provides a format by which harvest festivals and other community-oriented promotional activities can be organized. Gilroy, California, became "The Garlic Capital of the World" by sponsoring the Gilroy Garlic Festival.

The Festival promoted Gilroy garlic into minds and onto the tables of many consumers. There is not much an individual in the Rogue Valley can do to overcome the advantageous position Gilroy farmers have achieved. However, a co-op could challenge Gilroy for the title of "Garlic Capital of the World" and, by so doing, spark a competition which would benefit all concerned.

PACKAGING AND FACILITIES: Prices invariably plummet immediately after harvest. Better prices can be earned by storing harvested crops and selling at a later date. Storage, however, requires facilities and good facilities are often beyond the means of individual farmers.

By combining packaging and storage needs of its members, a co-op could build one facility for the entire Rogue Valley. This facility could reduce individual packaging and storage costs and, when combined with higher prices obtained through off-season sales, yield significant benefits for individual farmers.

GOVERNMENT SUPPORT: Small-scale farmers do not receive much direct support from the government. Consequently, many small farmers often feel that market's rules and regulations have been rigged against them. A co-op can help alleviate the sense of inferiority by securing the same kinds of support government affords big operators.

One way in which government supports big operators is by granting official status to their marketing orders, commissions and councils. As discussed earlier in this chapter, the programs are granted six kinds of authority. A determined effort by a co-op of Rogue Valley garlic growers could lead to the granting of a government sanctioned marketing order which would protect the Rogue Valley name from being used by outside producers.

INFORMATION: One often-voiced complaint of small-scale farmers is their lack of information. The reason for this feeling is not difficult to comprehend. For example, if an individual from the Rogue Valley was to sell garlic in San Francisco, he would have to rely on the San Francisco buyer's information. What kind of information would big-city buyers give to sellers just in from the countryside?

A co-op can alleviate the problem by serving as an information-gathering apparatus. Ten farmers can always gather more information than one. The information can then be passed to the co-op's manager, who can disseminate it by word-of-mouth, newsletter and seminar.

TAX BENEFITS: A marketing co-op is a nonprofit organization. Unlike a corporation, whose profits are taxed twice, a co-op pays no income taxes. Profits are distributed to members and members are taxed only once. A co-op, in other words, would allow individual garlic farmers in the Rogue Valley to enjoy the benefits of the corporate form of organization while also allowing them to escape the obvious limitations of double taxation.

Limitations of Selling Through a Cooperative

The effectiveness of a co-op, like the one being organized by the garlic growers in southern Oregon, is determined by the willingness of its members to cooperate. This fact exposes the co-op's principal limitation: Farmers are notoriously independent.

Many of the farmers in the Rogue Valley are refugees from big cities and many paid for their rural independence by sacrificing high-paying city jobs. Individuals often display a reluctance to submit to various kinds of authority which govern a co-op. This is not to say they will not pitch-in and help when needed. On the contrary, help is rarely more easily obtained than from a neighbor in the Rogue Valley. It is to say that individuals, when in a group, are sometimes unwilling to abdicate personal responsibility.

Because of the determined independence of individuals, co-op meetings can easily degenerate into a confrontation in which each member presents an argument for each point. Given this kind of absolute democracy, it becomes extremely difficult for the group to agree on compromises required to standardize a product. Co-ops are effective when individuals trade some of their independence for the good of the group. If individuals do not cooperate, their co-op will not function.

Establish a Cooperative

A co-op can be established where farmers in the same area share the same interests and when they are committed to working for the common good. Following are basic steps to organizing a co-op.

CONDUCT A FEASIBILITY STUDY: A feasibility study provides information for determining whether or not to proceed. First, it defines common interest and determines exactly what it will take to advance this interest. If common interest is a good market for garlic, the study determines where the market is and exactly how much garlic, of which grade, would be required to capture a share of the market. Second, it determines how much it will cost the co-op to achieve the objective and whether or not individual members are willing to pay the price.

ORGANIZE A LEGAL ENTITY: A co-op should be organized as a legal entity. The requirements for establishing a co-op vary from state to state.

California requires incorporation under the California Corporation Laws. The laws require filing articles of incorporation, bylaws and membership agreements. A small co-op can bypass some legal costs by using standardized articles and bylaws for incorporation. (The United States Department of Agriculture provides help and expertise for farmers interested in establishing a co-op.)

MetroFarm Market Comparison

	Traditional marketing system	Direct to retail	Direct to government institutions	Farmers markets	Roadside stands	Farm stands	U-Pick
Harvesting & marketing costs	2-3	3	3	3	2-3	2	1
Labor requirements: harvesting and marketing	1-2	2	2	2-3	2-3	2	1
Grading, packing & other marketing requirements	3	1-3	1-3	1-2	0	0	0
Requirements for licenses, permits fee payments, etc.	0	0	0-1	2-3	1-3	0-1	0-1
Establishment costs	0-1	1-3	0-1	0-2	2-3	1-3	1-2
Importance of produce variety in marketing	0	1-2	1-2	1-3	2-3	0-2	0-1
Importance of cooperation among farmers	0	0	0	0-1	0	0	0
Importance of proximity to customers	0	0-1	0-1	3	3	3	3
Volume limitations on the amount marketable through the alternative	0	2-3	2-3	2-3	2-3	2-3	2-3
Existence of market power advantages to large growers	3	1	1	0-1	0-1	0-1	0-1
Limitations on the type of crop marketable through the alternative	0	0-1	0-1	0-1	1-2	1-2	3
Limitations on the number of new entrants	0	2-3	2-3	2-3	2-3	2-3	2-3

0=NONE   1=LOW   2=MODERATE   3=HIGH

FIG 4.15 A survey of markets open to metrofarmers.

Voting is the mechanism by which a co-op develops its membership agreements. There are various methods for determining voting rights. A common method is the one member, one vote method where each member, regardless of how much he or she produces, has an equal say.

Membership agreements define how the co-op will grow and what services it will provide. They define the duties of the board of directors, the manager and other hired specialists; set meeting schedules and establish procedures for admitting new members; and determine how much each member will be assessed and how they will be paid.

BEGIN OPERATIONS: Finally, the membership, through its board of directors and manager, must get down to business. It must move into an office— or onto a kitchen table— and determine how best to achieve the objectives established in the feasibility study.

Review The Cooperative Sector

A marketing co-op has the potential to reduce or eliminate many economies of scale limitations which reduce the competitiveness of small-scale farms. To realize the benefit, however, individual farmers within the co-op must surrender some of their personal independence for the good of all.

REVIEW

The first step toward winning the competition for consumer dollars is to obtain information. The best information is obtained through a constant survey of the market. A constant survey will reveal the right crop, the right form, the right place, the right time, the right quality, the right quantity, the right price and the right standards and programs. Information will enable you to avoid strength and attack weakness, move from one market to the next and see changes coming in time to make necessary adjustments.

CHAPTER FOUR EXERCISES

i   Title a fresh sheet of paper with "The Survey" and place it in the "Production" section of your metrofarm workbook. Below the title list different kinds of information the market can provide. (See Figure 4.1.)

ii   Survey wholesalers in your market. List their names and companies on fresh sheets of paper and place them in "The Survey" subsection of your metrofarm workbook. Approach several wholesalers in a slack period of their work day, buy them a cup of coffee and ask for information. Below each name list where the market's supply is strong and where it is weak.

iii   Survey retailers in your market. List as many as possible. Select representative candidates from the list and approach them with your list of questions. Below each name list where the market's supply is strong and where it is weak.

iv   Survey direct markets. Visit as many farmers markets, roadside stands, farm stands and other direct markets as possible. Write the name of each business visited on a fresh sheet of paper for your metrofarm workbook. Engage direct marketers in conversation and identify their strengths and weaknesses. List the strengths and weaknesses in "The Survey" subsection of your workbook.

v   Survey your market for production and marketing cooperatives. Do other farmers in the area share your interests. Are they organized? Do they pool resources to gain benefits of economies of scale? List formal cooperatives in your metrofarm workbook. List the names of farmers who could be organized into a formal cooperative.

Evaluating and Controlling Land

Oh beautiful for spacious skies,
for amber waves of grain;
for purple mountain majesties,
above the fruited plain;
America! America!
God shed his grace on thee...

Samuel Ward
"America The Beautiful"

They came to take advantage of a new frontier. Land, plentiful and cheap, was incentive to grow. And grow they did. In 250 years, immigrants cultivated the wide-open spaces out of existence.

Land is not cheap anymore. Indeed, space sells for a premium and immigrants are an endangered species. Nevertheless, near population centers there still exists a frontier of vacant lots, scrub-covered hillsides, greenbelts, floodplains and other small parcels of unused land. It is a frontier filled with incentive to grow.

The following chapter presents a guide to taking advantage of this incentive. The first section consists of a guide to evaluating land as a possible metrofarm resource. The second section consists of a guide to controlling the land.

EVALUATE THE LAND

Farmland is judged by its capacity to generate income. There are environmental, geographical, political and economic factors which govern capacity.

In China these conditions are called "feng shui" (pronounced "feng shway;" literally, "wind and water").

The study of feng shui helps Chinese farmers integrate agricultural practices with factors which govern productive capacity of farmland. If a farm has good feng shui, it will likely succeed; if it has bad feng shui, it will likely fail. The concept of feng shui is rooted deep in Chinese culture and language. The south-facing side of a hill or valley, for example, is called "sheng bian" or "growth-side." In modern times the concept evolved to include factors of a man-

FIG 5.1 The land's ability to produce is rooted deep in the Chinese language. The south-facing side of a hill, for example, is called "sheng bian" or "growth side." Photo by Scott Murray.

made world as well as ones of nature. Feng shui experts are now paid huge commissions to evaluate Hong Kong office space for prospective tenants.

Why should an ancient technology of China be of interest? After all, the Goliaths of modern agriculture now possess the technological wizardry to enable one person to farm a thousand acres! The reason is simple: The ability to evaluate land has a profound effect on the productivity of small farms. Consider some very round figures: Small farms of China, which average one-half acre in size, typically lack the benefits of modern equipment or chemicals, and occupy only about one-tenth the arable land in the United States, sustain a population greater than exists in the United States, Russia and all of Western Europe combined!

This section will help you become an expert in the art and science of feng shui. It provides a guide to environmental, geographical, political and economic factors which govern the land's capacity to generate income.

Consider the Environment

Plants grow and develop in direct response to their environment. It follows some environments are more productive than others. Climate, water and soil are the environmental factors which determine the productive capacity of a metrofarm.

CLIMATE: Elements of climate which plants respond to include day and night temperatures of the air and soil, light intensity and period, relative humidity, wind and rain. The more closely elements match a crop's needs, as defined by the crop's genotype, the more value land will have as a metrofarm resource. Three levels of climate must be considered when judging the productive capacity of a small parcel of land.

Macroclimate: The overall condition of a broad geographical region is called macroclimate. The United States, for example, can be divided into 13 macroclimates, ranging from the cold and dry arctic of Northern Alaska to the warm and humid tropics of Southern Florida and Hawaii.

Mesoclimate: Intermediate variations within macroclimates are called mesoclimates. San Francisco and Los Angeles, though sharing the same macroclimate, have vastly different mesoclimates. Los Angeles is hot and dry; San Francisco is cool and humid.

Microclimates: Local variations of climate are called microclimates. By collecting and redistributing heat, concentrating light, redirecting wind or encouraging precipitation, local influences such as hills, buildings, bodies of water, degree and direction of slope and groves of trees, create climates unique to small areas. There are cool and humid microclimates in the Los Angeles mesoclimate, as well as hot and dry microclimates in the San Francisco mesoclimate.

Large-scale farms locate in areas where broad conditions of climate are favorable. Metrofarms, on the other hand, may operate wherever nature or man has constructed the proper small climate. Consider strawberries, by way of illustration: Strawberry plants produce red, sweet-tasting fruits in many temperate climates. However, scientific studies indicate the plants must ripen in temperatures of 50 degrees F for one week to produce berries with a good strawberry flavor. (Which explains why the first berries of spring are so delicious.)

To satisfy this genetic requirement, large-scale farms congregate in cooler mesoclimates like California's central coast. By selecting a proper microclimate, however, metrofarmers can grow strawberries in many locations, including the arid mesoclimate of Eastern Montana. Charles and Lois Heyn (Part IV, Conversations) parlayed this wisdom into a thriving U-Pick strawberry business near Billings, Montana.

Important information is obtained by evaluating the climate of a small parcel of land, including which crops may be grown most efficiently and the best way to align a production system. Research and practice are required before one can accurately read the climate. As the Chinese have learned, the ability to evaluate a microclimate is a prerequisite for generating maximum return from minimum space.

WATER: Living plants are approximately 90 percent water by weight. In the environment, water is a flow resource which becomes available only through precipitation: Today's precipitation can be used to grow today's crops or stored in reservoirs, snowpacks and aquifers to grow tomorrow's crops. Unlike credit, however, tomorrow's precipitation cannot be used to grow today's crops. Water is, therefore, a prime factor in determining the capacity of land to generate income.

By way of illustration, consider a high-contrast snapshot of farmland in Baja California Sur, Mexico. Southern Baja consists of mile after mile of bone dry hills sparsely covered with tenacious cactus and shrubs. It is said, somewhat jokingly, cattle must graze at 60 miles-per-hour to survive in the hills. However, between La Paz and San Jose Del Cabo, there exists a tiny verdant oasis called San Bartola, which springs out of the drab desert with an abundance of tropical vegetation and streets lined with fruit and vegetable stands. Agricultural enterprises of San Bartola are made possible by groundwater bubbling to the surface from aquifers. Whereas farmland in the vast bone dry hills is valued in terms of square kilometers, farmland in tiny San Bartola is valued in terms of square meters.

This snapshot of extremes is offered to emphasize one irrefutable fact: The productive capacity of land— and its value as a metrofarm resource— is governed by its supply of water.

SOIL: Technically speaking, soil is a resource which covers every parcel of land. However, some soil conditions are better able to support growth of crops than others. Following are five soil conditions to evaluate when measuring the productive capacity of land:

Utility: Soil must be usable. If it is covered by large stones and fallen trees or confined by steep slopes and deep ravines, then time and money will have to be invested to develop the soil into a productive resource.

Stability: Soil

FIG 5.2   The farmer asks, "What must I invest to maintain this fertility?" And, "Will my investment return a profit or a loss?"

must be stable. If it erodes with wind and rain when subjected to cultivation, then corrective measures such as cover crops, contour plowing or terracing will have to be employed.

Texture: Soil must have a good texture, or tilth. It must allow moisture and oxygen to reach plant root zones, store water for future plant needs and allow excess water to drain away. It must facilitate cultural practices such as tilling and weeding. Soil with poor texture must be built-up with organic matter or inert materials like vermiculite and sand.

Fertility: Soil should yield the 13 essential soil nutrients plants need. If the nutrients are not available, they must be added through an application of organic materials or fertilizers.

Chemistry: Soil must be chemically balanced. If it is too acidic or basic, then its *pH* must be corrected with amendments such as lime or organic matter. If soil has sizable concentrations of salts or minerals (*e.g.*, selenium), then it must be washed clean.

Plants live in the soil. If soil conditions are poor, conditions must be improved. The more acres of land a farmer controls, the more soil he must build and maintain. Consequently most large-scale farmers are confined to areas rich in topsoil, like Iowa or California's Central Valley.

The fewer acres a farmer has, the less soil there is to build and maintain. Metrofarmers therefore are not confined to land with deep topsoil. They may locate on small parcels with thin topsoil near valuable markets and then build soil up to ideal levels.

COMPETITION: Each environment contains organisms which compete for the energy in living crops. Competitors range the spectrum from microscopic bacteria to large mammals and include weeds, insects, birds, rodents and even people. One must either control competitors or lose a portion of the crop to them. The competitors in an environment are therefore a factor to be considered when evaluating the capacity of land to generate income.

By way of illustration, consider neighboring flower gardeners in the hills of Northern California. The first had situated his garden in tall grass on a hillside— a grasshopper paradise. The second situated her garden in a stand of tall man-zanita— a natural grasshopper shelter-belt. When the hoppers made their run at the luscious gardens in late summer, the first gardener tried everything to stop them, including repeated applications of the strongest pesticides on the market. The grasshoppers won. The second gardener fought off the hoppers which breached the manzanita shelter-belt with water from the garden hose. She won. As the gardeners learned, competitors can have a profound effect on yield. By evaluating which competitors are dominant in an environment, one can estimate the costs which must be paid to establish pest control.

Crops grow and develop in direct response to their environment. The

Environmental Checklist

S	NA	W	A. CLIMATE
			1. Climate of Area
			A. Length of Growing Season
			1. First frost
			2. Last frost
			B. Direction and Velocity of Prevailing Wind
			C. Average Rainfall
			D. Seasonal Temperature Extremes
			E. Light Intensity
			F. Relative Humidity
			2. Climate of Site
			A. Temperatures
			1. Day
			2. Night
			3. Soil
			4. Air
			B. WATER
			1. Source
			A. Rivers, Streams, and Other Natural Sources
			B. Irrigation Projects and Associations
			C. Depth and Gallon-Per-Minute Averages for Local or On-Site Wells
			2. Quality
			A. Concentration of Natural Minerals
			B. Concentration of Toxic Chemicals
			3. Delivery and Drainage Systems
			A. Irrigation Systems
			B. Drainage Systems (if necessary)

S=Strength   NA=Not Applicable   W=Weakness

FIG 5.3 A list of environmental factors to check when evaluating land for a metrofarm.

S	NA	W	**C. SOIL**
			1. Utility
			A. Size of Arable Parcels
			B. Natural or Man-Made Restrictions
			C. Proximity to Dwelling and Roads
			2. Stability
			A. Evidence of Wind Erosion
			B. Evidence of Water Erosion
			3. Texture
			A. Depth of Topsoil
			B. Air and Water Penetration Characteristics
			C. Water Retention
			D. Water Drainage (Depth of Hardpan)
			E. Organic Content
			F. Tillage Characteristics
			4. Fertility
			A. Natural Fertility
			B. Present Fertility
			C. Estimated Fertilization Requirements
			5. Chemistry
			A. *pH*
			B. Saline Content
			C. Mineral Concentrations
			D. COMPETITORS
			1. Diseases
			2. Insects
			3. Rodents
			4. Birds
			5. Small and Large Animals
			6. Others

FIG 5.3 Continued

environment is therefore an important factor in evaluating the capacity of land to generate income. In fact, a good environment can even negate the lack of an immediate market. The Rogue Valley of Southern Oregon is an example of such an environment.

Small farmers in the Rogue Valley grow *Alliums* like garlic and shallots for sale in big-city markets. Some even sell to the gourmet food sections of department stores in New York City. How is it possible to grow *Alliums* on small farms in Oregon and sell them to upscale retailers in New York City? Because the microclimate is so favorable, the soil so rich and the water supply so plentiful, that plants thrive and yield superior products. Many consumers, especially ones living and working in the grit of big cities, find the products to be an attractive value and are willing to pay the freight.

Consider the Geography, Politics and Economy

Though environmental factors determine a crop's ability to grow, other factors determine a farmer's ability to grow crops. Geographical, political and economic factors must also be considered when evaluating land's capacity to generate income.

Geographical Checklist

S	NA	W	Geographical Location
			1. MARKETS
			A. Processing Facilities
			B. Transportation Facilities
			C. Access to Local Markets
			D. Access to Regional Specialty Markets
			2. COMMUNITY
			A. Character of Neighbors
			B. Character of Town
			C. Availability of Amenities
			D. Neighborhood Associations and Agreements

S=Strength   NA=Not Applicable   W=Weakness

FIG 5.4 A list of geographical factors to check when evaluating land for a metrofarm.

GEOGRAPHICAL FACTORS: People in the business of buying and selling land say there are three factors to consider when judging value: "Location, location and location."

Since land is the physical foundation upon which a metrofarm is built, its geographical location is indeed a prime factor in determining its value. Two aspects of location should be considered: proximity to a good market and proximity to a good community.

Economic benefits of farming close to a good market provide important competitive advantages, including access to processing and transportation, access to the market's wholesale, retail and direct sectors and access to the opportunities of developing auxiliary retail enterprises such as a farm stand. The money is in the market and a good market, close at hand, can significantly increase the capacity of land to generate income.

Metrofarmers, by virtue of their business, are tied to the community. A good community, populated with friendly, cooperative neighbors, can provide important competitive advantages such as basic services and a comfortable place in which to live and work. A bad community, populated by unfriendly, uncooperative neighbors, can drive the cost of business to prohibitive heights. Real estate people are correct: "Location, location and location."

POLITICAL FACTORS: Individuals control land in partnership with the society in which they live. Society always withholds certain property rights, including the right to create and enforce laws, the right to tax and the right to buy when deemed in the public interest (*i.e.,* eminent domain).

Society's rights can have a profound effect on the capacity of land to generate income. The rights may include restrictive covenants placed on deeds, property easement restrictions, zoning laws, building codes, health codes, tax rates and business licensing requirements.

The more populated an area becomes, the more restrictions will be placed on the use of land. Costs of political restrictions must therefore be considered when determining value of land. Imagine, for one wild example, the political costs of establishing a hog or turkey farm midst the well-pruned landscape of an upscale suburban neighborhood!

ECONOMIC FACTORS: Farms are business enterprises and farmers, entrepreneurs. In business the measure of productive capacity is the margin between what something costs and what it will return. There are many economic factors associated with a parcel of land which will affect this margin.

A good place to begin an evaluation is with competition for the land. If demand is high, then the monthly costs of using the parcel will likely be high. If demand is low, one should be able to use the parcel for little cost. Many parcels of land near a city cannot be developed (*e.g.,* green belts) and may therefore be metrofarmed for a small monthly cost. Jeff Larkey (Part IV, Conversations) leased

Political Checklist

S	NA	W		
			1.	LEGAL DESCRIPTION OF PROPERTY
				A. Restrictive Covenants
				B. Easements
			2.	ZONING RESTRICTIONS
				A. Limits on Agricultural Use
				B. Building Codes
				C. Possibility of Variances
				D. Others
			3.	HEALTH AND SAFETY CODES
				A. Sewage Disposal Restrictions
				B. Agricultural Chemical Restrictions
				C. Others
			4.	BUSINESS CODES
				A. License Requirements (on-site sales, etc.)
				B. Business Licenses
				C. Others

S=Strength   NA=Not Applicable   W=Weakness

FIG 5.5 A list of political factors to check when evaluating land for a metrofarm.

prime bottom land one mile from the city's center for the cost of its taxes.

The land's permanent improvements also affect the margin: Is there a residence in which to live? Are there buildings or other improvements which can be used for production-related chores? Are utility costs reasonable? How much will it cost to pump water? To heat the greenhouse? Where are the nearest gas station and grocery store?

Economic factors like rent, improvements, utilities and services have a direct impact on business margin. Costs can add up over the course of a production cycle and, if not controlled, "nickel and dime" one out of business. Prudent entrepreneurs evaluate economic factors when judging the value of a parcel of land.

Geographical, political and economic factors have a direct impact on a farm's costs. If factors are unfavorable, costs of production may be too high. If factors are favorable, they can even negate the costs of a bad environment. By way

Economic Checklist

S	NA	W	
			1. COMPETITION FOR LOCAL REAL ESTATE
			A. Comparison of Local Asking Prices
			B. Comparison of Actual Sale Price
			C. Previous Sale Price of Parcel
			2. PROPERTY TAX RATES
			3. DOMESTIC IMPROVEMENTS
			A. Dwelling
			B. Roads and Fences
			C. Landscaping
			D. Others
			4. PRODUCTION IMPROVEMENTS
			A. Buildings (e.g., barn, packing shed)
			B. Fences
			C. Irrigation Facilities
			D. Access Roads to Field
			E. Others
			5. UTILITIES
			A. Availability
			B. Costs
			6. PROXIMITY TO SERVICES
			A. Distance to Supplies
			B. Distance to Market

S=Strength   NA=Not Applicable   W=Weakness

FIG 5.6 A list of economic factors to check when evaluating land for a metrofarm.

of illustration, consider Glie Farms Inc. Glie Farms grows 32 different culinary herbs on a desolate, rubble-strewn lot in New York City's South Bronx. Though this land might easily be considered one of the world's worst environments for farming, Glie recently projected sales of one million dollars per year, according to *The Wall Street Journal*. How is it possible to generate one million dollars in sales by farming a vacant lot in the South Bronx? Because New York City consumers, of which there are many, demand fresh culinary herbs and Glie Farms Inc. is close enough to satisfy their demand.

Glie Farms is not alone in farming small parcels of land in or near metropolitan areas. According to a recent Census of Agriculture, the most productive farms in the United States, in terms of dollar value of crops per acre, are ones in the Borough of the Bronx. The second most productive farms are those in the City and County of San Francisco, and they generated an average of $76,421 per acre. And so on.

CONTROL THE LAND

It takes time to grow a crop from seeds. It takes time to build a business from opportunity. Secure enough time; take control of the land. There are two ways to take control; purchase and lease. The way you take control will influence the selection of crops, the design of a production system and even your relations with the neighbors. The best way is the one which will enable you to generate the most income from the least amount of effort and risk.

This section will help you find the best way to control land. It defines ownership and leasehold, discusses benefits and limitations of each and outlines procedures for acquiring a title or lease.

Consider a Purchase

The most complete way to control land is to hold a fee simple title. A title grants exclusive right to use land and is subject only to rights withheld by society. As mentioned, society maintains the right to enact and enforce laws and taxes and purchase property when deemed necessary. The remaining rights may be exercised as the owner sees fit, including the sale, lease or bequeathment to other parties.

BENEFITS OF OWNING: The following benefits of ownership are considered from the perspective of business and do not address the emotional benefits of holding a fee simple title. And emotional benefits can be substantial ones.

Decision-Making: Ownership allows one the maximum amount of freedom to make decisions. The owner can do anything he or she wants with the land so long as it does not infringe upon rights withheld by society.

Equity: Ownership enables one to build long-term equity at the expense of short-term profit. By investing operating revenue into such improvements as buildings, cover crops, soil-conservation terraces and drainage systems, the owner can shelter personal income from taxes and improve the land's productivity and market value.

Crop Selection: Ownership allows the largest number of crop opportunities. The owner can invest in short term opportunities, like a crop of radishes or in very long-term projects, such as a "heritage" crop of black walnut trees. Black walnut veneer now sells for just under one cent per square inch; in 75 years, when the wild black walnuts have all been harvested, the veneer could sell for one dollar. By planting an acre or two of black walnut trees today, the owner's great grandchildren could harvest college educations a few decades down the road.

Inflation: Land is real property. When the value of cash erodes during periods of high inflation, the value of land tends to hold firm. Indeed, the owner can use inflation to pay for land. An annual inflation rate of ten percent, for example, reduces the principal cost of a land by ten percent a year. The principal cost of the mortgage then erodes into an inconsequential pennies-on-the-dollar sum in a very short period of time.

Long-Term Capital Gains: An expanding economy means more people living on the same amount of land. Resulting competition for resources tends to increase the real value of farmland, making it a good long-term investment. Consider the value of Chinese farmland, as described by David Fairchild in *The World Was My Garden*, "Thousands of little gardens crowded each other and covered every available inch of space. Land is so valuable that it must be kept busy producing vegetables every month of the year."

LIMITATIONS OF OWNING: If there are benefits to owning land, there are also limitations. Following are four limitations to ownership which should be considered before land is purchased as a metrofarm resource.

Operating Costs: Farmland is valued by its ability to grow crops. Metropolitan land, however, is often valued as a future housing development, industrial park or shopping center. The market value of prime metrofarm land is consequently very high and monthly interest and principal payments can easily starve a metrofarm business out of operating funds.

Flexibility: Though land may be purchased in a day or two, it often takes much longer to sell. Consequently, should a parcel prove unproductive for one reason or another, the owner may find it difficult to move on to greener pastures. This is especially true during periods of sharp recession, when the seller's high appraised value is confronted by the buyer's expectations of falling prices.

Community Responsibilities: Ownership consists of a bundle of rights and responsibilities, among which is the responsibility to help maintain the health of the community. This often means contributing valuable time and expense to such community organizations as road associations and school boards.

Long-Term Capital Loss: The way to make money is to buy low and sell high. When land is purchased at a high price, however, it will likely lose value during ensuing periods of deflation. Loss may be compounded when other "distressed" properties are put on the market, thereby exacerbating the erosion of prices.

OFFER TO BUY: Buying real estate is a complicated business, especially in metropolitan areas. One should use the technical assistance of a real estate professional and take the following steps in order to use professional services effectively.

Determine a Good Price: As mentioned above, the road to financial success is always: "Buy low and sell high!" But how does one determine what is low and what is high?

Land is valued by what it can produce today and by a vision of what it can produce in the future. The environmental, geographical, political and economic factors discussed earlier govern capacity. Land is also valued by the competition for local resources. The demand is evident in prices paid for comparable properties in the area.

Prices may therefore be evaluated by looking at the capacity of land to generate income and at competitive pressures to own the land. A "good" price occurs when capacity is great and demand is slight. A "fair" price occurs when capacity and demand have achieved an equilibrium. A "bad" price occurs when capacity is slight and demand great.

Arrange Payment: Most buyers finance the purchase of land with a promissory note to the current owner, family or friends, a financial institution or a government agency. A favorable loan may enable one to increase the productive capacity of land. For example, where inflation is high and interest rates low, a smart buyer borrows most of the selling price and lets inflation pay the mortgage. Many variables must be evaluated when determining the best way to finance a land purchase. The most desirable way is the one which leaves you with enough capital to operate the business.

Make an Offer: An offer is a legal bid. The offer should therefore be well-defined and contain the contingencies needed to protect the buyer's interests. For example, a buyer making an offer on a ten-acre parcel of land for a berry farm might well make the offer contingent upon obtaining a zoning variance for a roadside stand.

Search the Title: A title contains the legal description of real property. The smart buyer searches the title for liens, covenants, conditions, easements and water and mineral rights which might adversely affect business operations. A title search is how a buyer discovers exactly what is being purchased.

Close: When the preceding steps have been successfully completed, sit down with all of the principals and professionals and sign the papers.

Owning land is a good way to farm because it provides a stable foundation upon which to conduct business. This fact was demonstrated to me one summer when, as a respite from studies at the University, I took a job on a family-owned wheat and cattle ranch in Central Montana. After several weeks of working with me, the family left for a midsummer fishing trip in the mountains. I suddenly

Offer to Buy

☑	**LEGAL DESCRIPTION OF PROPERTY**	Give a complete and clear-cut description of what you are offering to buy.
☑	**NAMES OF BUYER AND SELLER**	List full legal names of all parties included in transaction.
☑	**PRICE**	Include such items as the amount of interest to be paid, additional payments for assuming an existing mortgage and total price.
☑	**PERSONAL PROPERTY**	Include the personal items you want included in the real property transaction (*e.g.*, tractors, trucks).
☑	**FORM OF DEED**	Specify which kind of deed you will accept from the owner (*e.g.*, warranty deed, quit claim deed).
☑	**DATE AND PLACE OF CLOSING**	List date and location where transaction will be consummated.
☑	**CONDITIONAL CLAUSES**	Specify conditions such as change in a zoning law or completion of financing.
☑	**OTHER CLAUSES**	Specify who is to pay closing costs; how taxes, rents, etc., are to be prorated; who is to pay for termite inspections, etc.
☑	**DEADLINE**	Specify how long the offer will last.

FIG 5.7 An offer to buy is a legal proposal which includes articles listed in this chart.

became the sole operator of a thousand-acre wheat and cattle ranch.

As I went about doing the ranch's work— bailing hay, checking cattle, mending equipment— I marveled at the security the family must have felt in leaving their business and home to a student. I then realized there was simply not much I, nor anybody else, could do to disrupt their security. The family owned their land and ownership is backed by the reign of law. There is nothing more secure than owning productive land.

Consider a Lease

A lease is a contract in which an owner grants temporary property rights to a tenant in exchange for a consideration called rent. This contract is enforceable in a court of law and allows a farmer to control land without owning it. There are five types of farm leases:

Share Lease: Tenant pays owner a share of the crop. Share lease is favored because it allows landowners to participate in management decisions and tenants, with small cash reserves, to share the risks of crop failure and price fluctuations.

Cash Lease: Tenant pays owner a fixed rent in cash or services. A cash lease is preferred because it guarantees the owner a steady income and it gives experienced tenants more flexibility to make decisions and more incentive to increase production. Cash rent is usually less than share rent because the tenant assumes all risks.

Share and Cash Lease: Tenant pays owner a fixed rent and a share of the crop. The combination lease is often used in situations where there are two activities. For example, the owner may lease a dwelling for a fixed rent and land for a share of the crop.

Crop Lease: Tenant pays owner a guaranteed amount of the crop as rent. The crop lease is favored because it guarantees the owner a fixed amount of product to sell and because it protects the tenant from sudden fluctuations in prices.

Management Contract Lease: Owner hires a professional manager to operate the farm. Owner may either pay the manager a fixed sum, in which case the owner assumes all risks, or a percentage of the crop, in which case risks are shared.

BENEFITS OF LEASING: Though owning land may be the most secure form of control, leasing also offers substantial benefits to those who cannot buy:

Operating Costs: Prime land located close to a metropolitan area is often valued as a future residential development or industrial park. Until the potential is realized, however, the owner must pay principal, interest and taxes on idle property. Idle property is often leased as an agricultural resource for a nominal monthly rent which allows the owner to generate income and the farmer to produce without the burden of high principal and interest payments.

Equity: A business, like a parcel of land, provides a means to accumulate equity. Harry and David, for example, focused their capital investments in production and marketing, instead of more land. As a consequence of their investments, Harry and David were able to build the Fruit of the Month Club into a successful nationwide mail order business. And when it came time to "cash in the chips," Harry and David's business was worth far more than the Oregon land upon which it was built.

Management Flexibility: Though it takes time to grow and develop a business, one can usually see within a relatively short period of time whether the business will grow and develop into a worthwhile enterprise. If prospects are not favorable, a lessor can move on to greener pastures with less effort and expense than an owner.

Expansion: Leasing may allow one to expand production without becoming burdened by the expense of huge land payments. For example, a berry farmer might double production with little extra expense simply by leasing the neighbor's vacant lot instead of buying it.

LIMITATIONS OF LEASING: Limitations of tenant farming are well-documented. Indeed, abuse of the owner-tenant relationship has sparked revolutions and destroyed countless acres of topsoil. It is therefore important to consider the limitations before taking a lease.

Equity: A lessor does not benefit from reinvesting operating revenues into long-term improvements in the land, nor from the appreciation of the land's value. And the capital gains from appreciating land values have traditionally been one of the principal economic benefits of farming.

Natural Resources: Tenants with short-term leases cannot invest in long-term improvements like soil conservation terraces and drainage systems. This fact defines an age-old problem of land tenure: Because of the duration and uncertainty of using someone else's land, tenants tend to cash-out resources such as topsoil for short-term profit.

Crop Selection: The number of crops which a tenant farmer can grow is limited by the length of the lease. A year-to-year lease, for example, would preclude developing a tree crop like apples which would not yield for several years. The tenant farmer's lease may therefore limit cropping possibilities to short-term enterprises such as annual ornamentals, vegetables and small fruits.

Management Flexibility: A lease is a legal contract between two consenting parties. This partnership adds one more level of complexity to the farmer's decision-making process. Since both owner and tenant contribute to the production of crops, both have a say in the management of business affairs. And sometimes owners and tenants do not agree.

OFFER TO LEASE: A lease is a legal contract which can be as simple as a word-of-mouth agreement between two individuals or as complex as a signed, witnessed and notarized multi-page document between two corporations. The more

complex the business arrangement between owner and tenant, the more complex
the lease. There are three points to cover when arranging any lease. They are the
legal requirements, rent and the agreement itself.

Satisfy Legal Requirements: As a contract, a lease must conform to certain
legal requirements in order to be valid. Legal requirements include the following:

- Owner and lessor must consent on all terms.
- Owner and lessor must be legally competent.
- The purpose of the lease must be legal.
- A lease agreement for longer than one year must be in written
  form.

Pay Rent: Rental rates are determined by supply and demand. An equitable
rent is one in which owner and tenant share rewards of production in the same
proportion as they share risks. The following computations will provide a fairly
good idea of what a fair rent might be:

- Determine contributions put at risk by each party. Include fixed
  labor, depreciation, taxes, insurance and conservation expenses;
  and variable costs like seeds, fertilizers, fuel, utilities, seasonal
  labor, chemicals and machine work.
- Assign each contribution a cash value.
- Add total cash value of all production inputs.
- Calculate percentage of contributions put at risk by each party and
  then base rent on that percentage.

Sign a Lease Agreement: A lease should protect the interests of owner and
tenant by defining each party's rights and responsibilities. It should be written,
signed, dated and witnessed. And if the lease is complex, it should be reviewed by
a competent legal expert.

In many metropolitan areas, the lease provides many individuals with an
effective way to control land. Gerd Schneider (Part IV, Conversations) is a good
example of a metrofarmer who has leased for many years. When interviewed for
this book, Gerd was asked why he had leased his land, instead of purchasing it. Gerd
answered, "This land is too expensive and my business is too important!"

Gerd's investment strategy prevented him from realizing significant capital
gains from the rapidly inflating California land values. However, being a lessor did
not prevent him from succeeding. Shortly after the interview Gerd was offered a
50 percent interest in a 16-acre nursery in exchange for his business goodwill and
technical know-how. (And highly inflated California real estate prices fell by 35
percent!) By continually reinvesting operating revenue into his business, instead of

Lease Agreement

 Names of owner and lessor.

 Date when lease begins and ends.

 Description of property to be leased.

 Amount of rent to be paid, including when and where it will be paid.

 A list describing those inputs furnished by the owner and those by the tenant.

 A description of how management decisions are to be made.

 A list of rights reserved by the owner, such as the rights of entry and inspection.

 Restrictions on the tenant, such as control of weeds, maintenance practices on buildings and dwellings, and soil conservation practices to be followed.

 How and when the lease will be terminated.

 How the tenant will be compensated for any improvements which will extend beyond the life of the lease.

 A description of rights held by heirs of the owner and tenant should either die or become incapacitated.

 A procedure for settling disputes.

FIG 5.8 A lease is an agreement in which an owner grants temporary rights to a tenant in exchange for rent.

the land, Gerd's equity grew from a one and one-quarter acre leasehold to a 50 percent interest in a 16 acre nursery business.

REVIEW

One way to win the competition for consumer dollars is to generate more production from less land. To succeed you must accurately evaluate environmental, geographical, political and economic factors which govern the productive capacity of land. The more skilled you become in evaluating feng shui, the better your chances become of generating more yield with less land. Another way to win is through the control you exert over the land. Some metrofarmers buy and some lease. Each form of control is governed by economic benefits and limitations. The more skilled you become in evaluating benefits and limitations, the better your chances become of generating more income from less land.

CHAPTER FIVE EXERCISES

i   Title a fresh sheet of binder paper with "Land" and place it in the "Production" section of your metrofarm workbook. Make copies of environmental, geographical, political and economic checklists included in this chapter (Figures 5.3 to 5.6) and include copies in the new subsection of your workbook.

ii   Shop for three attractive small parcels of land in or near your market. Any three parcels will be fine, as long as you think they have potential as a metrofarm resource. To save time, ask a real estate professional for a list of prospects.

iii   Use your checklists to evaluate each parcel of land. Check each item in the lists with *S* for *Strength*, *NF* for *Not a Factor* or *W* for *Weakness*. Carefully weigh strengths and weaknesses of each parcel. Select the best parcel and write a sentence describing why you gave each *S*, *NF* or *W*.

iv   Title a fresh sheet of paper with "Offer to Buy." Evaluate how much it would cost to buy the parcel of land you selected in the previous exercise. Include the downpayment, monthly payments, taxes, assessments, dues and other relevant costs. Write an offer to buy the parcel and make your offer contingent on receiving a zoning variance for a roadside stand.

v   Title a fresh sheet of binder paper with "Offer to Lease." Construct three five-year lease proposals for your parcel. Make one proposal a cash lease, one a share lease and one a crop lease.

Selecting Crops

He began to inspect them, one by one, with great concentration, separating the good from the bad. When he had set aside a large enough pile of good acorns he counted them out by tens, meanwhile eliminating the small ones or those which were slightly cracked, for now he examined them more closely. When he had thus selected one hundred perfect acorns he stopped and went to bed.

Jean Giono, *The Man Who Planted Hope and Grew Happiness*

Selection of a crop is the most important decision in farming. No other decision more thoroughly defines the farm enterprise; no other has more potential for generating profit or loss.

Selection is a complex and demanding task for metrofarmers because there are so many possibilities from which to select. Consider the metropolitan diet, which now consists of an eclectic blend of all the world's foods. We eat huevos rancheros for breakfast, snow peas with shiitake mushrooms for lunch and pasta with pesto sauce for dinner. We decorate our homes with orchids from Thailand, our yards with cactuses from South America and our land with dwarf peach trees from China. We have collected the best plant and animal crops the world has to offer and are hungry for more.

The diversity of metropolitan living stimulates demand for thousands of unique plant and animal products. Metrofarmers, unrestricted by the requirements of mass production and distribution, are ideally situated to supply this demand. They can grow fresh tomatoes for local markets, lemon basil for regional gourmet markets or ginseng for international rare commodity markets. They can establish stands of apple and pineapple guava. They can farm catfish in ponds, llamas in pastures or pheasants in cages. They can grow containers of juniper and roses or flats of floribunda.

The crop is your principal weapon in the competition for consumer dollars. This chapter presents a four-step guide to help you select the right crop. The first step consists of selecting the right category of crops; the second, of selecting the right crop; the third, of selecting the right variety and strain; and the fourth, of selecting the right seeds or starts.

This selection process will help you win the competition for consumer dollars. Be deliberate in taking the four steps and you will, like "the man who planted hope and grew happiness," be rewarded with many good nights of sleep.

SELECT THE CATEGORY

The first step toward selecting a crop consists of determining which category of crops is best. Begin this evaluation by defining your ability to invest. Are you a renter just starting out and in need of a quick turnover on your time, know-how and money? Or are you a landowner capable of making long-term investments? Do you have a full-time job elsewhere which requires most of your time and talents?

Next, read the descriptions of the crop categories offered below and ask two questions of each: How much time, know-how and money will it cost to farm this category? When will the investments be returned?

Finally, match your investment abilities with the category's investment requirements. If you are a renter in need of a fast turnover on your investments, eliminate the tree, shrub and vine crops because they will not yield for years. If you have another job which takes a lot of your time, eliminate the animal, bird and fish crops because they require constant attention. Match "haves" and "needs" to find which crop is best.

There are many crops from which to select. More than 1,000 are listed by category in the charts which accompany this chapter. In the charts each species represents a single crop. Apples are one crop and oranges, another. Each is listed by its common name (*e.g.*, apple) and by its Latin binomial (*e.g., Malus sylvestris*). This Latin binomial identifies a crop's genus and species. Many prefer the binomial to the common name, which often will vary from location to location.

Categories of Prospective MetroFarm Crops

(Refers to Figure 6.6)	(Refers to Figure 6.7)	(Refers to Figure 6.8)	(Refers to Figure 6.9)
CATEGORY I: **TREE, SHRUB AND VINE**	**CATEGORY II:** **ROW, BED AND FIELD**	**CATEGORY III:** **NURSERY CROPS**	**CATEGORY IV:** **ANIMAL, BIRD AND FISH**

CATEGORY I: TREE, SHRUB AND VINE

(Tree Crops)
A. Deciduous Fruits
B. Citrus Fruits
C. Subtropical Fruits
D. Nuts
E. Firewoods
F. Building Woods
G. Animal Fodders
H. Oils, Chemicals, Drugs, Spices

(Shrub and Vine Crops)
I. Small Fruits and Nuts
J. Grapes
K. Lubricants and Rubbers
L. Miscellaneous

CATEGORY II: ROW, BED AND FIELD

(Row and Bed Crops)
A. Vegetables
B. Culinary Herbs
C. Medicinal Herbs
D. Cut Flowers

(Field Crops)
E. Food Grains
F. Animal Fodders and Feed Grains
G. Fibers
H. Green Manures

CATEGORY III: NURSERY CROPS

(Flowering Crops)
A. Small Plants
B. Shrubs and Vines
C. Trees

(Foliage Crops)
D. Small Plants
E. Shrubs and Vines
F. Trees

(Other Nursery Crops)
G. Ground Covers
H. Trees for Bonsai
I. Trees for Christmas
J. Succulents
K. Exotic or Unusual Plants for Pots

CATEGORY IV: ANIMAL, BIRD AND FISH

A. Animal Crops
B. Bird Crops
C. Fish Crops
D. Miscellaneous Small Animal Crops

FIG 6.1 Match your ability to invest with the category's investment requirements.

Tree, Shrub and Vine Crops

This category consists of woody perennials which yield many different kinds of marketable products, including edible seeds and fruits, building and fuel woods, animal fodders, oils, chemicals, drugs and spices.

Tree, shrub and vine crops are long-term investments into which a considerable amount of time, know-how and money must be invested before a return is realized. Land must be cleared, contoured and graded; seedlings planted and watered; growing plants pruned, trained, protected, watered and fed until they reach maturity; and, finally, the crop must be harvested. Though start-up costs may be amortized for tax purposes over the 15 to 75-year life span of a well-maintained project, they must still be paid.

Most tree, shrub and vine crops are labor-

FIG 6.2 Tree, shrub and vine crops are long-term investments into which a considerable amount of effort must be placed before a return is realized.

intensive. Though many breakthroughs have been made in mechanization, no machine can match the intelligence of a person for the tasks of pruning, training and harvesting. Given the prohibitive costs of labor-saving devices, many small-scale farmers still rely on the labor of experienced hands to prune the right branch, train the right leader and harvest the ripe fruit.

Since large commercial growers cannot switch from one tree, shrub or vine crop to another without incurring a considerable loss in production, they often form growers' associations and cooperatives to protect themselves against glutted markets and low prices. Associations can have a profound impact on a competitive marketplace. Their government-backed marketing orders, for example, can go so far as to restrict the sale of certain farm products to specific size standards.

Where conventional cultivation and marketing practices are employed, tree, shrub and vine crops are typically not big income-producers on a per-acre basis. According to a recent report by the San Diego County Department of Agriculture, for example, fruit and nut trees in the County produced only $2,372 per acre, whereas cut flowers generated $48,382 per acre.

High-density metrofarm techniques could increase per acre yield significantly. One example of this technology would be the intensive training, pruning, watering and feeding of a deciduous fruit crop on a dwarf root stock. Though expensive in terms of labor, the techniques, when combined with other sound metrofarm strategies, could increase the total production and productivity of an acre significantly, thereby providing an edge in the competition for consumer dollars.

For an example of a tree crop enterprise, read Larkey in Part IV, Conversations.

Row, Bed and Field Crops

This category consists of annuals and perennials. Subcategories include vegetables, culinary and medicinal herbs, cut flowers, grains, fibers, animal feeds and lubricants.

Row, bed and field crops represent a relatively short-term investment. Effort invested in a crop of radishes, for example, might generate a return in as little as 21 days. In addition, land planted in vegetables one growth cycle might be planted in cut flowers the next. This allows for a considerable amount of flexibility in planning a crop enterprise.

Conventional technologies used in the large-scale production of row, bed and field crops rely on expensive equipment, fuels, pesticides, fertilizers and plant varieties. As the technologies are capital-intensive, most conventional farmers have specialized in a few staple crops like processing tomatoes, wheat, feed corn and soybeans and have increased the size of their farms to benefit from the resulting economies of scale. As a consequence, ever larger farms grow fewer and fewer crops.

Economies of scale give big operators an advantage in production of staples. The remaining row, bed and field crop opportunities, of which there are many, belong to metrofarmers. By combining high-density production techniques with sound metrofarm business strategies, you can take advantage of the opportunities. For an insight into three successful row, bed and field crop metrofarmers, read Heyn, Larkey and The Flower Ladies in Part IV, Conversations.

FIG 6.3 Row, bed and field crops are a relatively short-term investment which allow for a considerable degree of flexibility.

Nursery Crops

This category consists of the seeds and young plants which are, in themselves, marketable products. Subcategories include edible crops like vegetable starts and bare-root fruit trees, ornamental crops like house plants and lawn grasses and special occasion crops like Christmas trees.

Nursery crops can be short or long-term investments, depending upon the plants and the facilities required for their care. Jan LaJoie (Part IV, Conversations)

FIG 6.4 Nursery crops tend to be capital-intensive investments because of the technologies required to care for many young plants in confined environments.

propagates houseplants for sale to local grocery stores. Her investments, including those in several small greenhouses, generate a return in a few weeks. Joe Ellis, of Ellis Farms in Borrego Springs, California, propagates ornamental palms for retail nurseries and landscapers. His investments may take years to mature.

Nurseries tend to be capital-intensive businesses because of the greenhouses and related equipment required to breed, propagate and nurture young plants.

Nurseries also tend to be technology-intensive. One must know how to breed, propagate and force plants; how to grow them out in an unnatural environment like a greenhouse; and how to sell them through a marketing system which, more likely than not, is dominated by large integrated wholesalers. The technological skills are especially important because nurseries produce so many plants in so little space. One small mistake, such as an application of too much soluble nitrogen fertilizer, can result in a major financial loss.

Given a modest degree of technological proficiency, nursery crops provide a good metrofarm opportunity because they generate a large return from a small space. Even big nurseries operate on small parcels of land. In the San Diego County study previously cited, nursery crops generated an average of $30,176 per acre. However, Gerd Schneider and Jan Lajoie, two of the metrofarmers featured in Part IV, Conversations, produced considerably more on a per acre basis.

Animal, Bird and Fish Crops

This category consists of animals, birds and fish which either yield a product of economic value or are valued as products in themselves. Subcategories range from protein producers like beef, chicken and fish, to recreational or ornamental products like horses, peacocks and koi carp.

As investments, animal, bird and fish crops can be short-term, like a hutch of rabbits, or long-term, like a stable of race horses. The category tends to be money and technology intensive, especially where high-density production techniques are employed. Facilities must be constructed to breed, propagate and raise the crop. And a knowledgeable person must be present to ensure the production system—housing, feeds, water, medicines, etc.— is functioning smoothly.

The production of Category IV crops on small parcels of land is a burgeoning industry. In times past, cattle, hogs and chickens were crops produced on an open range. Today, they may be placed briefly on the open range but will likely be finished in factory-like feedlots located on small parcels of land. This same trend is true for fish.

Aquaculture has never proved a big money maker in the United States because of the abundant supply of wild fish. The supply, however, is rapidly falling prey to efficient fishermen and polluted water. Here, next to the sparkling waters of Monterey Bay, there now exists a burgeoning trade in farmed trout, which comes from several acres of concrete beds in Southern Idaho.

Animals, birds and fish as recreational and ornamental products will present metrofarmers with many good opportunities in the near future. Many city people enjoy pets. And the more crowded the environment becomes, the more valuable a pet becomes in the life of its owner.

The four crop categories described above are sometimes combined into one production system. For example, a stand of apples might be grazed by a herd of sheep or intercropped with beds of vegetables. Though diversity in conventional agriculture is fast fading away, metrofarmers are in a position to benefit. After all, the successful market farmers of China and Japan profited from diversity for centuries.

There are other categories of crops to be considered as well. There are green manure crops which improve the nutrient and humus content of soil; cover crops which protect soil from erosion; nurse crops which protect seedlings; supplementary and companion crops which increase production and protect other crops from insects; and catch crops which fill voids in the production rotation. These categories are discussed further in Chapter Eight.

Finally, there are farms which do not fit into any of the traditional farm categories. Near Carmel, California, a farm grows turf grass and ornamental trees. Its customers travel from points all over the world and pay a small fortune

FIG 6.5 Animal, bird and fish crops tend to be capital-intensive investments because of technologies required to manage confined environments, like the beds of this Idaho trout farm.

for the privilege of batting a little white ball around the parcel. Though Pebble Beach is one of the world's most successful agricultural enterprises, few consider it a farm at all.

SELECT THE CROP

As Napoleon the Pig remarked in George Orwell's *Animal Farm*, "All animals are equal, but some are more equal than others." The market decides which plant and animal crops are "most equal" in winning the competition for consumer dollars. It reveals the level of consumer demand, potential for future demand and the competition's strengths and weaknesses in supplying that demand.

The market presents low and high-risk selection opportunities. Low-risk crops consist of staples, such as apples and oranges, which have an established production technology and a stable market. Though a relatively safe investment, staple crops are subject to the strenuous pressures of competition and therefore tend to yield a smaller return. High-risk crops consist of exotics, like cherimoyas and pawpaws, which may be highly profitable because they are new or because the normal swing of supply and demand has, for one reason or another, left them in scarce supply. Though promising a high rate of return, exotics are subject to rapid fluctuations in price and can also return a high rate of loss.

After you have selected a crop from the appropriate category, subject it to a thorough scrutiny to determine whether it is indeed a good opportunity. Following is a series of ten checklists which will help with your evaluation. Though it will cost time to successfully complete the check, it will be time well spent, as demonstrated by the following analogy.

Airplane pilots conduct a series of inspections when flying their machines into the sky. There are preflight inspections, prestart inspections, engine run-up inspections, takeoff inspections, cruise inspections and landing inspections. Each inspection is organized into formal checklists to make certain nothing goes unnoticed. Though completing the system of checklists takes time, it helps the pilot conduct a thorough inspection every time. Recently the pilot of a small single-engine airplane, who was also a full-time pilot with a major airline, took off on a cross-country flight. He did not, for one reason or another, complete his landing checklist and arrived gear-up at his destination.

The crop you have selected is like your airplane. The checklists which follow will help you make certain everything is in order before you take off with your crop. The time you invest in completing the checklists will not be wasted. To the contrary, the effort will save you from arriving gear-up at your destination.

Category I: Tree, Shrub and Vine Crops

Trees
A. DECIDUOUS FRUITS

(common name)	(botanical name)
Apple	*Malus sylvestris*
Apricot	*Prunus armeniaca*
Cherry, Sweet	*Prunus avium*
Cherry, Sour	*Prunus cerasus*
Jujube	*Zyziphus jujube*
Kafir Plum	*Harpephyllum caffrum*
Nectarine	*Prunus persica (v.nectarina)*
Pawpaw	*Asimina triloba*
Peach	*Amygdalus persica*
Pear	*Pyrus communis*
Persimmon	*Diospyros kaki*
Plum	*Prunus domestica*
Plum, Japanese	*Prunus salicina*
Pomegranate	*Punica granatum*
Quince	*Cydonia oblonga*
Quince, Chinese	*Chaenomeles sinensis*

Trees
B. CITRUS FRUITS

Calamondin	*Citrus mitus*
Grapefruit	*Citrus paradisi*
Kumquat	*Fortunella japonica*
Kumquat, Chinese	*Fortunella margarita*
Lemon	*Citrus limonia*
Lime	*C. aurantifolia*
Mandarin	*C. reticulata*
Orange	*C. sinensis*
Orange, Satsuma	*C. nobilis*
Orange, Seville	*C. aurantium*
Pummelo	*C. maxima*

Trees
B. CITRUS FRUITS (CONT.)

Tangelo	tangerine and mandarin X
Tangerine	*Citrus nobilis*
Tangor	mandarin and orange X

Trees
C. SUBTROPICAL FRUITS

Avocado	*Persea americana*
Banana	*Musa paradisica*
Barbados Cherry	*Malpighia glabra*
Capulin Cherry	*Prunus capuli*
Carob	*Ceratonia siliqua*
Cherimoya	*Annona cherimoya*
Citron	*Citrus medica*
Date	*Phoenix dactylifera*
Fig	*Ficus carica*
Guava	*Psidium guajava*
Guava, Strawberry	*Psidium cattleianum*
Jaboticaba	*Myrciaria cauliflora*
Litchi	*Litchi chinesis*
Longan	*Euphoria longana*
Loquat	*Eriobotrya japonica*
Limequat	Lime and Kumquat X
Mango	*Mango indica*
Mangosteen	*Garcinia mangostana*
Natal Plum	*Carissa grandiflora*
Olive	*Olea europaea*
Papaya	*Carica papaya*
Passion Fruit	*Passiflora incarnata*
Pineapple Guava	*Feijoa sellowiana*
Prickly Pear	*Opuntia tuna*
Sapote	*Diospyros ebenaster*
Surinam Cherry	*Eugenia uniflora*

FIG 6.6 A sample of tree, shrub and vine crops to consider for metrofarming.

Trees

D. Nuts

Almond	*Prunus amygdalus*
Cashew	*Anacardium occidentale*
Chestnut	*Castanea dentata*
Hazelnut	*Corylus americana*
Hickory	*Carya ovata*
Macadamia	*Macadamia ternifolia*
Pecan	*Carya pecan*
Pinyon	*Pinus cembroides (v edulis)*
Pistachio	*Pistacia vera*
Walnut	*Juglans regia*

Trees

E. Firewood

Birch, Yellow	*Betula lutea*
Birch, Sweet	*Betula lenta*
Eucalyptus	*Eucalyptus (80+ species)*
Juniper, Rocky Mountain	*Juniperus scopulorum*
Juniper, Western	*Juniperus occidentalis*
Oak	*Quercus (60+ species)*
Pine	*Pinus (60+ species)*
Pinyon	*Pinus cembroides*
Poplar	*Populus (40+ species)*
Sycamore	*Platanus occidentalis*

Trees

F. Hardwoods

Alder	*Alnus rubra*
Ash	*Fraxinus (6 species)*

Trees

F. Hardwoods (cont.)

Basswood, American	*Tilia americana*
Basswood, White	*Tilia heterophylla*
Birch, Yellow	*Betula lutea*
Birch, Sweet	*Betala lenta*
Black Cherry	*Prunus serotina*
Black Walnut	*Juglans nigra*
Boxelder	*Acer negundo*
Buckeye, Ohio	*Aesculus glabra*
Buckeye, Yellow	*Aesculus octandra*
Chestnut	*Castanea dentata*
Dogwood, Flowering	*Cornus florida*
Dogwood, Pacific	*Cornus nuttallii*
Elm	*Ulmus (6 species)*
Hickory	*Carya (8 species)*
Holly	*Ilex opaca*
Laurel	*Umbellularia californica*
Maple	*Acer (5 species)*
Magnolia	*Magnolia grandiflora*
Mulberry	*Morus rubra*
Oak	*Quercus (28 species)*
Pecan	*Carya pecan*
Persimmon	*Diospyros virginiana*
Yellow Poplar	*Liriodendron tulipifera*
Sweetgum	*Liquidambar styraciflua*

Trees

G. Softwoods

Sycamore, American	*Platanus occidentalis*
Cedar, Alaska	*Chamaecyparis nootkatensis*
Cedar, Port-Orford	*Chamaecyparis lawsoniana*
Cedar, Western Red	*Thuja plicata*

FIG 6.6 Category I: Tree, Shrub and Vine Crops (continued).

Trees
G. Softwoods (cont.)

Douglas Fir	*Pseudotsuga taxiflora*
Fir	*Abies* (8 species)
Hemlock, Eastern	*Tsuga canadensis*
Hemlock, Western	*Tsuga heterophylla*
Larch	*Larix occidentalis*
Pine	*Pinus* (18 species)
Redwood	*Sequoia sempervirens*
Spruce	*Picea* (6 species)

Trees
H. Animal Fodder

Beech	*Fagus grandifolia*
Carob	*Ceratonia siliqua*
Honey Locust	*Gleditsia triacanthos*
Mulberry	*Morus alba*

Trees
I. Oils, Chemicals, Drugs & Spice

Camphor-Tree	*Cinnamomum camphora*
Cedar	*Thuja occidentalis*
Chestnut	*Castanea dentata*
Chinese Tallow	*Sapium sebiferum*
Eastern Redcedar	*Juniperus virginiana*
Eucalyptus	*Eucalyptus globulus*
Larch	*Larix occidentallis*
Laurel	*Laurus nobilis*
Maple	*Acer saccharum*
Sassafras	*Sassafras albidum*
Sweetgum	*Liquidambar styraciflua*
Tung	*Aleurites fordii*

Shrubs and Vines
J. Small Fruits & Nuts

Barberry	*Berberis vulgaris*
Blackberry	*Rubus allegheniensis*

Shrubs and Vines
J. Small Fruits & Nuts (cont.)

Blueberry, low bush	*Vaccinium pennsylvanicum*
high bush	*V. corymbosum*
Boysenberry	*Rubus villosus*
Cranberry	*Vaccinium macrocarpon*
Currant	*Ribes sativum*
Elderberry	*Sambucus canadensis*
Gooseberry	*Ribes hirtellum*
Juneberry	*Amelanchier stolonifera*
Kiwi	*Actinidia chinensis*
Loganberry	*Rubus loganobaccus*
Pomegranate	*Punica granatum*
Peanut	*Arachis hypogaea*
Prickly Pear	*Opuntia vulgaris*
Raspberry	*Rubus ideaeus*
Strawberry	*Fragaria chiloensis*

Shrubs and Vines
K. Grapes

European or Californian	*Vitus vinifera*
Eastern or Concord	*Vitus labrusca*
Southern or Muscadine	*Vitus rotundifolia*

Shrubs and Vines
L. Lubricants, Rubbers

Guayule	*Parthenium argentatum*
Jojoba	*Simmondsia californica*

Shrubs and Vines
M. Miscellaneous

Tea	*Thea sinensis*
Hop, American	*Humulus americanus*
Hop, European	*Humulus lupulus*

FIG 6.6 Category I: Tree, Shrub and Vine Crops (continued).

Category II: Row, Bed, and Field Crops

Row and Bed Crops
A. VEGETABLES

Amaranth	*Amaranthus hypochondriacus*
Arrowhead, Chinese	*Sagittaria sagittifolia*
Arrowroot	*Maranta arundinacea*
Artichoke	*Cynara scolymus*
Artichoke, Jerusalem	*Helianthus tuberosus*
Artichoke, Chinese	*Stachys sieboldii*
Asparagus	*Asparagus officinalis*
Bean	*Glycine, Phaseolus, Dolichos, Vigna*
Bean, Adzuki	*Phaseolus angularis*
Bean, Asparagus	*Vigna sesquipedalis*
Bean, Fava	*Vicia faba*
Bean, Goa (Winged)	*Psophocarpus tetragonolubus*
Bean, Kidney	*Phaseolus vulgaris*
Bean, Lima	*Phaseolus limensis*
Bean, Mung	*Phaseolus aureus*
Bean, Soybean	*Glycine max*
Bean, Yard-long	*Vigna sesquipedalis*
Beet	*Beta vulgaris*
Bitter Melon	*Momordica charantia*
Black-eyed Pea	*Dolichos sphaerospermus*
Broccoli	*Brassica oleracea (v. italica)*
Broccoli, Chinese	*Brassica alboglabra*
Burdock	*Arctium lappa*
Butter Bur	*Petasites hybridus*
Cabbage	*Brassica oleracea (v. capitata)*
Cabbage, Bok Choy	*Brassica chinensis*
Cabbage, Chinese	*Brassica pekinensis*
Cabbage, Flowering	*Brassica oleracea*
Carrot	*Daucus carota (v. sativa)*

Row and Bed Crops
A. VEGETABLES (CONT.)

Cassava	*Manihot esculenta*
Cauliflower	*Brassica oleracea (v. capitata)*
Celery	*Apium graveolens*
Chard	*Beta vulgaris (v. cicla)*
Chick-pea (garbanzo)	*Cicer arietinum*
Chayote	*Sechium edule*
Corn, Sweet	*Zea mays (v. rugosa)*
Cucumber	*Cucumis sativus*
Daikon	*Raphanus sativus*
Dandelion	*Taraxacum officinale*
Endive	*Cichorium endivia*
Eggplant	*Solanum melongena (v.esculentum)*
Garland Chrysanthemum	*Chrysanthemum coronarium*
Fuzzy Gourd	*Benincasa hispida*
Jicama (Yam Bean)	*Pachyrhizus erosus*
Kale	*Brassica oleracea (v. acephala)*
Kohlrabi	*Brassica caulorapa*
Leek	*Allium porrum*
Lentil	*Lens esculenta*
Lettuce, Head	*Lactuca sativa (v. capitata)*
Lettuce, Leaf	*Lactuca sativa (v. crispa)*
Lotus root	*Nelumbo nucifera*
Malanga	*Xanthosoma sagittifolia*
Mushroom, Chanterelle	*Cantharellus cibarius*
Mushroom, Common	*Agaricus campestris*
Mushroom, Enokitake	*Flammulina velutipes*
Mushroom, Morel	*Morchella esculenta*
Mushroom, Shiitake	*Lentinus edodes*

FIG 6.7 A sample of row, bed and field crops to consider for metrofarming.

Row and Bed Crops	
A. VEGETABLES (CONT.)	
Muskmelon	*Cucumis melo*
Muskmelon, Cantaloupe	*Cucumis melo (v. cantalupensis)*
Muskmelon, Honeydew	*Cucumis melo (v. indorus)*
Mustard	*Brassica juncea*
Mustard, Chinese	*Brassica juncea*
Mustard (Peppergrass)	*Brassica japonica*
Okra	*Hibiscus esculentus*
Okra, Chinese	*Luffa acutangula*
Onion	*Allium cepa*
Parsnip	*Pastinaca sativa*
Pea	*Pisum sativum*
Pea, Goa	*Psophocarpus tetragonolobus*
Pea, Snow	*Pisum sativum*
Pepper, Sweet	*Capsicum frutescens (v. grossum)*
Pepper, Chili	*Capsicum frutescens*
Pickling Melon, Chinese	*Cucumis conomon*
Potato	*Solanum tuberosum*
Potato, Sweet	*Ipomoea batatas*
Pumpkin	*Cucurbita pepo*
Pumpkin, Chinese	*Cucurbita pepo*
Purslane	*Portulaca oleracea*
Radish	*Raphanus sativus*
Radish, Chinese	*Raphanus sativus*
Rhubarb	*Rheum rhaponticum*
Rutabaga	*Brassica napobrassica*
Scallion	*Allium fistulosum*
Shallot	*Allium crepa*
Spinach	*Spinacia oleracea*
Squash, Summer	*Cucurbita pepo (v. melopepo)*

Row and Bed Crops	
A. VEGETABLES (CONT.)	
Squash, Winter	*Cucurbita maxima*
Sunflower	*Helianthus annuus*
Tarro	*Colocasia esculenta*
Tomatillo	*Physalis ixocarpa*
Tomato	*Lycopersicon esculentum*
Turnip	*Brassica rapa*
Water chestnut	*Eleocharis dulcis*
Water cress	*Nasturtium officinale*
Watermelon	*Citrullus vulgaris*
Winter Melon	*Benincasa hispida*
Yam, Chinese	*Dioscorea batatas*
Yam, White	*Dioscorea alata*

Row and Bed Crops	
B. CULINARY HERBS	
Angelica	*Angelica archangelica*
Anise	*Pimpinella anisum*
Basil, Sweet	*Ocimum basilicum*
Borage	*Borago officinalis*
Burnet	*Poterium sanguisorba*
Caper	*Capparis spinosa*
Caraway	*Carum carvi*
Cardamom	*Elettaria cardamomum*
Celantro (Coriander)	*Coriandrum sativum*
Celeriac (Celery Root)	*Apium graveolens (v. rapaceum)*
Chervil (Beaked Parsley)	*Anthriscus cerefolium*
Chives	*Allium schoenoprasum*
Chives, Chinese	*Allium tuberosum*
Cicely, Sweet	*Myrrhis odorata*
Cummin	*Cuminum cyminum*
Dill	*Anethum graveolens*

FIG 6.7 Category II: Row, Bed and Field Crops (continued).

Fennel	*Foeniculum vulgare*
Fenugreek	*Trigonella foenumgraecum*
Garlic	*Allium sativum*
Garlic Chives	*Allium tuberosum*
Garlic, Elephant	*Allium ampeloprasum*
Garlic, Society	*Tulbaghia violacea*
Ginger	*Zingiber officinale*
Horehound	*Marrubium vulgare*
Horseradish	*Armoracia rusticana*
Japanese Parsley	*Cryptotaenia japonica*
Lovage	*Levisticum officinale*
Majoram, Sweet	*Majorana hortensis*
Mint	*Mentha* (10+ species)
Peppermint	*Mentha piperita*
Spearmint	*Mentha spicata*
Nasturtium	*Tropaeolum magus*
Oregano	*Origanum vulgare*
Parsley	*Petroselinum crispum*
Pepper, Cayenne	*Capsicum frutescens (v. longum)*
Pepper, Chili	*Capsicum frutescens*
Radicchio (Leaf Chicory)	*Cichorium intybus*
Rocket (Arugula)	*Eruca sativa*
Rosemary	*Rosmarinus officinalis*
Saffron	*Crocus sativus*
Sage	*Salvia officinalis*
Sage, Pineapple	*Salvia elegans*
Salsify (Oyster Plant)	*Tragopogon porrifolius*
Savory, Summer	*Satureia hortensis*
Savory, Winter	*Satureia montana*
Sesame	*Sesamum orientale*

Sorrel	*Rumex acetosa*
Tarragon	*Artemisia dracunculus*
Thyme	*Thymus vulgaris*
Truffle, White	*Tuber magnatum* (v. pico)
Truffle, Black	*Tuber melanos* (v. porum)
Tumeric	*Curcuma domestica*
Watercress	*Nasturtium officinale*
Woodruff, Sweet	*Asperula odorata*

Row and Bed Crops
C. MEDICINAL HERBS

Aconite (Monkshood)	*Aconitum napellus*
Aloe Vera	*Aloe perryi*
Angelica	*Angelica archangelica*
Bee Balm	*Monarda didyma*
Betony	*Stachys officinalis*
Calamus root	*Acorus calamus*
Camomile	*Anthemis nobilis*
Catnip	*Nepeta cataria*
Comfrey	*Symphytum uplandicum*
Cannabis (illegal)	*Cannabis indica, sativa*
Crocus, Autumn	*Colchicum autumnale*
Echinacea	*Echinacea pallida*
Fennel	*Foeniculum vulgare*
Foxglove	*Digitalis purpurea*
Garlic	*Allium sativum*
Ginseng	*Panax schinseng*
Goldenseal	*Hydrastis canadensis*
Larkspur	*Delphinium ajacis*
Lemon Grass	*Cymbopogon citratus*
Lemon Verbena	*Lippia citriodora*
Licorice	*Glycyrrhiza glabra*

FIG 6.7 Category II: Row, Bed and Field Crops (continued).

Row and Bed Crops		Row and Bed Crops	
C. Medicinal Herbs (Cont.)		**D. Cut Flowers (Cont.)**	
Lobelia	*Lobelia inflata*	Bells of Ireland	*Molucella laevis*
Loveage	*Levisticum officinale*	Blue Lace	*Didiscus coeruleus*
Mandrake	*Mandragora officinarum*	Butterfly Flower	*Schizanthus*
		Calliopsis	*Coreopsis* (20+ species)
Mescal	*Lophophora Willimsii*	Candytuft	*Iberis umbellata*
Monkshood	*Aconitum columbianum*	Carnation	*Dianthus Caryophyllus*
Mugwort	*Artemesia vulgaris*	Celosia, Cockscomb	*Celosia cristata*
Mullein	*Verbascum thapsus*	Celosia, Plumed	*Celosia plumosa*
Mustard	*Brassica nigra*	China Aster	*Callistephus chinensis*
Passion Flower	*Passiflora incarnata*		
Penny Royal	*Hedeoma pulegioides*	Chinese Lantern	*Physalis alkenkengi*
Plantain	*Plantago major*	Chrysanthemum	*Chrysanthemum (60+ species)*
Poppy (illegal)	*Papaver somniferum*		
Rabbit Tobacco	*Gnaphalium obtusifolium*	Cleome	*Cleome spinosa*
		Coreopsis	*Coreopsis lanceolata*
Rosehips	*Rosa rugosa*	Cornflower	*Centaurea cyanus*
Rosemary	*Rosmarinus officinalis*	Cosmos	*Cosmos* (4 species)
Rue	*Ruta graveolens*	Cyclamen	*Cyclamen indicum*
Sage	*Salvia officinalis*	Daffodil	*Narcissus* (20+ species)
Tarragon	*Artemesia dracunculus*	Dahlia	*Dahlia* (8 species)
Tobacco	*Nicotiana tabacum*	Daylily	*Hemerocallis fulva (20+ species)*
Thyme	*Thymus vulgaris*		
Valerian	*Valeriana rubra*	Delphinium	*Delphinium elatum (100+ species)*
Wormwood	*Artemesia frigida*		
Yellowroot	*Xanthorhiza simplissima*	Dutch Iris	*Iris* (200+ species)
		Feverfew	*Chrysanthemum Parthenium*

Row and Bed Crops	
D. Cut Flowers	
		Forget-Me-Not	*Myosotis oblongata*
		Foxglove	*Digitalis gloxiniaeflora*
Acroclinium	*Helipterum roseum*	Freesia	*Freesia* (7 species)
Alstroemeria	*Alstroemeria (10+ species)*	Gaillardia	*Gaillardia grandiflora*
		Gladiola	*Gladiolus* (50+ species)
Anemone	*Anemone coronaria (70+ species)*	Globe Amaranth	*Gomphrena haageana*
		Gloriosa Daisy	*Rudbeckia hirta*
Aster	*Aster* (150+ species)	Godetia	*Godetia* (10+ species)
Azalea	*Rhododendron Azalea*	Hollyhock	*Althaea rosea*
Baby's Breath	*Gypsophila elegans*		

FIG 6.7 Category II: Row, Bed and Field Crops (continued).

Honesty (Money Plant)	*Lunaria annua*
Iceland Poppy	*Papaver nudicaule*
Iris	*Iris* (150+ species)
Larkspur	*Delphinium ajacis*
Lavender	*Lavandula vera*
Lily	*Lily Martagon* (50+ species)
Lupine	*Lupinus* (40+ species)
Marguerite	*Chrysanthemum frutescens*
Marigold	*Tagetes* (4 species)
Matricaria	*Matricaria* (5 species)
Mexican Sunflower	*Tithonia* (3 species)
Monarch of the Veldt	*Venidium fastuosum*
Monkshood	*Aconitum napellus*
Painted Daisy	*Pyrethrum roseum*
Phlox	*Phlox drummondi* (50+ species)
Pot Marigold	*Calendula officinalis*
Ranunculus	*Ranunculus asiaticus*
Rose	*Rosa borboniana* (200+ species)
Scarlet Sage	*Salvia splendens*
Shasta Daisy	*Chrysanthemum leucanthemum*
Snapdragon	*Antirrhinum majus*
Star Flower	*Scabiosa stellata*
Statice	*Limonium* (70+ species)
Stock	*Mathiola incana*
Strawflower	*Helichrysum bracteatum*
Sunflower	*Helianthus* (20+ species)
Sweet Pea	*Lathyrus odoratus*
Sweet Sultan	*Centaurea imperialis*
Sweet William	*Dianthus barbatus*

Tritoma (Red Hot Poker)	*Kniphofia* (10+ species)
Tulip	*Tulipa Gesneriana* (60+ species)
Velvet Flower	*Salpiglossis sinuata*
Xeranthemum	*Xeranthemum annum*
Yarrow	*Achillea filipendulina*
Zinnia	*Zinnia elegans* (10+ species)

Field Crops
E. FOOD GRAINS

Amaranth	*Amaranthus gangeticus*
Barley	*Hordeum vulgare*
Buckwheat	*Fagopyrum esculentum*
Corn	*Zea mays*
Corn, Sweet	*Z. mays* (v. *rugosa*)
Corn, Popcorn	*Z. mays* (v. *everta*)
Kafir	*Holcus sorghum* (v. *caffrorum*)
Millet, Broomcorn	*Panicum miliaceum*
Millet, Pearl	*Pennisetom glaucum*
Oats	*Avena* (6 species)
White	*A. sativa*
Red	*A. byzantina*
Rice	*Oryza sativa*
Rye	*Secale cereale*
Sorghum, Grain	*Sorghum vulgare*
Sorghum, Sweet	*Sorghum vulgare*
Soybean	*Glycine max*
Safflower	*Carthamus tinctorius*
Sunflower	*Helianthus*
Triticale	Wheat X Rye
Wheat	*Triticum aestivum*
Wild Rice	*Zizania*

FIG 6.7 Category II: Row, Bed and Field Crops (continued).

Field Crops
F. ANIMAL FODDER & FEED GRAINS

Alfalfa	*Medicago*
Purple Blossom	*M. sativa*
Variegated	*M. media*
Yellow Blossom	*M. falcata*
Barley	*Hordum vulgare*
Birdsfoot Trefoil	*Lotus corniculatus*
Bluegrass, Kentucky	*Poa pratensis*
Bromegrass	*Bromus inermis*
Buffalograss	*Buchlow dactyloides*
Clover	*Trifolium* (10+ species)
Alsike	*T. hybridum*
Crimson	*T. incarnatum*
Red	*T. pratense*
White	*T. repens*
Comfrey	*Symphytum uplandicum*
Corn, Field	*Zea mays* (v. *indentata*)
Fescue, Meadow	*Festuca elatior*
Fescue, Tall	*Festuca arundicanacea*
Lespedeza	*Lespedeza striata*
Lovegrass, Weeping	*Eragrostis curvula*
Millet, Pearl	*Pennisetum glaucum*
Millet, Foxtail	*Setaria italica*
Oats	*Avena sativa*
Orchardgrass	*Dactylis glomerata*
Rye	*Secale cereale*
Ryegrass, Perennial	*Lolium perenne*
Sorghum (Milo)	*Sorghum vulgare*
Sudangrass	*Sorghum vulgare* (v. *sudanese*)
Sweetclover, White	*Melilotus alba*
Sweetclover, Yellow	*M. officinalis*
Timothy	*Phleum pratense*

Field Crops
F. ANIMAL FODDER & FEED GRAINS (CONT.)

Vetch	*Vicia sativa*
Wheatgrass	*Agropyron trachycaulum*

Field Crops
G. FIBER

Cotton	*Gossypium hirsutum*
Flax	*Linum usitatissimum*
Hemp (illegal)	*Cannabis sativa*
Ramie (Silk Plant)	*Boehmeria nivea*

Field Crops
H. GREEN MANURES (LEGUMES)

Alfalfa	*Medicago* (10+ species)
Purple	*M. sativa*
Variagated	*M. media*
Yellow	*M. falcata*
Bean	*Phaseolus, Dolichos, Vigna*
Clover	*Trifolium*
Alsike	*T. hybridum*
Crimson	*T. incarnatum*
Red	*T. pratense*
White	*T. repens*
Lentil	*Lens esculenta*
Lespedeza	*Lespedeza striata*
Lupine	*Lupinus* (40+ species)
Pea	*Pisum sativum*
Peanut	*Arachis hypogaea*
Soybean	*Glycine max*
Sweetclover, White	*Melilotus alba*
Sweetclover, Yellow	*M. officinalis*
Vetch	*Vicia sativa*

FIG 6.7 Category II: Row, Bed and Field Crops (continued).

Category III: Nursery Crops

A. FLOWERING PLANTS FOR BEDS OR POTS (CONT.)

African Violet	Saintpaulia ionantha
Anthurium	Anthurium andreanum (20+ species)
Astilbe	Astilbe arendsii (20+ species)
Azalea	Azalea indica, A.amoena
Begonia	Begonia (100+ species)
Summer	B. semperflorens
Winter	B. Gloire de Lorraine
Bellflower	Campanula (100+ species)
Browallia	Browallia elata and speciosa
Calceolaria	Calceolaria hybrida
Camellia	Camellia japonica
Campanula	Campanula isophylla
Celosia	Celosia cristata (10+ species)
Cineraria	Senecio cruenta
Christmas cactus	Zygocactus truncatus
Chrysanthemum	Chrysanthemum carinatum (3 species)
Crocus	Crocus (40+ species)
Cyclamen	Cyclamen persicum (15 species)
Erica	Erica gracilis (60+ species)
Evening Primrose	Oenothera speciosa
Exacum	Exacum (5 species)
Felicia	Agathea amelloides
Flame Violet	Episcia (5 species)
Flamingo Flower	Anthurium scherzerianum
Foxglove, Mexican	Allophyton mexicanum

A. FLOWERING PLANTS FOR BEDS OR POTS (CONT.)

Fuchsia	Fuchsia (30+ species)
Gardenia	Gardenia veitchi
Geranium	Pelargonium zonale (50+ species)
Gloxinia	Sinningia speciosa
Hyacinth	Hyacinth (8 species)
Impatiens	Impatiens balsamina
Iris	Iris reticulata (200+ species)
Jerusalem Cherry	Solanum capsicastrum
Kalanchoe	Kalanchoe blossfeldiana (40+ species)
Lantana	Lantana montevidensis (10+ species)
Lavender, Fernleaf	Lavandula dentata
Lily	Lilium longiflorum (50+ species)
Lipstick Vine	Aeschynanthus
Marigold	Tagetes patula
Narcissus	Narcissus (30+ species)
Orchid, Epiphytics	Brassavola
	Cattleya
	Dendrobium
	Laelia
	Miltonias
	Odontoglossum
	Oncidium
	Phalaenopsis
	Vanda
Orchid, Terrestrials	Calanthe
	Cymbidium
	Cypridpedium
Oxalis	Oxalis hedysaroides (30+ species)

FIG 6.8 A sample of nursery crops to consider for metrofarming.

A. FLOWERING PLANTS FOR BEDS OR POTS (CONT.)	
Paris Daisy	*Chrysanthemum frutescens*
Peace Lilly	*Spathiphyllum (4 species)*
Pepper, Christmas	*Capsicum annuum*
Petunia	*Petunia hybrida*
Poinsettia	*Euphorbia pulcherrima*
Polyanthus	*Primula polyantha*
Pregnant Onion	*Ornithogalum caudatum*
Primose	*Primula sinensis (110+ species)*
Rose, Miniature	*Rosa chinensis (v. minima)*
Rouge Berry	*Rivina humilis*
Strawberry, Alpine	*Fragaria vesca (v. monophylla)*
Torenia	*Torenia fourneri*
Wallflower	*Cheiranthus cheiri*
Yellow Calla	*Zantedeschia elliottiana*

B. FLOWERING SHRUBS FOR BEDS, POTS OR LANDSCAPES	
Acacia	*Acacia pubescens (100+ species)*
Ardisia	*Ardisia crispa*
Bird of Paradise	*Strelitzia reginae*
Bougainvillea	*Bougainvillea glabra (7 species)*
Calamondin	*Citrus mitis*
Chorizema	*Chorizema cordatum*
Clerodendron	*Clerodendron thompsonae*
Cleveland Cherry	*Solanum capsicastrum*
Coralberry	*Symphoricarpos orbiculatus*

B. FLOWERING SHRUBS FOR BEDS, POTS OR LANDSCAPES (CONT.)	
Cytisus	*Cytisus canariensis*
False Goatsbeard	*Astilbe japonica*
Flowering Currant	*Ribes aureum*
Forsythia	*Forsythia intermedia*
	F. suspensa
Fuchsia	*Fuchsia hybrida (20+ species)*
Glossy Privet	*Ligustrum lucidum*
Heather	*Calluna vulgaris*
Hibiscus, Chinese	*Hibiscus rosasinensis*
Hydrangea	*Hydrangea hortensis (20+ species)*
Lilac	*Syringa (20+ species)*
Japanese	*S. amurensis*
Peking	*S. pekinensis*
Maple, Flowering	*Abutilon megapotamicum*
Rose	*Rosa (200+ species)*
Scarlet Sage	*Salvia splendens*
Service-berry	*Amelanchier laevis*
Spurge Laurel	*Daphne laureola*
Sweet Pepperbush	*Clethra alnifolia*

C. FLOWERING TREES FOR POTS OR LANDSCAPES	
Acacia	*Acacia (100+ species)*
Bailey	*Acacia baileyana*
Sweet	*A. farnesiana*
Weeping	*A. pendula*
Alder	*Clethra barbinervis*
Ash	*Sorbus (20+ species)*
Mountain	*Sorbus alnifolia*
European	*S. aucuparia*
Beebee-tree	*Evodia daniellii*

FIG 6.8 Category III: Nursery Crops (continued).

Beech	*Fagus grandifolia*
Birch, Gray	*Betula populifolia*
Bottlebrush	*Callistemon, Melaleuca*
Buckeye, Sweet	*Aesculus octandra*
Buckthorn, Wooly	*Bumelia lanuginosa*
Cape Chestnut	*Calodendrum capense*
Catalpa	*Catalpa* (7 species)
Chinese	*Catalpa ovata*
Common	*C. bignoniodes*
Northern	*C. speciosa*
Chaste Tree	*Vitex lucens*
Cherry, Flowering	
Black	*Prunus serotina*
Cornelian	*Cornus mas*
Fuji	*Prunus incisa*
Oriental	*P. serrulata*
Rosebud	*P. subhirtella*
Chestnut, Chinese	*Castanea mollissima*
Coral Tree	*Erythrina* (10+ species)
Cockspur	*E. crista-galli*
Kaffirboom	*E. caffra*
Naked	*E. coralloides*
Crabapple, Flowering	*Malus*
Cutleaf	*M. toringoides*
Flowering	*M. floribunda*
Tea	*M. hupehensis*
Crape Myrtle	*Lagerstroemia indica*
Desert Willow	*Chilopsis linearis*
Dogwood	*Cornus* (20+ species)
Chinese	*Cornus Kousa*
Flowering	*C. florida*
Dovetree	*Davidia involucrata*

Eastern Redbud	*Cercis canadensis*
Ehretia, Heliotrope	*Ehretia thyrsiflora*
Ficus	*Ficus carica*
	F. elastica
	F. retusa
Firewheel Tree	*Stenocarpus sinuatus*
Flame Tree	*Brachychiton acerifolium*
Australian	*B. acerifolius*
Pink	*B. discolor*
Franklin-tree	*Franklinia alatamaha*
Fringe-tree	*Chionanthus*
Chinese	*C. retusus*
White	*C. virginicus*
Goldenrain-tree	*Koelreuteria paniculata*
Haw, Black	*Viburnum prunifolium*
Hawthorn, Washington	*Crataegus phaenopyrum*
Hazelnut	*Corylus americana*
	C. cornuta
Jacaranda	*Jacaranda acutifolia*
Jerusalem Thorn	*Parkinsonia aculeata*
Linden	*Tilia* (10+ species)
Crimean	*T. euchlora*
Largeleaf	*T. platyphyllos*
Littleleaf	*T. cordata*
Locust, Black	*Robinia pseudoacacia*
Magnolia	*Magnolia* (20+ species)
Anise	*M. salicifolia*
Evergreen	*M. grandiflora*
Saucer	*M. soulangiana*
Sweetbay	*M. virginiana*

FIG 6.8 Category III: Nursery Crops (continued).

C. FLOWERING TREES FOR POTS OR LANDSCAPES (CONT.)

Maple	*Abutilon hybridum*
Michelia	*Michelia doltsopa*
Olive, Russian	*Elaeagnus angustifolia*
Orchid Tree	*Bauhinia* (10+ species)
Bauhinia	*B. variegata*
Brazilian	*B. forficata*
Hong Kong	*B. blakeana*
Peach	*Prunus persica*
Pear	*Pyrus calleryana*
Callery	*P. calleryana*
Evergreen	*P. kawakamii*
Photinia	*Photinia villosa*
Plum, Flowering	*Prunus* (80+ species)
	P. japonica
	P. maritima
	P. nana
Quince	*Cydonia japonica*
Raisin-tree, Japanese	*Hovenia dulcis*
Scholar-tree, Chinese	*Sophora japonica*
Silk-tree	*Albizia julibrissin*
Silk-Cotton Tree	*Bombax malabaricum*
Silverbell	*Halesia carolina*
Smoke-tree	*Cotinus coggygria*
Snowbell, Japanese	*Styrax japonica*
Snow-in-Summer Tree	*Melaleuca linariifolia*
Sorrel Tree	*Oxydendrum arboreum*
Stewartia	*Stewartia* (7+ species)
Japanese	*S. pseudo-camellia*
Korean	*S. koreana*
Strawberry Tree	*Arbutus unedo*
Sugar Plum Tree	*Lagunaria patersonii*

C. FLOWERING TREES FOR POTS OR LANDSCAPES (CONT.)

Surinam Cherry	*Eugenia uniflora*
Tipu Tree	*Tipuana tipu*
Tree of Heaven	*Ailanthus altissima*
Tulip-tree	*Liriodendron tulipifera*
Yellowwood	*Cladrastis lutea*

D. FOLIAGE PLANTS FOR BEDS OR POTS

Arrowhead Plant	*Syngonium podophyllum*
Baby's Tears	*Helxine Soleirolii*
Browallia	*Streptosolen jamesonii*
Caladium	*Caladium humboldtii* (4 species)
Calathea	*Calathea* (30+ species)
Chinese Evergreen	*Aglaonema* (8+ species)
Coleus	*Coleus blumei* (10+ species)
Columnea	*Columnea hirta*
Copper Leaf	*Acalypha wilkesiana macafeana*
Donkey's Tail	*Sedum morganianum*
Echeveria	*Echeveria secunda*
Fern	(6,000+ species)
Bear's Paw	*Polypodium aureum* (v.*mandaianum*)
Bird's Nest	*Asplenium nidus*
Holly	*Cyrtomium falcatum*
Miniature	*Polystichum tsus-simense*
Rabbit's Foot	*Davallia fejeensis*
Table	*Pteris cretica*
Fittonia	*Fittonia Verschaffeltii*
Freckleface	*Hypoestes*

FIG 6.8 Category III: Nursery Crops (continued).

D. FOLIAGE PLANTS FOR BEDS OR POTS (CONT.)

Gold Dust Plant	*Aucuba*
Hawaiian Ti	*Cordyline* (10+ species)
Haworthia	*Haworthia fasciata*
Kalanchoe	*Kalanchoe beharensis*
Kangaroo Vine	*Cissus antartica*
Lily-Turf	*Liriope Muscari*
Moses In The Cradle	*Rheo*
Myrtle	*Myrtus communis*
Night Jessamine	*Cestrum nocturnum*
Pellionia	*Pellionia daveauana*
Peperomia	*Peperomia obtusifolia*
Piggyback Plant	*Tolmiea Menziesii*
Pilea	*Pilea*
Artillery Fern	*P. serpillacea*
Creeping Charlie	*P. nummularifolia*
Moon Valley	*P. repens*
Pothos	*Pothos nitens, P. aureus*
Prayer Plant	*Maranta massangeana*
Purple Passion	*Gynura aurantiaca*
Rosary Vine	*Ceropegia woodii*
Shrimp Plant	*Beloperone guttata*
Snake Plant	*Sansevieria trifasciata*
Sonerila	*Sonerila margaritacea*
Spider Plant	*Chlorophytum vittatum*
Spreading Clubmoss	*Sellaginella kraussiana*
Sweet Flag	*Acorus calamus variegatus*
Tahitian Bridal Veil	*Gibasis*
Umbrella Plant	*Cyperus diffusus*

D. FOLIAGE PLANTS FOR BEDS OR POTS (CONT.)

Wandering Jew	*Tradescantia fluminensis* *Zebrina pendula*
Waxplant	*Hoya bella*
Zebra Plant	*Aphelandra squarrosa*

E. FOLIAGE SHRUBS FOR BEDS, POTS OR LANDSCAPES

Acalypha	*Acalypha marginata, A. obovata*
Agave	*Agave filifera*
Alternanthera	*Telanthera amoena*
Aralia	*Fatsia japonica*
Aralia, False	*Dizygotheca elegantissima*
Asparagus Fern	*Asparagus densiflorus*
Bamboo	*Bambusa* (30+ species)
Banana	*Musa ensete* (10+ species)
Beech, European	*Fagus sylvatica*
Caladium	*Caldium bicolor*
Cherry Laurel	*Prunus laurocerasus*
Coffee Tree	*Coffea arabica*
Croton	*Codiaeum variegatum*
Daphne	*Daphne cneorum* (20+ species)
Deutzia	*Deutzia crenata* (20+ species) *D. discolor* *D. gracilis*
Dieffenbachia	*Dieffenbachia amoena*
Dracaena	*Dracaena godseffiana* *D. sanderiana*

FIG 6.8 Category III: Nursery Crops (continued).

E. FOLIAGE SHRUBS FOR BEDS, POTS OR LANDSCAPES (CONT.)

Elephant's Foot	*Beaucarnea recurvata*
Euonymus	*Euonymus japonicus*
Fatshedera	*Hedera X Aralia japonica*
Ferns	*(6000+ species)*
Boston	*Nephrodium macrophyllum*
Maidenhair	*Adiantum trapeziforme*
Staghorn	*Platycerium bifurcatum*
Ficus	*Ficus benjamina, F. elastica*
Hawthorn	*Crataegus oxycantha*
Hemlock	*Tsuga heterophylla*
Honeysuckle	*Lonicera fragrantissima*
	L. standishii
Iron Plant	*Aspidistra lurida*
Ivy	
English	*Hedera helix*
German	*Senecio mikaniodides*
Grape	*Cissus rhombifolia*
Red	*Hemigraphis colorata*
Swedish	*Plectranthus australis*
Jade Plant	*Crassula argentea*
Japanese Andromeda	*Pieris japonica*
Jasmine	*Jasminum sambac*
Lonicera	*Lonicera japonica*
Norfolk Island Pine	*Araucaria excelsa*
Palm	*(200+ species)*
	Kentia belmoreana
	K. fosteriana
	Areca lutescens

E. FOLIAGE SHRUBS FOR BEDS, POTS OR LANDSCAPES (CONT.)

E. FOLIAGE SHRUBS FOR BEDS, POTS OR LANDSCAPES (CONT.)

	Latania borbonica
Philodendron	*(18+ species)*
Elephant's Ear	*Philodenron domesticum*
Monstera	*P. pertusum*
Heartleaf	*P. cordata*
Silverleaf	*P. sodiroi*
Pittosporum	*Pittosporum (20+ species)*
Pleomele	*Pleomele reflexa*
Podocarpus	*Podocarpus (10+ species)*
Polyscia	*Polyscias balfouriana (5 species)*
Rhododendron	*Rhododendron (200+ species)*
Rose Acacia	*Robinia hispida*
St. John's-Wort	*Hypericum calycinum*
Screw Pine	*Pandanus*
Shadbush	*Amelanchier laevis*
	A. oblongifolia
	A. ovalis
Silk Oak	*Grevellea robusta*
Smoke Tree	*Cotinus coggygria*
Umbrella Tree	*Brassaia actinophylla*
Vinca	*Vinca major*
Yew, English	*Taxus baccata*

F. FOLIAGE TREES FOR POTS OR LANDSCAPES

Alder	*Alnus (20+ species)*
Black	*A. glutinosa*
Italian	*A. cordata*
White	*A. rhombifolia*

FIG 6.8 Category III: Nursery Crops (continued).

Amur Cork	*Phellodendron amurense*
Araucaria	*Araucaria* (8 species)
Bunya-Bunya	*A. bidwillii*
Monkey Puzzle	*A. araucana*
Norfolk Island	*A. excelsa*
Avocado	*Persea americana*
Bastard Sandalwood	*Myoporum laetum*
Birch	*Betula alba* (20+ species)
Camphor	*Cinnamonum camphora*
Carob	*Ceratonia siliqua*
China Fir	*Cunninghamia lanceolata*
Chinese Pistachio	*Pistacia chinensis*
Chinese Tallow	*Sapium sebiferum*
Cypress	*Cupressus* (10+ species)
Arizona	*C. glabra*
Italian	*C. sempervirens*
Monterey	*C. macrocarpa*
Dawn Redwood	*Metasequoia glyptostroboides*
Elm	*Ulmus* (16 species)
Eucalylptus	*Eucalyptus polyanthemos* (80+ species)
European Hornbeam	*Carpinus betulus*
False Cypress	*Chamaecyparis*
Fir	*Abies* (30+ species)
Noble	*A. procera*
Nordmann	*A. nordmanniana*
White	*A. concolor*
Japanese Cedar	*Cryptomeria japonica*
Japanese Zelkova	*Zelkova serrata*
Juniper	*Juniperus* (30+ species)

Eastern Red	*J. virginiana*
Rocky Mountain	*J. scopulorum*
Hackberry	*Celtis occidentalis*
Honey Locust	*Gleditsia* (6 species)
Incense Cedar	*Calocedrus decurrens*
Katsura	*Cercidiphyllum japonicum*
Larch	*Larix leptolepis*
Madrone	*Arbutus menziesii*
Maple	*Acer* (60+ species)
Hedge	*A. campestre*
Trident	*A. buergeranum*
Japanese	*A. japonicum*
	A. palmatum
Norway	*A. platanoides*
Sugar	*A. saccharum*
Oak	*Quercus* (60+ species)
Blue	*Q. douglasii*
Buckleberry	*Q. vaccinifolia*
Canyon	*Q. chrysolepis*
Northern Red	*Q. rubra*
Water	*Q. nigra*
White	*Q. alba*
Olive	*Olea europaea* (5 species)
Oriental Arborvitae	*Platycladus orientalis*
Port Orford Cedar	*Chamaecyparis lawsoniana*
Peppermint Tree	*Agonis flexuosa*
Pine	*Pinus* (60+ species)
Aleppo	*P. halepensis*
Austrian	*P. nigra*
Bishop	*P. muricata*
Canary Island	*P. canariensis*
Japanese Black	*P. thunbergiana*

FIG 6.8 Category III: Nursery Crops (continued).

Japanese Red	*P. densiflora*
Monterey	*P. radiata*
Poplar	*Populus*
Carolina	*P. canadensis*
Japanese	*P. maximowiczii*
Narrow-leafed	*P. angustifolia*
Redwood	*Sequoia sempervirens*
Sassafras	*Sassafras albidum*
Sissoo	*Dalbergia sissoo*
Sour Gum	*Nyssa sylvatica*
Spruce	*Picea* (35+ species)
Black	*P. mariana*
Colorado	*P. pungens*
Red	*P. rubens*
White	*P. glauca*
Sumac	*Rhus glabra*
Sweet Bay	*Laurus nobilis*
Sweet Gum	*Liquidambar styraciflua*
Tea Tree	*Leptospermum laevigatum*
Tree Ferns	*Alsophila australis*
	Cyathea dealbata
	Dicksonia antarctica
	Lomaria gibba
Willow	*Salix alba tristis*

G. PLANTS FOR GROUNDCOVER

Lawn Grasses (cont.)

Bent grass	*Agrostis* (8 species)
Colonial	*A. tenuis*
Creeping	*A. palustris*
Redtop	*A. alba*
Fescue	*Festuca* (6 species)

Red	*F. rubra*
Korean lawn-grass	*Zoysia japonica*
Meadow grass	*Poa* (8 species)
Kentucky blue	*Poa pratensis*
Rough-stalked	*P. triviais*
Wood	*P. nemoralis*
Rye, perennial	*Lolium perenne*
Saint Augustine grass	*Stenotaphrum secundatum*

Ornamental Grasses

Aira	*Aira capillaris*
Brachypodium	*Brachypodium sylvaticum*
Brome Grass	*Bromus* (9 species)
Canary Grass	*Phalaris canariensis*
Eragrostis	*Eragrostis* (10+ species)
Finger Grass	*Chloris* (6 species)
Jobs Tears	*Coix Lacryma-Jobi*
Hair Grass	*Deschampsia flexuosa*
Palm Grass	*Setaria palmifolia*
Paspalum	*Paspalum dilatatum*
Quaking Grass	*Briza maxima*
Squirrel-tail Grass	*Hordeum jubatum*
Sweet Vernal Grass	*Anthoxanthum odoratum*
Wire Grass	*Eleusine indica*
Witch Grass	*Panicum capillare*

Ornamental Grasses

Bearberry	*Arctostaphylos Vuaursi*
Clover, Dutch White	*Trifolium repens*
Ivy, English	*Hedera Helix*

FIG 6.8 Category III: Nursery Crops (continued).

G. Plants for Groundcover (cont.)
Other Groundcovers (cont.)

Mazus	*Mazus japonicus*
Moneywort	*Lysimachia Nummularia*
Periwinkle	*Vinca minor*
Sandwort	*Arenaria verna*
Saxifragas	*Saxifragas decipiens*
Sedum	*Sedum* (7 species)
Spurge, Japanese	*Pachysandra terminalis*
Thrift	*Armeria maritima*
Yarrow	*Achillea Millefolium*

H. Trees for Bonsai

Arborvitae	*Thuja occidentalis*
Barberry, Crimson	*Berberis thunbergii*
Boxwood, Wintergreen	*Buxus koreana*
Juniper	*Juniperus*
Chinese	*J. chinensis*
Japanese	*J. procumbens*
Maidenhair-tree	*Gingko biloba*
Maple	*Acer*
Trident	*A. buergerianum*
Amur	*A. ginnala*
Pine	*Pinus*
Austrian	*P. nigra*
Japanese Black	*P. thunbergii*
Japanese White	*P. parviflora* (v. *pentaphylla*)
Limber	*P. flexilis* (v. *columnaris*)
Mugho	*P. mugho* (v. *mughus*)
Scotch	*P. Sylvestris*
Quince, Flowering	*Chaenomeles japonica*
Spruce	*Picea*

H. Trees for Bonsai (cont.)

Black Hill	*P. glauca* (v. *densata*)
Dwarf Albert	*P. glauca* (v. *conica*)
Saghalin	*P. glehni*
Succulent	*Crassula arborescens*

I. Trees for Christmas

Arizona Cypress	*Cupressus arizonica*
Fir	*Abies*
Alpine	*A. lasiocarpa*
Balsam	*A. balsamea*
Douglas	*Pseudotsuga taxifolia*
Red	*Abies procera*
White	*A. concolor*
Juniper	*Juniperus*
Hemlock	*Tsuga heterophylla*
Pine	*Pinus*
Jack	*P. banksiana*
Pinyon	*P. edulis*
Red	*P. resinosa*
Scotch	*P. sylvestris*
Virginia	*P. virginiana*
White	*P. monticola*
Redcedar	*Juniperus virginiana*
Spruce	*Picea*
Black	*P. mariana*
Colorado Blue	*P. pungens*
Norway	*P. abies*
Red	*P. rubens*
White	*P. glauca*

J. Succulents for Beds or Pots
Cactus

Column	*Cereus peruvianus*
	C. tetragonus

FIG 6.8 Category III: Nursery Crops (continued).

J. Succulents for Beds or Pots (cont.)		**K. Exotic or Unusual Plants for Pots**	
Cactus (cont.)		*Insect-Eating Plants*	
Easter Lily	*Echinopsis multiplex*	British Butterwort	*Pinguicula vulgaris*
Globe	*Echinocereus reichenbachii*	Pitcher Plant	*Nepenthes khasiana*
	E. rigidissimus	Australian	*Cephalotus follicularis*
Golden Barrel	*Echinocactus grusonii*		*Darlingtonia californica*
Jungle	*Epiphyllum oxypetalum*		*Heliamphora nutans*
Lobivia	*Lobivia*	Trumpet Leaf	*Sarracenia*
Mammillaria	*Mammillaria*	Sundew	*Drosera rotundifolia*
Notocactus	*Notocactus ottonis*	Venus Flytrap	*Dionaea muscipula*
	N. rutilans		
	N. haselbergii	*Touch-Sensitive Plants*	
Old Man	*Cephalocereus senilis*	Artillery	*Pilea muscosa*
Rattail	*Aporocactus flagelliformis*	Neptunia	*Neptunia oleraceae*
		Oxalis	*Oxalis sensitiva*
Other Succulents		Sensitive	*Mimosa pudica*
			M. sensitiva
Euphorbia	*Euphorbia* (100+ species)	Telegraph	*Desmodium gyrans*
Brain Cactus	*Euphorbia lactea*		
Candelabra	*E lactea*	*Unusually-Shaped Plants*	
Crown-of-Thorns	*E. milii*	Dutchman's Pipe	*Aristolochia indica*
Pencil	*E. tirucalli*	Japanese Fern Balls	*Davallia mariesii*
Living Stone	*Lithops*	Monarch of the East	*Sauromatum giganteus*
Redbird Cactus	*Pedilanthus*	Moonflower	*Calonyction aculeatum*
Silver squill	*Scilla violacea*	New Zealand Lobster	
Crassula arborescens	*Sanseveria zeylanica*	Claw	*Clianthus puniceus*
Echeveria globosa	*Sanseveria laurenti*	Pineapple	*Aechmea rhodocyanea*
Echinocactus	*Sedum acre*		*Bilbergia nutans*
Echinocereus	*Sedum sarmentosum*		*Cryptanthus bivittatus*
Mammilaria	*Sempervivum globiferum*		
Saxifraga sarmentosa	*Sempervivum tectorum*		

FIG 6.8 Category III: Nursery Crops (continued).

Category IV: Animal, Bird and Fish Crops

A. ANIMAL CROPS

Alpaca (fiber, recreation)	*Lama pacos*
Cat (recreation)	*Felis domesticus*
Cervidae (hides, antlers, meat)	
deer, mule	*Odocoileus hemionus*
deer, white-tail	*O. virginianus*
elk	*Cervus canadensis*
red deer	*C. elaphus*
reindeer	*Rangifer tarandus*
Chinchilla (fur, recreation)	*Chinchilla*
Cow (milk, hides, meat, breeding stock)	*Bos taurus*
Cow, Brahmin	*Bos indica*
Dog (recreation, work)	*Canis familiaris*
Fox (fur)	*Vulpes fulva*
Goat (milk, cheese, meat, recreation)	*Capra hircus*
Goat (mohair)	*Capra angorensis*
Guinea Pig (laboratory, recreation)	*Cavea porcellus*
Hamster (laboratory, recreation)	*Cricetus cricetus*
Hog (meat, hides)	*Sus scrofa*
Horse (breeding stock, recreation, work)	*Equus caballus*
Llama (recreation, work, wool)	*Lama glama*
Mink (fur)	*Mustela vision*
Mice (laboratory)	*Mus musculus*
Rabbit (meat, fur, recreation)	*Lepus cuniculus*
Rat (laboratory)	*Rattus norvegicus*
Sheep (meat, wool)	*Ovis aries*

B. BIRD CROPS

Chicken (eggs, meat, ornamental)	*Gallus gallus*
Duck (eggs, meat, ornamental)	*Anas platyrhynchos*
Game Birds (meat, game-preserve stock)	
Partridge, Hungarian	*Perdix perdix*
Pheasant, Chinese	*Phasianus colchicus*
Ruffed Grouse	*Bonasa umbellus*
Quail	*Coturnix coturnix*
Geese (meat, feathers, ornamental)	*Anser anser*
Guinea Fowl (ornamental)	*Numida meleagris*
Peacock (ornamental)	*Pavo*
Pigeon (squab, recreation, messengers)	*Columba livia*
Turkey (meat)	*Meleagris gallopavo*

C. FISH AND CRUSTACEAN CROPS

Abalone	*Haliotis*
Bass	
Largemouth	*Micropterus salmoides*
Striped	*Roccus saxatilis*
White	*Lepibema chrysops*
Bluegill	*Lepomis macrochirus*
Carp (edible and ornamental)	
Common, American	*Cyprinus carpis*
Bighead, Chinese	*Aristichthys noblis*
Grass, Chinese	*Ctenopharyngodon idella*
Israeli	*Cyprinus carpio*
Silver	*Hypothalmichthys molitrix*

FIG 6.9 A sample of animal, bird and fish crops to consider for metrofarming.

C. FISH AND CRUSTACEAN CROPS (CONT'D)	
Catfish	
Blue	*Ictalurus furcatus*
Channel	*Ictalurus punctatus*
White	*Ictalurus catus*
Clam (pearls, meat)	
Butterfly, American	*Plagiola securis*
Pearly, Chinese	*Anodonta woodiana*
Pearly, Chinese	*Hyriopsis cumingi*
Crayfish	
European	*Astacus*
American	*Cambarus*
Mussel, Fresh-water (Pearls)	*Anodonta woodiana*
	Hyriopsis cumingi
Oyster	*Ostreidae* (fam.)
Perch, Yellow	*Perca flavescens*
Shrimp	
Giant (fresh-water Malaysian)	*Macrobrachium rosenbergii*
Small	*Penaeid*
Tilapia	

C. FISH AND CRUSTACEAN CROPS (CONT'D)	
Blue	*Tilapia aurea*
Java	*Tilapia mossambica*
Nile	*Tilapia nilotica*
Trout	
Brook	*Salvelinus fontinalis*
Brown	*Salmo trutta*
Golden	*Salmo aqua-bonito*
Rainbow	*Salmo gairdnei*

D. MISCELLANEOUS SMALL ANIMAL CROPS	
Bees (honey, pollination)	*Apis mellifera*
Bull Frog (food)	*Rana catesbeiana*
Crickets (insectivorous food)	*Gryllus*
Mealworms (insectivorous food)	*Tenebrio*
Earthworm (fish bait and fish food, castings)	*Lumbricidae* (fam.)
Praying Mantis (beneficial garden insect)	*Mantis religiosa*
Snail (Escargot)	*Helix aspersa*
Salamander, Tiger (Laboratory, fish bait)	*Salamandra*

FIG 6.9 Category IV: Animal and Fish Crops (continued).

Climate Check

The first thing to check is whether the crop can be grown successfully in your climate. Following are five questions which will help you make this determination.

- Has this crop been successfully grown in your neighborhood before? If not, why not?

- Is the length of your growing season sufficient to meet the needs of this crop?
- Would an early or late season frost destroy this crop?
- Will the crop tolerate high temperature extremes of your growing season?
- Is there too much or too little rainfall for this crop in your area?

Market Check

Next check whether the crop will find a profitable market. The items in this checklist require a thorough survey of the market, as discussed in Chapter Four.

- What is the current per-unit cash value of this crop in the market?
- How does current per-unit value compare with the product's value last year? Ten years ago?
- What do experts expect this crop's per-unit value to be next year? Ten years from now?
- Can you detect any current conditions, such as drought, disease, pesticide contamination, etc., which will create abnormal swings in supply and demand for this crop? How will conditions affect the crop's per-unit value?
- Are there any restrictions, such as marketing orders, which set quality and quantity standards for this crop?
- How many market sectors can you reach with this crop?
- Are there wholesalers in your area who handle this crop? Will they give you a contract? How much are they willing to pay?
- How far must you transport this crop to market? How much will transportation cost?

Parcel Check

Check the crop's compatibility with your parcel of land. Following are seven items which will help you evaluate compatibility.

- Will the crop have a value-per-unit sufficient to support the cost-per-acre of your land?
- Is the crop compatible with your tenure? (*i.e.,* Long-term

investments, like orchards, require long-term tenure, like ownership.)

- Is there enough tillable land to support a market-sized production of this crop.
- Are the texture, *pH* and nutrient characteristics of your soil sufficient to meet needs of this crop. How much will it cost to make corrections?
- Which competitors— weeds, insects, diseases— will threaten the crop in your area? How much will it cost to control pests?
- How much water will this crop require? How much will it cost to irrigate?
- Are there government rules and/or regulations which restrict production of this crop in your area?

Production Check

Check requirements for establishing production. The six items which follow will help you evaluate whether you have the ability to produce the crop.

- Will the crop require a beginning, intermediate or advanced level of technical expertise?
- Do you have access to a sufficient amount of technical information to produce this crop for market?
- Does the crop present any special problems, such as a difficult pollination procedure, which have prevented it from being produced commercially in your area before?
- What are the competition's strengths and weaknesses in producing the crop for your market? Which high-density techniques could be employed to produce the crop on your small parcel of land?
- Which techniques could be employed to combat the crop's insect, disease and weed competitors?
- Can you develop the ability to manage production of this crop?

Rotation Check

The objective of this check is to determine whether the crop under consideration can be grown with other crops in rotation to increase production efficiency.

- Will this crop be compatible with other crops you have selected?
- Can you use the same equipment used in the production of other crops?
- Is the crop susceptible to soilborne diseases or nutrient deficiencies which might result from previous crops?
- Will this crop create any problems, such as soilborne diseases, for the next crop in the rotation?
- Can you sell this crop to the same markets as other crops in your rotation or will it require developing a new market?

Labor Check

The objective of this check is to determine what kind of labor requirements must be satisfied to produce the crop.

- How much extra labor will you need to hire in order to successfully produce the crop?
- Will your hired labor need to be technologically proficient? (*e.g.*, pruners, equipment operators, etc.)
- Can you reduce the need for hired labor through mechanization?
- Where will you secure the hired labor and how much will it cost?

Equipment Check

The objective of this check is to determine what kind of equipment will be required to produce the crop and how much that equipment will cost.

- Will production of the crop require the use of special equipment?
- Is this equipment available?
- How much will the equipment cost to buy or rent?

Infrastructure Check

The objective of this check is to determine what kind of investments you will have to make in a production system in order to produce the crop.

- Will the crop require use of special structures like greenhouses, packing sheds, storage facilities or equipment barns?

- Will the structures require special features, like ventilation systems or heated beds?
- Will the crop require construction of additional roads, fences or irrigation systems?
- How much will the necessary infrastructure developments cost?

Harvest Check

The objective of this check is to determine the costs of harvesting and processing the crop into a marketable product.

- How many times must the crop be harvested each season to obtain a profitable yield?
- Will additional labor be required for the harvest?
- How would adverse conditions, such as frost or rain, affect the harvest?
- What kind of special handling will the crop require after being harvested? How much will this handling cost?
- Are there any federal or state standards for packaging this crop for the market?
- Will the crop reach market during a period of oversupply, thereby reducing its per-unit value?

Bottom-Line Check

The objective of this check is to determine whether the crop will return a profit.

- What are projected costs for producing this crop?
- Where will you obtain money to cover the costs?
- What are the terms for obtaining this money and how much will it cost?
- What are the projected sales?
- What is the projected net profit?

You will have to invest a considerable amount of time to check each item in the lists. Though most of your answers will be mere projections, they will help you develop a good understanding of the crop and its potential before you invest your precious resources in production.

SELECT THE VARIETY

Variety is a class subordinate to species. Though sharing basic characteristics of species, each variety differs in one or more major features. There are, for example, about 7,000 varieties of the apple (*Malus sylvestris*) species. Some apple

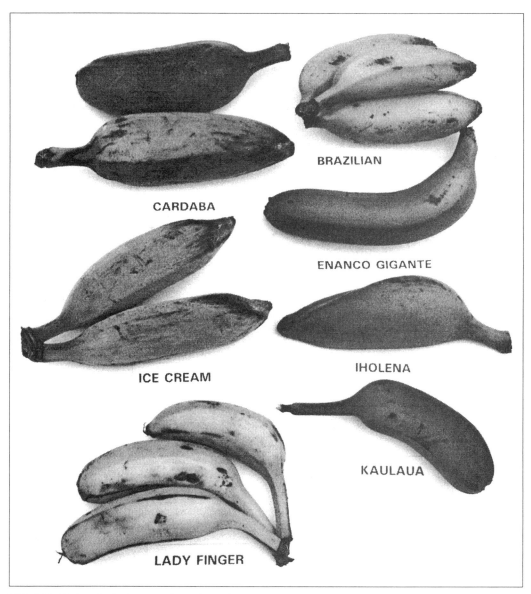

FIG 6.10 Variety is the degree of genetic variation within a species. Some crop varieties are economically viable while others are not. Featured are some of the viable banana varieties grown at the Seaside Banana Gardens near Ventura, California.

varieties are good to eat, others are not. There are hundreds of varieties cultivated for their sweet taste; one such variety is called the Delicious apple.

Strain is a class subordinate to variety. The longer a variety, like the Delicious, has been subjected to cultivation, the more likely it will have generated different strains. Though sharing basic characteristics of variety, each strain differs in one or more secondary features. There are Red Delicious apples and there are Golden Delicious apples. The more each strain is bred for its favorable characteristics, the more likely it will collect additional classifications (*e.g.*, "XYZ Corporation's Super 'D' Red Delicious apples").

The line between variety and strain is often indistinct and confusing. Some people call the strain a variety and others call the variety a strain. The important thing to remember is variety and strain are terms used to describe degrees of genetic variation. Some variations are good and some, bad. As farmer, you have the responsibility to decide which is best. For this reason, you must develop a good working knowledge of genetic variations in your crop species.

It would be impossible to list all varieties and strains of each crop in this text. Develop information sources of your own, including published material like seed catalogs, magazines and textbooks, as well as personal contacts such as local farmers, county extension agents, land grant colleges and, of course, representatives of seed producers.

Become an aggressive collector of information. If you elect to grow day lilies as a crop, then join the Hemerocallis Society and learn who is growing the different varieties of day lilies, which ones are best for your purposes and where you can obtain them.

The best strategy for selecting a variety is to go with the staples and continually experiment with newly developed variations. This strategy guards against two common mistakes. First is buying the enthusiastic advertisements contained in a seed catalog. While the exciting new variety advertised may indeed have a good feature, such as an early maturation date, it may also have a bad feature, such as a vulnerability to a disease or a poor taste. The other mistake is getting left behind. When consumers find an exciting new taste or color, the old staples may get left on the shelves in the rush for the new.

A good example of the conservative side to this strategy can be heard in the talk of Charles Heyn, who "tried it (the new hybrid) for several years before I decided to pay the extra money." The liberal side is well-represented by Jeff Larkey, with "oddities like purple broccoli, white and purple radishes, golden zucchini, golden beets, golden bell peppers and yellow tomatoes."

Following are checklists to help you evaluate the potential of different varieties and strains. Use the checklists as a research guide and become the expert in your crop's genetic variations before you place your time, know-how and money at risk.

Production Check

Determine how the variety will respond to your production efforts.

- Does this variety have a demonstrated history of success in your location? Describe this history.
- Will this variety respond well to your climate? To your soil?
- Will this variety reach maturity before your growing season ends?
- Can this variety tolerate the extreme weather conditions of your area such as frost or drought?
- Can this variety resist insects and diseases common to your location?
- How does the variety's yield compare with others?

Market Check

Determine how the market will respond to the variety.

- Will this variety mature early or late enough to enable you to avoid the harvest-time glut of supply in the market?
- Does this variety produce exotic or unique features, such as a different color or size, which might enable you to capture a market niche?
- How does the nutrient content of this variety compare with other varieties?
- How does the variety taste? Does it compare favorably with others?
- Will this variety conform to the prevailing market standards for product size and color?
- How does this variety stand-up to post-harvest processing? Shipping? Does it have a good shelf-life?

SELECT SEEDS

There are two ways to obtain seeds or living starts for production. You may grow them yourself or buy them from others. However, because of the work it takes to produce viable, true-to-type seeds and starts, most commercial farmers buy them from professional growers. Since prospects for the entire season are contained in tiny bundles of genes called seeds it is extremely important to select the right seeds from the right sources. Following is a guide to making this selection.

Select Viable Seeds

The crop will be no better than the seeds you plant. Good seeds are clean, free of dirt, chaff and other bits of extraneous material. They have embryos which are alive and ready to grow. Good seeds and starts do not harbor soilborne diseases and pests and, in certain cases, are certified as being free of diseases. Good seeds are true to type. They are good representatives of the variety and will grow as advertised. The best seeds are expensive to produce and often command the top price.

Select Viable Sources

Viable sources are the seed growers with a reputation for integrity. Each has a list of satisfied customers who will attest that the source's seeds grow as advertised. Good sources will reveal the origin of their seeds and furnish a lot number so you can keep an accurate record of the variety's performance. Good sources have trial grounds where each variety is tested for its ability to grow and yield as advertised.

There are many reputable seed companies and professional growers from which you can purchase seeds and starts. The three examples listed below represent a good cross-section of the seed industry. One is a mainstream, corporate-owned seed company; one, a small specialized seed company; and one, a society of like-minded individuals.

BURPEE SEEDS: This company is representative of the world's large, corporate-owned seed companies. It has offices in Pennsylvania, Iowa and California; it has trial grounds on both coasts and several locations in between. Burpee produces popular varieties of many different vegetables, flowers and trees.

SHEPHERD'S SEEDS: This company is as small as Burpee is large. It is headquartered in the home of Renee Shepherd in Felton, California. Renee travels to Europe and finds unique varieties of European garden produce, including such exotics as white sweetcorn, lemon basil and savoy cabbage. She tests the interesting varieties in plots next to her office in Felton. Varieties which prove good are then packaged under the Shepherd label for redistribution.

THE SEED SAVER'S EXCHANGE. This is an organization of like-minded individuals working toward preserving heirloom and endangered varieties of garden vegetables. It was organized by Kent Whealy and is located in Princeton,

Missouri. The Seed Saver's Exchange is a swap meet which operates through a catalog. People from all over the world list their names and the varieties of seeds they have in possession. Other people list their names and varieties of seeds for which they are looking.

REVIEW

The crop is your weapon in the competition for consumer dollars. Deciding which crop can win this competition is consequently one of the most important decisions you will make as a metrofarmer. There is a simple, four-step process to making a good selection:

First, select the right category. This selection will help you stay within the confines of your ability to invest. Second, select the right crop from your category. The market will provide high and low-risk selection opportunities and each must be subjected to a thorough business analysis. Third, select the right variety. Genetic variation may help you grow outside of the normal growing season and region, sell into the market at a favorable time and earn a favorable price by providing consumers with exciting new colors, shapes and tastes. Fourth, select the right seeds. Your entire season is focused in the tiny bundles of genes called seeds, so invest the time and effort it takes to select viable seeds from viable sources.

You must select the right crop to win the competition for consumer dollars. If you select the right crop you will, like "the man who planted hope and grew happiness," be rewarded with many good nights of sleep.

CHAPTER SIX EXERCISES

i   Title a fresh sheet of paper with "Crop Enterprise" and insert the sheet into the "Production" section of your metrofarm workbook. Write one paragraph describing your capacity to invest and the technical expertise you can bring to bear in a crop enterprise.

ii   Compare your ability to invest with each crop category's investment requirements. Select the category which matches your ability to invest.

iii Examine the crops listed in your category (Figures 6.6 to 6.9). On fresh sheets of binder paper, separate the high-risk, high-return crops— like the cherimoya— from the low-risk, low-return crops— like the apple. Decide which direction you wish to take. Survey the market for each crop. Ask about the level of consumer demand, the potential for future demand and the competitions' strengths and weaknesses in supplying the demand. Select the best crop or crops on your list.

iv   Title a fresh sheet of paper with the name of the crop you have selected. Below list names of the most popular varieties, like Red and Golden Delicious. Next, list varieties which offer a difference, like Winesap and Gravenstein. Finally, subject each variety to the "Production" and "Market" checks found in the chapter. Select the best variety.

v   On a fresh sheet of paper list as many sources for your crop as you can find. Call or write each source and ask for a catalog and a list of references. Call or write several references and ask, "Would you buy seeds from this company again?" List the best sources in your metrofarm workbook.

Organizing the Business

Spring came; he worked on his patch of ground, and planted potatoes. His live stock multiplied; the two she-goats each had twins, making seven in all about the place. He made a bigger shed for them, ready for further increase, and put a couple of glass panes in there too. Ay, 'twas lighter and brighter now in every way.

Knut Hamsun, *Growth of the Soil*

A well-organized business, like a good tractor, has many uses: It can haul in start-up financing, process information, plow through government red tape, cultivate returns and stow equity. Consider, for example, how a well-organized business was used to haul in start-up financing:

Mark Twain called them a "rare and curious luxury." Cherimoyas (*Annona cherimoya*) are a subtropical fruit which have a delicious pineapple-coconut flavor and a custard-like texture. These rare and curious luxuries recently sold for up to seven dollars per pound in Los Angeles and San Francisco markets.

Seeing opportunity, a business imported a hardy cherimoya varietal from Spain with the intent of establishing production in California. This business devised two tactics to guard against overproduction. First, it reached an agreement with the Spanish Government to limit importation of cherimoya varietals to 1,000 acres. Second, it devised a gibberellin formula which would enable orchardists to produce seedless cherimoya fruit.

A financial projection on an acre of cherimoya production showed it to be a fascinating prospect. Trees planted 109 to the acre would yield from 600 to 1600 pounds of fruit, with an approximate net return of $2.50 per pound. So, $2.50 per pound X 600 pounds per tree X 109 trees per acre would produce $163,500 in annual net profit per acre.

A farmer would need the right environment, know-how and about $30,000 per acre to invest in this cherimoya enterprise. In addition, the farmer would also have to survive for five years without a return on this investment.

Farm start-up costs are often financed with personal time and money borrowed from family savings, a friend or a financial institution. But what happens when an extraordinary opportunity comes along for which conventional financing is not available?

A young California couple had the right environment and know-how. They did not, however, have the $30,000 per acre to place at risk in this cherimoya prospect and so organized a limited partnership business. Following is a paragraph from the prospectus which described their offering:

> We intend to plant 8 acres of Cherimoyas, 109 trees per acre, at our citrus ranch. We are offering the opportunity for investors to take part in this exciting project as limited partners. The cost, $27,350 per acre, includes planting costs, irrigation system, trees, bulldozer and backhoe work, cover crop frost protection and all fertilizer, maintenance, cultural costs and insurance for five years. These costs are deferred over six payments. The trees are expected to begin bearing a commercial crop within five years. The investor will receive 50% of all harvest proceeds after deduction of harvesting costs estimated at $1,000 per acre, such as picking, boxes, refrigeration, Association assessments etc., for 15 years. At this time the General Partners will buy out the investors' share of the business at its appraised value. In addition to the tax shelter benefits during the initial 5 year period, the return to the investor prior to the buy out is conservatively projected to total over $993,000 per acre.

Rare and curious luxuries, like the cherimoya, are a high-risk business in which a relatively small investment, like $30,000, might generate a relatively high return, like $993,000. The investment might also return nothing.

If general partners had to rely on their personal savings, they would have had to forego the cherimoya opportunity. If limited partners had to rely on their personal time, land and know-how, they, too, would have had to forego the opportunity. A well-organized business enabled the young couple and their limited partners to plant eight acres of cherimoyas and get a huge jump on their competition.

A well-organized business is a weapon in the competition for consumer dollars. This chapter presents two steps to help you organize. The first step is to establish ownership; the second is to establish management. If you master your business, as David mastered his slingshot, you will have the ability and confidence to face down the Goliaths which dominate the competition for consumer dollars.

ESTABLISH OWNERSHIP

There is more than one way to own a business. You can own it yourself or share ownership with a partner, limited partners or fellow shareholders.

No single form of ownership is best for all metrofarmers. The best way is the one which enables you to generate a maximum return for minimum cost. To find this way, examine benefits and limitations of each form of ownership to determine which one will best allow you to achieve your strategic business objectives.

Consider a Sole Proprietorship

The sole proprietorship gives legal right of exclusive ownership to one person. This person organizes the business, contributes all its resources, makes all its decisions, assumes all its risks and pays personal income tax on all its rewards. The sole proprietorship business begins and ends at the will of the proprietor and terminates at his or her death.

BENEFITS OF A SOLE PROPRIETORSHIP: There are two benefits which make this form of ownership the most widely used in the business of farming.

Management Flexibility: The principal benefit of a sole proprietorship lies in its simplicity. One person provides all resources, assumes all risks, makes all decisions and receives all rewards. No organization is more readily managed than a simple sole proprietorship.

Accumulation of Equity: A second benefit is the building of equity in long term capital resources such as business, land, buildings and equipment at the expense of short term wages. This benefit enables one to avoid paying high income taxes when the business is young (and gives credence to the saying, "Farmers live poor but die rich.")

LIMITATIONS OF A SOLE PROPRIETORSHIP: Three limitations to sole proprietorship should be evaluated when considering this form of ownership.

Investment Opportunities: Many metrofarms require a substantial amount of developmental capital, especially where land values are highly inflated or where capital-intensive production systems are required. To take advantage of opportunities, the sole proprietor must provide capital or borrow it from others. Personal resources of the sole proprietor, in other words, limit the number of opportunities available to the business.

Progressive Income Taxes: Profits from a sole proprietorship are taxed as personal income. This is a progressive tax. Progressive does not mean the more you make, the more you pay; it means the more you make, the higher percentage you pay. If you barely eke out a profit, you might pay only 10 percent; but if you become a smashing success, you might pay 50 percent!

Progressive taxes cause two kinds of unproductive business activity. One, called "new-paint disease," is the unnecessary purchase of new equipment simply to prevent generating enough profit to qualify for a higher tax bracket. The second, lethargy, is the why-bother attitude propagated when any additional effort is taken by government and redistributed elsewhere. Both afflictions limit how much profit is saved against the eventuality of hard times.

Inheritance Taxes: It is said there are only two certainties: death and taxes. In farming this is true because when a sole proprietor dies, the farm business dies, too. Though the farm may be carried on by a family member, the business is legally a new entity. This transfer of equity is often called the "family-farm cycle."

The family-farm cycle works as follows: A farmer begins production with much labor and little cash. Over time, cash is accumulated and substituted for labor until such time as the business achieves its most profitable ratio of cash and time. Upon the farmer's reaching retirement age, equity has likely grown dramatically through individual efforts and inflation. When a farmer dies the transfer of equity frequently precipitates a cash crisis, forcing heirs to mortgage the farm to pay inheritance taxes. The family farm cycle then begins anew, with much labor and little cash.

ESTABLISH A SOLE PROPRIETORSHIP: There are no legal restrictions to organizing a sole proprietorship. And, with the exception of tax papers, fictitious name statements and business licenses, there are no papers to file with government. Business begins when the proprietor begins business.

Because of its simplicity, the sole proprietorship is the form of ownership most often used in farming. The Heyn and LaJoie families (Part IV, Conversations) provide good examples of sole proprietorship. Heyn's Country Gardens and The Plant Carrousel have been organized into effective instruments of competition, enabling the Heyn and LaJoie families to win their fight for consumer dollars.

Consider a Partnership

A general partnership is a legal agreement in which two or more individuals are granted ownership, liability and profit rights in equal proportions. A limited partnership consists of a general partner, who operates the business and whose liability is unlimited and a limited partner(s), who does not participate in the operation of the business and whose liability is limited to the amount invested.

Partnerships are like sole proprietorships in that liability, profits and losses are distributed directly to partners. However, partnerships differ from sole

Three Forms of Ownership

	Sole Proprietor	Partnership	Corporation
NATURE OF ENTITY	Single individual	Association of two or more individuals	Legal entity separate from shareholders
LIFE OF BUSINESS	Terminates on death	Agreed terms; but terminates at death of a partner	Perpetual or fixed term of years
LIABILITY	Personally liable	Each partner liable for all partnership obligations	Shareholders not liable for corporate obligations
SOURCE OF CAPITAL	Personal funds or loans	Partners' contribution or loans	Contribution of shareholders for stock; sale of stock; bonds and other loans
MANAGEMENT DECISIONS	Proprietor	Agreement of partners	Shareholders elect directors who manage business
LIMITS ON BUSINESS ACTIVITY	Proprietor's discretion	Partnership agreement	Articles of incorporation and state corporation law
TRANSFER OF INTEREST	Terminates proprietorship	Dissolves partnership; but new partnership may be formed	Transfer of stock does not affect continuity of business — may be transferred to anyone if no restrictions
EFFECT OF DEATH	Liquidation	Liquidation	No effect on corporation. Stock passes by will or inheritance
INCOME TAXES	Income taxed to individual—50% deduction for long term capital gains.	Partnership pays no tax. Each partner reports share of income or loss, capital gains and losses as an individual.	Corporation pays tax on net income before distribution of dividends. No 50% deduction for capital gains.

FIG 7.1 Three ways to own a business. From *Business Organization for Modern Farms*, A. Doyle Reed. Leaflet #2358, Division of Agricultural Sciences, University of California, 1978.

proprietorships in two important ways:

First, general partners share an unlimited personal liability for all the partnership's debts. If the business cannot pay its bills for one reason or another, all general partners must pay out of their own pockets. If one partner is involved in a business-related accident for which there is no insurance, all partners are liable for damages. If one partner makes a bad decision, all partners are liable for terms of the decision.

Second, partners have a fiduciary responsibility to work in each other's best interests. This responsibility means it is illegal to cheat a partner, to withhold information, to use partnership assets and property without a partner's knowledge and consent or to steal the partnership's assets.

BENEFITS OF A PARTNERSHIP: There are four important benefits to partnership which should be evaluated when considering this form of business ownership.

Investment Opportunities: Whereas a sole proprietorship is limited by resources of an individual, a partnership combines the resources of its partners. This larger resource base can then be used to control better property, build a more elaborate production system (*e.g.*, greenhouses, drip irrigation) and buy more productive seeds and fertilizers.

Management and Labor Costs: A well-organized partnership reduces management and labor costs by pooling time and talents of its partners. Each partner can specialize in a different aspect of the business, thereby providing for a more efficient use of resources.

The Flower Ladies (Part IV, Conversations) are a good example of these economies. Though all three general partners are capable of doing all of the partnership's tasks, each has assumed a specialty: One watches over the gardens, one over arranging flowers and one over soothing brides.

Tax Benefits: A partnership is not a legal entity which earns profit or pays taxes. Instead, each partner's share of the profit or loss is taxed as personal income. This makes it possible to create attractive investment opportunities for limited partners with high taxable incomes. As the young couple with the cherimoya prospect demonstrated at the beginning of this chapter, limited partners can write off short-term losses and hope for long-term gains without incurring undue risk.

Management Flexibility: Though not as flexible as a sole proprietorship, a partnership may still allow for a considerable degree of management flexibility. This is especially true where general partners are capable of working closely, as is the case with The Flower Ladies. And with the exception of limited partnerships, which are regulated by state law, no government agencies monitor the business affairs of partnerships.

LIMITATIONS OF A PARTNERSHIP: Three limitations to the partnership form of ownership should be considered.

Conflict Resolution: The principal limitation of a partnership is the difficulty with which conflicts between general partners are resolved. Unless clearly defined procedures are established, small inconsequential disagreements can easily be fanned into major conflagrations by stressful conditions of competition.

Personal Liability: Individual partners are liable for actions and decisions of their partners. Though each partner's fiduciary responsibility to the partnership protects against theft and fraud, it does not prevent one partner from making a bad decision for which the remaining partners become personally liable.

Equity Transfer: Transferring a partnership's equity from one principal to another or from one principal to an outside investor is difficult. The reason for this difficulty is the partnership must literally be reformed each time a general partner dies, withdraws or enters into the business. At this time all accounts must be settled and the only way to determine the true value of the partnership's equity is to sell the entire business. This makes it very difficult for principals to enter and leave a general partnership.

ESTABLISH A PARTNERSHIP: A general partnership agreement can be spoken, written or even unspoken if partners show a decision to share control, profit and risk. The partnership begins when partners reach an agreement.

Though it is not mandatory for a partnership agreement to be written, it helps the organization function more smoothly. Written agreements clarify details and force partners to think things through before becoming legally bound to each other. Many state business codes require that partnerships which last longer than one year must have a written partnership agreement. A well-planned partnership agreement should contain the following articles:

Identification: List the name of the partnership, its location and names of partners.

Duration of Partnership: Estimate how long it will take for the business to generate a profit and agree to remain in business for at least this period of time.

Description of Business: List the partnership's objectives in terms which will enable the business to expand. Limit its scope to specific objectives. (*i.e.,* Do not use broad terms like "To farm the land.")

Contributions: List each partner's contributions and loans to the partnership and show each partner's share of total assets. Describe what happens if these contributions and loans are not made and provide for future contributions and loans.

Profits and Losses: Describe how profits and losses are to be divided and when they are to be distributed. List salaries of operating partners and how they are to be distributed.

    Records and Accounting: Describe accounting and banking
        procedures and where the books are to be located.

    Management: Describe the decision-making process for various
        aspects of the farm operation. List who will make these
        decisions and how other partners will be informed.

    Reformation of Partnership: Describe procedures for reforming
        the partnership if a partner leaves. Describe how departing
        partner's assets and liabilities will be transferred. Describe
        how a new partner can buy into the partnership.

    Dissolution of Partnership: Describe a procedure for terminating
        the partnership. Describe how assets and liabilities will be
        distributed.

Pogonip Farms and The Flower Ladies (Part IV, Conversations) provide good examples of how partnerships can be used to improve the competitive position of a metrofarm. Jeff Larkey and his fellow Pogonip farmers organized a partnership to gain control of prime land. The Flower Ladies organized a partnership, and friendship, to gain efficiencies through the specialization of labor and management.

Consider a Corporation

Corporate farming is not limited to the giants of agribusiness. It is often used by individuals, families and small groups of associates as well. In fact, the corporation is being used with increasing frequency to rescue many small farms from the treadmill effect of the family-farm cycle described earlier.

A corporation, like a person, is an entity which can buy and sell land, grow crops, make contracts, sue and be sued in court and pay taxes. Having no flesh and blood, however, a corporation cannot exist apart from the laws which give it the authority to exist. Though corporate law varies from state to state, the basic characteristics of the corporate entity are generally similar.

Ownership is controlled through paper certificates, called "stocks," which are given to individuals in exchange for capital. Each stock certificate, called a "share," gives its owner one vote in the corporate decision-making process, one share in the profits (called a "stock dividend") and one share in the corporation's assets and liabilities when the corporation is terminated.

Shareholders meet once a year to elect a board of directors. Directors, in turn, meet regularly to make policy decisions, such as which crops to plant, which land to buy or lease and when to pay dividends. Directors also appoint corporate officers (president, secretary and treasurer) to implement policy and manage day-to-day operations.

The corporation is governed by two sets of rules. The first, called the "shareholder's agreement," is a legal contract which describes the relationship and responsibilities of shareholders. The second, called "corporate bylaws," governs the day-to-day business operations conducted by corporate officers.

There are regular corporations, which may include shareholders from the general public; closed corporations, which are limited to 10 or fewer shareholders and closed to the public; and subchapter S corporations, which are taxed as limited partnerships. Of these, the closed corporation and the subchapter S are most frequently used in farm operations because they can be readily used by one person, one family or a small group of associates.

BENEFITS OF A CORPORATION: Many farmers are intimidated by the amount of paper work required to establish a corporation and consequently decide to operate with a simple sole proprietorship. Consider the following benefits before making your decision.

Investment Opportunities: A corporation can raise operating funds by selling additional stock or by borrowing against equity. By creating a larger reservoir of capital, the business can take advantage of many additional opportunities, such as better land, a more efficient production system, a more detailed marketing campaign and so on. These opportunities may not be available to sole proprietors or to general partners with limited resources.

Equity Transfer: Sole proprietors and partners own a business until they sell it or die. The business is then terminated and proceeds become subject to taxes.

A corporation, on the other hand, is an entity which continues to function regardless of the status of its shareholders. Personal considerations such as divorce and death affect shareholders, but not the corporation.

This feature of the corporate form of organization facilitates transfer of equity from parents to children. Children may earn shares gradually by investing their time in the farm business. This gradual accumulation of equity eases enormous tax burdens which come at the end of the family-farm cycle. This feature also enables parents to more readily divide equity between children who stay and work and ones who move on to other occupations. By organizing their business as a corporation, farm families may step off the treadmill of the family-farm cycle.

Personal Liability: The liability of corporate shareholders is limited to their investment. Since creditors cannot reach through a corporate entity to attach personal assets, potential shareholders might be more inclined to invest. This limitation of personal liability makes it easier for a farmer to generate additional capital.

Fringe Benefits: A corporation can provide fringe economic benefits for shareholders who are also employees. Benefits include medical insurance, tax-sheltered retirement plans, membership in social organizations and housing. While these benefits are individual expenses in a sole proprietorship or partnership, they may be counted as a business expense in a corporation. This feature of the corporate form of organization can provide a farmer with basic amenities and reduce the farm's tax burden.

Tax Options: Corporate profits are taxed twice: first as corporate income and then as shareholder income on wages and dividends. Though it might seem income taxed twice would be more than income taxed once, such is not always the case. By carefully surveying for the best ratio between flat corporate tax and progressive personal tax, a corporate farmer can sometimes reduce the overall tax burden.

LIMITATIONS OF A CORPORATION: There are four limitations to the corporate form of organization to be considered.

Management Flexibility: A sole proprietor is free to make decisions as he or she sees fit. Corporate directors and officers, on the other hand, are restricted by the corporation's charter and government regulation. (This limitation is mitigated in most closed and subchapter S family corporations because shareholders, directors and officers can confer and make decisions while doing breakfast dishes.)

Organizational Costs: It costs time and capital to organize and maintain a corporation. Shareholders must write a corporate charter, file with the state and pay its fees, conduct regular meetings, establish accounting procedures and pay corporate and personal taxes. Costs of satisfying these requirements can become significant if professional services of attorneys and accountants are employed.

Another organizational cost is the amount of resources which must be expended in dealing with disagreeable shareholders. Though this cost may be insignificant when business is good, it can grow quickly into a major cost when business is down.

Tax Burdens: As outlined in the benefits section, corporate income is taxed twice: once as corporate income and again as shareholder income. Though the manager of a closely held family corporation may actually be able to reduce taxes by finding the best ratio of flat corporate and progressive personal taxes, getting taxed twice will likely cost more than getting taxed once.

This double taxation can be avoided if all shareholders elect to pay taxes on their share of the total corporate income. This legal mechanism, called a subchapter S corporation, is similar in organization to limited partnership.

Personal Liability: There are several ways in which the corporate form of organization increases one's vulnerability.

First, by incorporating the farm, one loses protection of federal bankruptcy laws, which protect farm sole proprietorships and partnerships from being placed into involuntary bankruptcy by creditors. Second, if all one's personal assets are placed into the corporation, then all assets become subject to the corporation's risk. And finally, by acting as a corporate director or officer, one becomes liable for corporate activities and subject to being sued together with the corporation.

ESTABLISH A CORPORATION: Corporate law varies from state to state. Each state has a proscribed procedure for forming a corporation. In basic terms, incorporation is a two-step process in which shareholders write articles of incorporation and then file the articles with appropriate governmental agencies.

Articles of incorporation are a corporation's flesh and blood. These articles should be well-considered and include the following items:

Corporate Name: List the name of the corporate body and confirm no similar corporate name exists.

Identification: State the purpose for which the corporation was formed and business the corporation proposes to do.

Location: Name the county where the corporation's offices will be located.

Directors: List names and addresses of the corporation's board of directors. (Usually three directors are required by state law.)

Shares: State the total number of stock certificates of each class (*i.e.*, common and preferred) to be issued and monetary value of each.

Shareholder Agreement: List rights and responsibilities of shareholders. This article is similar to a partnership agreement in both purpose and importance.

The corporate form of organization enables one to step off the treadmill of the family-farm cycle. It allows one to take advantage of many previously unobtainable business opportunities. It limits personal liability of investors, provides many attractive fringe benefits and increases the number of tax options. As a consequence, the closed and subchapter S corporations are becoming an increasingly popular way to organize one's farm business.

Select the Ownership

Sole proprietorship, partnership or corporation? No single way to own a business is best for all farmers. To find your way, evaluate each of the following demands on ownership and decide which way will help you generate maximum return for minimum cost.

PROVIDE CAPITAL: A sole proprietor provides all capital or borrows against equity. A partnership does likewise or takes in additional partners. A corporation borrows against equity or sells additional shares.

MANAGE RISK: A sole proprietor assumes all risk, collects all rewards and is liable for all claims against the business. General partners share risks and rewards equally, but are individually responsible for all claims against the partnership. Limited partners share risks, rewards and liabilities only in proportion to their investments. Corporate stockholders share risk, reward and liability in proportion to their investment. Stockholders who serve as officers also share corporate liability.

CONTROL VOLUME: Sole proprietorships and general partnerships are simple to establish and inexpensive to maintain and are therefore suitable for small-volume businesses. Limited partnerships and corporations, on the other hand, are complex and require the expensive professional services of attorneys and accountants. They are more suitable for large-volume businesses.

MAINTAIN FLEXIBILITY: An individual (or married couple) manages a sole proprietorship. General partners manage general and limited partnerships. A chain-of-command, from stockholders to board of directors to officers, manages a corporation.

TRANSFER EQUITY: A business is dissolved and proceeds become subject to taxes when a sole proprietor or general partner sells or dies. On the other hand, business continues unabated when a corporate shareholder sells or dies.

MANAGE TAXES: Personal income from a sole proprietorship is taxed progressively: The more you make, the higher the rate you must pay. Income from a general partnership is divided among partners and then taxed as personal income. Corporate profits are taxed twice: first at a flat corporate rate and then at a progressive personal rate. Since tax rates and schedules continually change, only a thorough understanding of the current tax situation will reveal which way is best for managing taxes.

ESTABLISH MANAGEMENT

There are many decisions to be made in business. Some, like which crops to grow, are big. Others, like whether or not to take a day off, are small. Someone must make decisions and the someone is the manager.

There are three steps to an efficient manager's decision-making: First, the business objective is recognized; second, information relating to achieving this objective is collected and processed into alternative courses of action; and third, a course of action is selected which will generate maximum return for minimum resources.

To become a proficient manager, learn how to recognize the objective, collect and process the information and write the plan. This ability will enable you to control your metrofarm's business, instead of allowing the business to control you.

Define the Management Objective

Some think profit is what is left at the end of the year ("If I'm lucky!"). This thinking will not win the competition for consumer dollars.

Profit is the reward for putting time, money and know-how at risk in a business. While a reward is never guaranteed, it will most likely end up in the hands of those who plan for it. Every management decision, from big ones like which crops to grow to small ones like whether to take a day off, affects the possibility of generating profit.

Management's objective, therefore, is to generate profit on the time, money and know-how placed at risk in the business. To achieve this objective, the manager must identify financial requirements of each business decision, raise necessary capital, manage cash flow and protect assets.

IDENTIFY FINANCIAL NEEDS: The manager plans how the metrofarm will survive and grow by identifying how much capital the business will need and when it will be needed.

RAISE MONEY: There are two kinds of money: equity and debt. Equity is personal assets like savings, gifts, inheritances or wages from an outside job. Debt is assets borrowed from family, friends or financial institutions. Equity plus debt constitute the primary source of external money in a business. Internal money is generated through profit on the sale of products.

The manager plans how to satisfy money requirements of the metrofarm by locating sources of money, evaluating which source provides the best terms and determining how and when this money will be repaid.

MANAGE MONEY FLOW: Money is the "lifeblood" of business. The flow of money through a farm business is managed by forecasting sales, margins and expenses; budgeting to meet the timely payment of obligations; evaluating financial progress; and taking corrective actions when actual results differ from budgeted objectives.

PROTECT ASSETS: The manager plans how to protect assets which have been acquired by the business. This plan includes establishing procedures for eventual transfer of equity (estate planning); maintaining adequate insurance coverage; providing adequate employee training; and giving attention to opportunities to diversify.

Each management decision must help generate a profitable return on the time, money and know-how invested in the business. Successful managers focus on this objective with the intensity of David staring down mighty Goliath on the field of battle.

Process Information

Some farmers would rather shovel manure for two hours than keep records for one. This reluctance explains why, as agricultural economists often note, more farmers fail because of poor management than because of poor production.

Keeping accurate business records is a primary responsibility of management. There are three reasons why this responsibility must be fulfilled.

First, records provide an accurate picture of the farm's financial condition. They reveal its current status, why it is in this condition, and how this condition compares to the competition. In addition, accurate records tell exactly where a farm profits and where it forfeits.

Second, accurate records tell how much income must be relinquished to government in taxes.

Third, records provide solid information for making financial projections. Most projections are calculated with probable costs and returns. The soundness of projections depends on the accuracy of information upon which they are based. Many factors, including weather conditions, crop diseases and severe price fluctuations, cannot be calculated with certainty. A set of accurate records mitigates uncertainties by telling the manager exactly what has happened in the past.

Successful managers establish the right accounts, identify and adopt the best record-keeping system and then create sound financial statements.

ESTABLISH ACCOUNTS: Accurate record-keeping begins with timely posting of transactions to proper accounts. There are three types of accounts to establish: personal, capital and operational.

Personal Accounts: Personal and business transactions often become intermingled in metrofarming. The electricity bill, for example, may include charges for a greenhouse fan and a kitchen fan. While the greenhouse fan is a business expense, the kitchen fan is a personal one. This is true for the revenue side as well. Proceeds from the greenhouse are business, while proceeds from an outside job are personal. This explains why an account must be established where personal transactions may be segregated from business transactions.

Capital Accounts and Depreciation Schedules: Capital is property. It may exist as cash, equipment, know-how, livestock or real estate. Many production systems require a capital outlay which will last over a number of years. Examples include greenhouses, tractors and irrigation systems. The purchase of equipment and facilities is recorded in a capital account and in an accompanying depreciation schedule.

Consider, by way of example, the purchase of a new greenhouse: If the entire price of this greenhouse is charged to current operations, it would severely distort the farm's profit or loss picture for a year. The greenhouse is therefore listed as a capital outlay and its cost spread over its estimated life in the depreciation

Farm Record Checklist

Information wanted	Check your needs ✓	What each system provides:			
		Check book	Farm record book	Double-entry accounting	Enterprise accounting
Income tax—cash basis (Farm Schedule 1040F)		Yes	Yes	Yes	Yes
Farm profit—accrual basis		No	Yes	Op.*	Yes
Personal return—Form 1040, including nonfarm incomes, capital gains, etc.		No	Yes	Yes	Yes
Farm profit statement for management, using inventories and true depreciation		No	Yes	Op.	Yes
Net worth or financial statement		No	Yes	Yes	Yes
Reference record of financial transactions		No	Yes	Yes	Yes
Owner's investment and withdrawals		No	Yes	Yes	Yes
Investment and depreciation record, listing individual items		No	Yes	Su.**	Yes
Inventory of crops, livestock, and supplies on hand annually		No	Yes	Su.	Yes
Payroll record		No	Su.	Su.	Yes
Social security tax records		Su.	Su.	Su.	Yes
Crop and livestock production records		No	Yes	Op.	Yes
Joint ownership accounts		No	No	Yes	Yes
Accounts receivable and payable		No	No	Yes	Yes
Enterprise profit statement and analysis		No	No	No	Yes
Operating costs of service units		No	No	No	Yes
Gasoline use for tax refunds		Op.	Op.	Op.	Yes

* Optional record or statement that may be made available if desired, but is frequently not part of the usual system.
** A supplemental record that may be kept if desired or needed.

FIG 7.2 Four ways to keep records. From *Financial Records for California Farmers*, A. Doyle Reed. Leaflet #2709. Division of Agricultural Sciences, University of California, 1975.

schedule. Capital accounts with accompanying depreciation schedules help make financial sense of long-term transactions.

Operational Accounts: Operational transactions are those required for the day-to-day operation of the farm. When money is borrowed to buy a new greenhouse, the principal payment is posted to a capital account. Interest payment, on the other hand, is posted to an operations account. When money is expended to run the greenhouse, the transaction is also recorded in an operational account.

There are two methods for recording operational transactions, cash or accrual. In the cash method, transactions are recorded when paid or received. For example, if seed is purchased this year for sowing next, the transaction would be recorded as this year's operational expense. In the accrual method, transactions are recorded in the year in which they are expended; *i.e.*, seeds purchased this year would be charged to next year or whenever they are sown.

Most small farms use the cash method because of its simplicity. Most large corporate farms and limited partnerships use the accrual method because they are required to do so by government.

SELECT A RECORD-KEEPING SYSTEM: A record-keeping system is a group of records designed to provide an accurate picture of one's personal and business affairs. As each farm is unique, so too is each farm's record-keeping system. There are four basic systems to consider. The general rule: The more complex the farm, the more complex its record-keeping system.

Check Book System: The simplest way to keep records is by using a check book. All personal, capital and operating transactions can be recorded on the check stub and the deposit slip. At the end of the year, the transactions can be separated into their proper account, tallied and then used to calculate profit or loss.

Invoices, receipts, asset schedules, employees' earnings records and production records should be kept to augment the check book.

The check book is easy to use, takes little time and is not expensive. Beyond supplying basic data for calculating tax, however, the check book does not provide much useful information to the manager. Furthermore, the time saved by not recording information on a regular basis may have to be paid back when all the year's transactions must be separated and tallied at tax time.

Farm Record-Book System: A farm record book is a published book of forms designed to provide the essential farm records in one package. The most popular record books, like the *California Farm Record Book*, are published by land grant colleges and made available through local farm advisors. There are many other record books and many have been packaged into computer software programs. Some record books are worthwhile, others are not. A good record book will contain the following forms:

- a cash journal to keep a running account of all personal, capital and operational expenses and revenues.
- a page for mapping the farm and its crops.
- crop and livestock production records.
- inventories for land, crops, supplies and livestock.
- depreciation schedules for crops, buildings and improvements, equipment and vehicles.
- miscellaneous accounts receivable and payable.
- a Social Security tax record for employees.

A good farm record book system provides management with information with which it can make sound business decisions. In addition, the system is simple and requires no special training. Its limitation is the time required to dutifully enter each transactions.

Double-Entry Ledger System: In double-entry record-keeping, each transaction is first entered into a cash journal like the one found in a farm record book. The amount is later posted to a ledger or book of accounts as a debit to one account and a credit to another (thus the double entry). When tallied, total debits equal the total credits.

Double entry record-keeping is used when accuracy must be proven and is therefore used by most partnerships and corporations or by any large business which has employees who handle cash flow. The system is highly technical and is usually kept by trained specialists who design each ledger or book of accounts to fit the farm's operations.

There are three benefits to using the double-entry system. First, more segregations of income and expense are possible in the unlimited number of accounts which can be constructed. Second, credit transactions can be tracked more efficiently. And third, accuracy can be audited and proved.

The limitation stems from the system's complexity. Unless the manager has technical training in accounting, the books are usually maintained by a professional.

Enterprise System: This record-keeping system begins where the double-entry one quits. It uses an account, or group of accounts, for each crop. For example, the farm's cabbage crop has one set of accounts, while its cut flowers crop has another set. Each enterprise is credited with its direct costs, its share of the general overhead and its sales.

Enterprise accounting provides management with detailed information about which crop wins and which loses. Enterprise system, in short, provides management with the most comprehensive information.

The limitation of the enterprise system is complexity. Establishing accounts for each enterprise may require more accounting expertise than a farmer has and more agricultural expertise than an accountant has. And once established, the

Balance Sheet

(also called Net Worth or Financial Statement)

ASSETS

a. Personal Assets	
b. Capital Assets	
c. Operating Assets	+

Total Assets

LIABILITIES

a. Personal Liabilities	
b. Capital Liabilities	
c. Operating Liabilities	+

Total Liabilities

NET WORTH

a. Total Assets	
b. Total Liabilities	-

Net Worth

FIG 7.3 A balance sheet is based on the principle that assets equal liabilities plus net worth. It shows how solvent a business is at a specific point in time, how its assets have grown and what will happen if it borrows money.

system requires more time because transactions for each crop enterprise must be posted to separate accounts.

DEVELOP FINANCIAL STATEMENTS: The good manager uses information from records to create financial statements which allow for control of liquidity, leverage and profit.

Liquidity means having enough cash to meet obligations in a timely manner. Leverage is maintaining a reasonable and safe mix between the farm's

Income Statement

(also called Profit Or Loss Statement)

REVENUE

a.   Personal Revenue   _____

b.   Capital Revenue   _____

c.   Operating Revenue   +   _____

Total Revenue   _____

EXPENDITURES

a.   Personal Expenses   _____

b.   Capital Expenses   _____

c.   Operating Expenses   +   _____

Total Expenditures   _____

PROFIT OR LOSS

a.   Total Revenue   _____

b.   Total Expenditures   -   _____

Profit or Loss   _____

FIG 7.4 An income statement is based on the principle that net profit is equal to revenue minus expenses. It shows how much profit or loss has been earned and provides documentation for securing credit.

equity and debt capital. And profitability means earning a good return on this capital.

Financial statements used to gain and maintain control are the balance sheet, income statement and cash-flow statement. Information taken from the metrofarm's record-keeping system provides the raw materials needed for creating these financial statements.

Balance Sheet: The balance sheet lists all of the metrofarm's assets and liabilities at a specific point in time. It is based on the fundamental accounting equation of assets equal liabilities plus net worth.

Cash Flow Statement

BEGINNING CASH BALANCE

a. Checking Account

b. Other Cash Accounts   +

Total Beginning Balance

REVENUE FOR PERIOD

a. Personal Revenues

b. Capital Revenues

c. Operating Revenues   +

Total Revenues

TOTAL CASH AVAILABLE

a. Total Beginning Balance

b. Total Revenues   +

Total Cash Available

EXPENDITURES FOR PERIOD

a. Personal Expenses

b. Capital Expenses

c. Operating Expenses   +

Total Cash Expenditures

ENDING CASH BALANCE

a. Total Cash Available

b. Total Cash Expenditures   -

Ending Cash Balance

FIG 7.5 A cash flow statement is based on the principle that cash available must be equal to or greater than cash required. It tells how much cash is available, how much is required and how much must be obtained to cover expenses.

The balance sheet shows how solvent a farm is at a specific point in time; tells how much additional debt it can service and how much additional collateral it can pledge; and reveals how equity has grown. The balance sheet helps a manager think like a banker when applying for credit.

Income Statement: The income statement examines the farm's net profit or loss for one year by subtracting all costs from all revenues.

The income statement tells how the metrofarm performed during the year and how this performance compares with previous years. It aids in preparation of taxes, helps in analyses of return on capital and is used as a supporting document when seeking additional debt capital.

Cash Flow Statement: A cash flow statement examines cash flowing into the business and cash flowing out. It does not examine non-cash needs, nor the need for cash in future years.

The cash flow statement helps project how much cash will be needed for the year. And cash needs must equal cash sources. If cash needs exceed cash flow, then the difference must be taken from savings or borrowed. Cash flow statements aid in establishing credit because it tells when cash will be needed and when it will be paid back.

Write a Business Plan

The objective is to control the metrofarm's business without being controlled by it. To achieve control, the manager must establish realistic financial objectives and a set a pre-determined course of action. Culmination of this process is the writing of a business plan.

This business plan is written for the metrofarm's owner, manager and lender (even if all positions are occupied by the same person). The plan tells where the business is, where it is going, how it is going to get there and what it is going to do when it reaches the objective.

Demonstrate you are in control of your business. Write a business plan and include the articles discussed below:

SUMMARIZE THE BUSINESS: Write an introductory paragraph giving an overview of your metrofarm's name, its business and products and its location. Include more detailed information in the following sections.

> *Name of Business:* Give the business name and, if necessary, include details about previous uses of the name, filing of a fictitious business name statement, and results of a name search with the Secretary of State (or equivalent) and Federal Trademark Register.

Location and Facilities: Identify the metrofarm's location and describe how it relates to the rest of the community. Describe why this particular location will support the success of the business and its products. Identify your current facilities and describe how they will help satisfy the farm's production requirements. Identify plans for improving facilities and provide reasons for making the improvements.

Product, Market and Competition: Provide an introductory description of your metrofarm's product, its market and your competition for consumer dollars. Expand this description in following paragraphs.

Describe the product. Start with a simple description anyone can understand and provide a detailed analysis of why your product is unique and what makes it better than the competition's product.

Describe the market. Give a simple overview of the market. Who buys the product and what do they do with it? Give specifics about your product and why consumers will buy it rather than the competition's product.

Describe the competition. Start with a simple overview of the competition and then give specific details about the competition in your market. Provide an in-depth analysis of their competitive strengths and weaknesses.

Ownership: Describe how you have organized the ownership of the business. Are you a sole proprietorship, partnership, or corporation? Describe any plans you might have for changing this organization and why.

Objectives: Provide a broad description of how you see the business growing. Include changes in your product line, business location, organization, and personnel. Project growth in expenses and income.

ANALYZE THE BUSINESS: Write an introductory description of your metrofarm business. Include more detailed information in the following sections.

Production: Describe how you grow and process crops into marketable products. Include the processes, materials, suppliers and lenders which make production possible. Do not include proprietary information.

Technology: Compare your production technology with the competition's. Describe what makes your technology more efficient and/or productive than the competition's.

Equipment: Describe the equipment required for your metrofarm. Provide, if applicable, informed projections on how additional or upgraded equipment could improve productivity.

Business Strategies: Identify your major strengths and weaknesses. Describe how specific factors such as location, product, technology and market trends will enable you to avoid the competition's strength and attack its weaknesses.

DESCRIBE MANAGEMENT: Write an introductory paragraph describing management of your metrofarm business. Provide details in the following sections.

Experience: Introduce the principal manager(s) of your metrofarm. Describe relevant experience, training and education. Where applicable, include a formal written resume.

Labor: Describe specific personnel requirements of your metrofarm. Include skills needed, pay rates and plans for additional hiring. Identify paid consultants and other temporary personnel.

Operations: Provide a description of how your business is managed. Include, if applicable, a flow-chart showing who is doing what and to whom they report. Describe management of all farm operations, including production, processing, marketing and sales.

PROVIDE FINANCIAL INFORMATION: Write an introductory paragraph describing financial management of your metrofarm. Provide details in the following sections.

Balance Sheet: Show all of your company's assets, liabilities and equity. Use this balance sheet to demonstrate that your business has the ability to support its projected growth.

Income Statement: Show how effectively your resources have been used and how well financial management of the business has been performed. Demonstrate how successful your farm has been during the past year.

Cash Flow Statement: Summarize how cash has moved in and out of the business during the past year. Demonstrate when the business has excess cash and when it is in deficit.

Loan Budget: Give an overview of all loans and outstanding financial obligations. Include their length and terms. Demonstrate how you can service these obligations.

Business Plan

I.   SUMMARY OF BUSINESS

    A.   Name of business

    B.   Location and description of business facilities, condition, and planned improvements

    C.   Product, market, and competition

        1.   Description of product(s)

        2.   Description of market in general and your approach in particular

        3.   Description of competition

    D.   Ownership: Legal structure and documents

    E.   Goals

II.   ANALYSIS OF BUSINESS

    A.   Manufacturing/production (if relevant): Description of process, timing and supplier

    B.   Technology: Compare your technology with competitors'

    C.   Capacity: Compare your capacity to produce or serve relative to competitors'

    D.   Business equipment

    E.   Major strengths and weaknesses

    F.   Key success factors

FIG 7.6 A business plan shows a realistic financial objective has been established and a pre-determined course of action set.

III. **MANAGEMENT**

    A.   Experience of owners and key personnel

    B.   Staffing plans

    C.   Operating plans

IV. **FINANCIAL DATA**

    A.   Pro forma balance sheet(s)

    B.   Profit and loss projections

    C.   Cash flow projections

    D.   Loan Budget

    E.   Insurance

    F.   Description of collateral

    G.   Tax returns

FIG 7.6 Business Plan (continued).

Insurance: List the type, amounts and providers of all insurance policies. Include specific problem areas such as worker's compensation and customer liability.

Collateral: List all personal assets pledged as collateral against loans. Include a description of each item, its value and how its ownership is identified.

Tax Returns: If applicable, as is the case when applying for a loan, include copies of your personal tax returns for the past several years.

Writing this plan is the culmination of establishing management control over the business. It describes, in black and white, how you will control the financial operations of your metrofarm without allowing the operations to control you.

REVIEW

A well-organized business is a weapon in the competition for consumer dollars. There are two steps to organizing a business: First, establish ownership; second, establish management.

You may own a business as a sole proprietor, partner or shareholder. No one way is best for all metrofarmers. Each way has its benefits and limitations. You must carefully weigh these factors and then decide which way will enable you to generate maximum return for minimum cost.

There are many decisions to be made in the management of a metrofarm business. Some, like which crops to grow, are big; others, like whether or not to take a day off, are small. Each decision affects your ability to generate profit. Your ability to make good decisions is governed by how well you recognize the objective, keep records and plan.

Organization will enable you to control your business without being controlled by it. This control will give you the ability and confidence to face down the Goliaths which dominate competition for consumer dollars.

CHAPTER SEVEN EXERCISES

i   Begin a subsection entitled "Ownership" within the "Production" section of your metrofarm workbook. Title one page "Sole Proprietorship;" another, "Partnership;" and a third, "Subchapter S Corporation." On each page list the principal strengths and weaknesses of each respective form of business ownership.

ii   Research laws of your state which govern formation of limited partnerships. Write a limited partnership proposal to generate enough capital to produce and market the crops selected in Chapter 6 exercises. Make certain your financial proposal conforms to current laws of your state. (A local library should be able to provide good information sources.)

iii   Begin another subsection and title it "Management." Title the first page of this subsection, "Management's Objective." Write a one sentence description of the manager's business objective. Next write a one sentence description of how the manager achieves this objective.

iv   Imagine your household as a business enterprise. Select one of your major household purchases— car, house or television— and write a depreciation schedule for the item. Next, write a balance sheet, an income statement and a cash-flow statement on the business of running your household. Use actual figures.

v   Write a business plan for your household enterprise. Include all items listed in the "Write the Business Plan" section of this chapter. Use this business plan to demonstrate how you control your finances.

Establishing Production

With our broad fields, our machinery and few people, their system appears to us crude and impossible, but cut our holding to the size of theirs and the same stroke makes our machinery, even our plows, still more impossible, and so the more one studies the environment of these people, their numbers, what they have done and are doing, against what odds they have succeeded, the more difficult it becomes to see what course might have been better.

F. H. King, *Farmers of Forty Centuries: Permanent Agriculture in China, Korea, and Japan*

The objective is to survive competition and prosper. Typical obstacles include insufficient size and a shortage of capital. Obstacles can be overcome and objective achieved by establishing an efficient system of production.

Metrofarm production systems are complex units formed of many different elements. The design of each system is determined by the metrofarm's crops, size, topography, climate and proximity to market. Since no two metrofarms operate in exactly the same conditions, no two metrofarm production systems are exactly the same.

Though your metrofarm will be as unique as a fingerprint, it must overcome two production objectives shared by all metrofarms: First, it must yield the maximum number of products in the minimum amount of space. In lieu of the industry standard of three tons per acre, establish a production system which will yield nine tons per acre! Second, it must yield maximum return for minimum cost. In lieu of the industry standard of a $12 product for $10, establish a production system which will yield a $15 product for $5!

This chapter will help you establish production. It begins by offering a precedent for a system which has survived, successfully, for 4,000 years. It then offers a guide to drafting a system design, building soil, applying fertilizer, managing water, controlling pests, forcing growth and development, managing equipment and labor, providing facilities and scheduling crops.

RECOGNIZE THE PRECEDENT

The safe way is one which has proved successful in the past. Precedent reveals if it can be done and how it can be done. There is a precedent for the profitable farming of small parcels of land and it has a 4,000 year history of success.

In 1900, F.H. King, a former chief of the U.S. Department of Agriculture, studied the farm production systems of China, Korea and Japan. King's objective was to find ways to improve farm productivity in the United States. The essence of what he learned can be gleaned from a paragraph recorded in his book, *Farmers of Forty Centuries*:

> The man walking down the row with his manure pails swinging from his shoulders informed us on his return that in his household there were twenty to be fed; that from this garden of half an acre of land he usually sold a product bringing in $400. The crop was cucumbers in groups of two rows thirty inches apart and twenty-four inches between the groups. The plants were eight to ten inches apart in the row. He had just marketed the last of a crop of greens which occupied the space between the rows of cucumbers seen under the strong, durable, light and very readily removable trellises. On May 28 the vines were beginning to run, so not a minute had been lost in the change of a crop. On the contrary this man had added a month to his growing season by over-lapping his crops, and the trellises enabled him to feed more plants of this type than there was room for vines on the ground. With ingenuity and much labor he had made his half acre for cucumbers equivalent to more than two. He had removed the vines entirely from the ground; had provided a travel space two feet wide, down which he was walking, and he had made it possible to work about the roots of every plant for the purpose of hoeing and feeding. Four acres of cucumbers handled by American field methods would not yield more than this man's half acre, and he grows besides two other crops the same season. The difference is not so much in activity of muscle as it is in alertness and efficiency of the grey matter of the brain.

FIG 8.1 The small farms of Asia, like the ones pictured on the Japanese island of Kyushu, offer 4,000 years of precedent for the profitable farming of small parcels of land.

King's observation contains two important revelations: First, a one-half acre farm produced enough income to sustain twenty people. There are, in contemporary agriculture, thousand acre farms which do not sustain two people! Second, high-density production is based on the "alertness and efficiency of the grey matter of the brain," rather than on large amounts of money. The farmer, as King has observed, is the single most important ingredient in production.

Give the farmer of ancient China a lift into the here and now. Lease him a small parcel of land near a market. Lend him a few dollars for seeds, an old pickup truck and a few other inexpensive labor-saving devices. In a few years, he will likely be earning a substantial income on his own metrofarm. Sound incredible? The West Coast is well-peopled with this kind of success story. Consider, for example, the Chino family of Rancho Santa Fe, California:

After Junzo and Hatsyo Chino were released from the Japanese internment camps of World War II, they began growing vegetables and flowers on small plots in the Los Angeles and San Diego areas. They generated enough income to raise eight children— most of whom became doctors and lawyers— and to purchase some 56 acres of prime farmland near Rancho Santa Fe. Today the farm, which is valued at many millions of dollars, is run by the Chinos' son Tom, who is a former cancer research specialist.

FIG 8.2 The Vegetable Shop, near Rancho Sante Fe, California, traces its beginnings to 1945, when Junzo and Hatsyo Chino were released from a Japanese internment camp and began growing vegetables and flowers on small parcels of land near Los Angeles. The Chinos grew enough to raise eight children and buy 56 acres of what is now some of the most valuable land on the West Coast.

Every day during the sweet corn season a line of luxury cars pulls into the Chinos' parking lot where people stand in line up to an hour waiting to buy. When asked why she was willing to stand in line, one silver-haired matron said, "Because the corn is sweeter and more tender than any I've ever eaten. It's worth the wait."

Many imagine the farmer of ancient China walking down the road with manure pails swinging and think: *'FARMER \ 4a: an ignorant rustic: YOKEL, BUMPKIN b: a clumsy stupid fellow: DOLT.'* Others think of the "alertness and efficiency of the grey matter of the brain" and see, as so many immigrants have seen, the opportunity available in small parcels of unused land. Look to the individuals featured in Part IV, Conversations. They began farming with small sums of money, established a good production system, and then invested time and ingenuity to make their systems work.

Farmers have relied on the capital of time and know-how for thousands of years to generate big incomes from small parcels of land. Recognize the precedent. See your resources as the "farmers of forty centuries" would see your resources. Use the alertness and efficiency of your "grey matter" to establish a farm which will waste not a single moment nor a square inch in winning the competition for consumer dollars.

DRAFT A DESIGN

Begin by designing a production system which takes maximum advantage of environmental resources. The objective is to use each resource like a welder's torch uses gas and oxygen to force a soft orange flame into one of white-hot intensity.

EVALUATE LAY: Natural elements of a farm's environment, such as its elevation, direction and degree of slope and prevailing winds are called "lay."

Elements, as discussed in previous Chapters, govern the crop's rate of growth, development and yield. A south-facing slope, for example, receives more direct sunshine than one facing north. Crops planted on a south-facing slope will therefore tend to grow faster, longer and yield more than will crops planted on a north-facing slope. (This explains why the Chinese call the south-facing side of a hill "sheng bian," or growth side.)

Favorable features of lay, such as a south-facing slope, can help increase yields in areas where the overall climate is less than generous. Similarly, unfavorable features, such as a north-facing slope, can reduce productivity in the best of environments. Know the lay of your land. List each element of the environment which will have an effect on your crop's growth, development and yield.

EVALUATE LAYOUT: Man-made elements of a farm's environment, including its roads, wells, irrigation ditches, drainage systems and storage buildings are called "layout."

Elements of layout affect costs of production. For example, if a production site is located at the top of a hill and its water source at the bottom, pumping costs will be more than if the site were located at the bottom and the water at the top. Know the layout of your land. Identify each man-made element which will have an effect on how much it will cost to grow your crop.

SKETCH A DESIGN: Lay affects yield; layout affects efficiency. Sketch a system which takes maximum advantage of the positive elements in the environment and which avoids negative ones. The steps which follow will help you sketch a good design.

Draw the Map: Sketch a large, detailed map of your land on a piece of paper. An aerial photograph will help locate all the details but if one is not available, use road maps and personal observation.

Identify Elements of Lay and Layout: Find the elements of lay and layout which will affect yield and costs and note the elements on the map. Include all pertinent features outlined in Chapter Five, Evaluating and Controlling Land. This step will help you pay attention to important things, such as the path of the sun during its extremes of winter and summer solstice, prevailing winds, direction and degree of slope, roads and water sources.

Label the Elements: Label each element good (+) or bad (-). This labeling

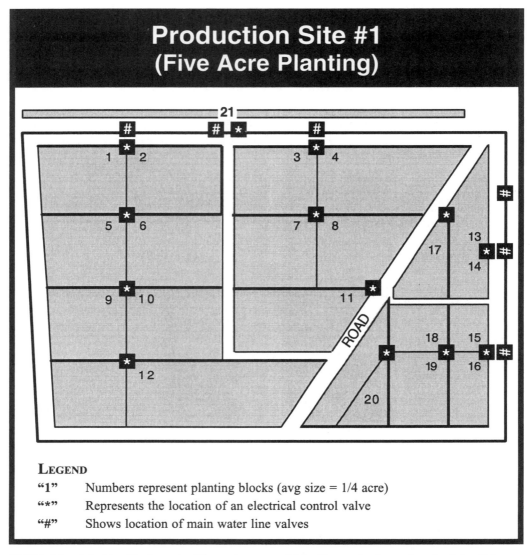

FIG 8.3 Maps help identify elements of the environment which affect production and productivity. This map identifies sources of water.

will require an evaluation of how each element of lay will affect yield and how each element of layout will affect costs.

   Lay It Out: Lay out a system which takes maximum advantage of positive elements and which avoids negative ones. For example, if your crop requires a long period of sunlight, situate the production site so it faces south and tracks the path of the sun. Also, draft plans to improve the situation. For example, if a large stand of trees creates a dangerous frost pocket, plan to cull a few trees so air can flow more freely.

Draft a design which masters your metrofarm's feng shui. It will cost time and effort but the investments will be returned many times over through increased yields and reduced costs.

BUILD SOIL

Soil is the foundation upon which crops grow and develop: It supports, nourishes, provides water to and regulates the metabolic rate of plant crops. It provides the plant nourishment required by animal, bird and fish crops.

A primary obstacle to farming a small parcel of land is lack of space with which to grow enough crops. Two kinds of soil mediums are often used by metrofarmers to overcome this obstacle. One consists of biologically-active ingredients blended into native soil. This medium is often called "organic." The other consists of biologically-inert ingredients. We will call this medium "synthetic."

Consider Organic Soil

Organic soils are topsoils with relatively large amounts of biologically-active ingredients, such as manures and composts, which provide nutrients, texture and a balanced *pH*. Building a naturally thin layer of topsoil into an artificially thick one enables plants to find nutrients and moisture by sending roots down a short distance rather than a long one. Consequently, many more plants can be grown on a much smaller parcel of land.

BENEFITS OF ORGANIC SOIL: In addition to the topsoil benefit listed above, organic soils offer metrofarmers many other ways to increase production and productivity. Following are discussions of some important ways.

Product Value: Crops grown in organic soils may qualify as "organic produce," and organic products often sell for a premium over those grown with synthetic fertilizers and pesticides.

Premium prices are paid because many consumers perceive organic products to be of higher quality. This perception is continually reinforced by news accounts of food products contaminated with various dangerous substances. California's Department of Health Services, for example, recently reported 106 of the 1,218 food and water samples it tested contained traces of pesticides and other potentially dangerous chemicals.

How consumers perceive organic products can generate substantial premiums for the farmer. At this writing, for example, I am selling an organic allium product into the wholesale market for $1.85 to $2.00 per pound, while the same

CONVENTIONAL SOILS   HIGH DENSITY SOILS

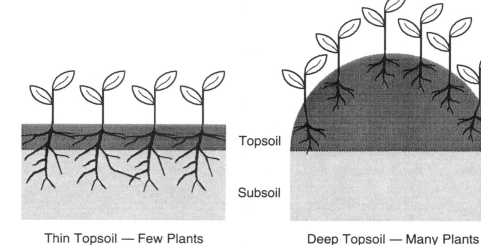

Topsoil

Subsoil

Thin Topsoil — Few Plants   Deep Topsoil — Many Plants

FIG 8.4 With large scale soil practices, plants must send their roots out for a sufficient supply of nutrients and water. With high density soil practices, plants' roots find all the nutrients and water they need close to the plant.

variety grown conventionally is selling for less than $1 per pound.

Water and Irrigation: Water is an expensive resource and providing it to crops increases the costs. Organic soil can reduce water and irrigation costs.

The humic acids released by decaying organic matter improve air and water penetration in heavily-textured soils like clay. The colloidal properties of humus improve the moisture-holding capacity of lightly-textured soils like sand by binding soil particles together into a mass capable of holding water. Government studies claim, and practicing farmers confirm, that organic soils can reduce water requirements by thirty to 300 percent.

Cost savings can be substantial in areas where water is an expensive resource. I am currently observing a controlled experiment in which a highly refined blend of organic materials has been applied to golf course turf grass. To date the blend has enabled the greenskeeper to reduce irrigation time on the computer-operated sprinkler system by a full fifty percent.

Fertilizers: There are three ways in which organic soils can reduce fertilizer costs.

First, organic matter provides an abundance of negatively-charged colloidal size particles called "humus" which are capable of holding and exchanging essential nutrients. Humus, in other words, is the glue which binds soil particles together, preventing nutrients from leaching away or escaping back into the atmosphere. The colloidal property is especially important because the major nutrient nitrogen is highly mobile and will be lost if not held in place.

Second, humic acids affect the formation of metal-organic complexes, thus stabilizing soil micronutrients which otherwise might not be available in sufficient quantities. Humic acids, in other words, dissolve soil particles which would otherwise keep micronutrients locked up and unavailable.

Third, organic matter releases the essential nutrient, iron, as a result of the vigorous microbial activity which occurs during decomposition.

By holding essential nutrients in place and by freeing other nutrients, organic soils encourage vigorous growth and development while reducing or even eliminating the need for frequent applications of expensive fertilizers.

Soil Conditioning: A well-designed organic soil is very forgiving. Decaying organic material contained in soil acts as a buffering agent by decreasing the tendency for abrupt *pH* changes when acid or alkaline-forming substances are added to the soil. How can this reduce costs?

Organic soils allow you to make mistakes. You can apply too much synthetic nitrogen or too little water and be forgiven by the buffering activity of decomposition. Soils lacking organic matter do not have the capacity to forgive and one small mistake, like applying too much synthetic nitrogen, can increase costs, reduce yields and may even destroy the crop.

Pest Control: Pests compete with you for the energy contained in your living crops. How can organic soils help you win this competition?

Weeds steal sunshine, water and nutrients. Crops planted extremely close together in organic soils can out-compete the few weeds which might sprout up between them by taking all the sunshine. In addition, weeds growing in well-textured organic soils are much easier to pull than weeds growing in hardpan clays. Organic soil, in other words, can reduce costs of weeding, eliminate need for expensive herbicides and reduce fertilizer and water costs.

Organic matter also helps reduce infestations of harmful soil pests, like nematodes, by encouraging natural predators, like nematode-trapping fungi. And finally, organic decomposition provides a balanced source of slow-release nutrients, thus promoting the kind of slow, steady and healthy growth which resists insect and disease pests which prey on weak crops. Organic soils, in other words, can reduce or even eliminate the need for expensive pesticides.

Erosion Control: A raindrop is like a miniature bomb which blasts particles of unprotected soil into the air and down the river. How can organic soil help reduce soil erosion costs? Colloidal properties of decomposing organic matter literally glue soil particles together, giving the particles capacity to resist the erosive effects of wind and rain.

Land Costs: The cost of controlling a parcel of land often provides the most difficult obstacle to establishing production. This is especially true where land

Selected Soil-Building Amendments

FOR HEAVY CLAY SOILS	FOR LIGHT SAND SOILS
Alfalfa Hay	Animal Manures
Clodbuster	Cocoa Bean Hulls
Compost	Compost (Mushroom)
Granite Dust	Green Manures (Fava Bean)
Green Manures (Rye)	Ground Bark
Gypsum	Leaf Mold
Rice Hulls	Peat Moss
Sand	Rice Hulls
Sawdust	Sawdust Sedge Peat
Straw	Vermiculite
Vermiculite	Woody Peat

FIG 8.5 Soil is amended to improve its moisture retention, drainage, aeration and nutrient-holding characteristics.

values are highly inflated, as they tend to be near large metropolitan areas. How can an organic soil medium help overcome this obstacle?

A well-designed organic soil can be built into almost any small parcel of unused land for a relatively small cost. This gives you the opportunity to establish production in marginal locations such as scrub-covered hillsides, vacant lots, green belts, power transmission corridors and so on. The otherwise unusable parcels of land can often be controlled for much less than a prime parcel of perfectly flat land which has been zoned for a shopping center or office complex.

Farm Appearance: The appearance of a farm becomes economically significant when the farm is used as a marketing facility. Consumers visiting the farm are in the market for more than a farm product; they also seek farm ambience. A well-designed organic soil can enhance ambience far better than an industrial grade soil labeled with a "Danger: Poison" sign.

A good example of how a well-designed organic soil can enhance a farm's

Soil *pH* Scale (1 to 14)

	3	4	5	6	7	8	9	10	11	
	Extremely Acidic		*Acidic*	*Weakly Acidic*	*Neutral*	*Weakly Alkaline*	*Alkaline*		*Extremely Alkaline*	

1 TO 4.3	4.4 TO 5.4	5.5 TO 6.4	NEUTRAL	7.4 TO 8	ABOVE 8
Injurious to most crops; promotes harmful bacteria; only a few mosses and algae can grow.	Azalea Blueberry Cranberry Heather Holly Rhododendron	Bentgrass Buckwheat Cotton Cucumber Field Bean Millet Oats Parsley Peanuts Potato Raspberry Rye Strawberry Tobacco Tomato Vetch	Most Crops	Alfalfa Asparagus Barley Beets Broccoli Cabbage Carnation Cauliflower Celery Dahlias Lettuce Muskmelon Onion Parsnip Spinach Sweet Clover	Soil too alkaline for most crops, very few have specialized enough to survive.

FIG 8.6 A soil's acidity or alkalinity, which is expressed in terms of *pH*, determines whether certain nutrients will be released to growing plants.

Selected Amendments for Controlling Soil *pH*

FOR RAISING *pH* OF ACID SOILS (Wet Climate Soils)	FOR LOWERING *pH* OF BASIC SOILS (Dry Climate Soils)
Calcitic Limestone	Animal Manures
Dolomitic Limestone	Composts
Hydrated Lime	Green Manures
Basic Slag	Crushed Eggshells
Calcium Silicate Slag	Peat Moss
Ground Shells	Redwood Mulch
Marble Dust	Sawdust
Marl	Leaf Mulch
Mud (Sugarcane and beet)	Gypsum
Paper Mill Sludge	Sulphur
Wood Ashes	Sulfuric Acid

FIG 8.7 Soil *pH* can be amended to satisfy the specific needs of individual crops.

appearance can be seen at Heyns' Country Gardens (Part IV, Conversations). People from nearby Billings, Montana, make reservations in January for the privilege of picking Heyns' strawberries in summer. Since consumers spend a lot of time on their hands and knees, a good soil plays an important part in the attractiveness and profitability of Heyns' business.

LIMITATIONS OF ORGANIC SOIL: Be aware of the costs of establishing organic soils. Start-up and management are two to watch.

Start-up: It costs time to learn the chemical characteristics of organic materials; time and money to gather, handle and compost materials into a usable form; time and money to arrange soil into rows, beds or boxes; and time to allow soil decomposers to convert organic materials into available nutrients and humus. Although organic soils improve and become less costly with use, start-up costs must be paid when you may least be able to afford them.

Management: Organic soils must be recharged with periodic applications of nutrient-rich plant and animal matter. (The more you take out of a soil, the more you must return.) Therefore the manager of an organic soil must have a good understanding of nutrient characteristics and handling requirements of organic materials. It costs time to develop this know-how.

ESTABLISH ORGANIC SOIL: There are two factors to consider when establishing an organic soil: content and form.

Select the Right Ingredients: A well-designed organic soil contains a sufficient and available supply of nutrients, a good texture and a neutral or near-neutral *pH*. There is no one way to build this organic soil; there is only your way. An elementary recipe with which to begin establishing an organic soil is as follows:

First, locate the most cost-efficient organic materials in your neighborhood. Second, analyze characteristics of the materials. For example, the chicken manure your neighbor is trying to give away may contain too much salt or be too hot (*i.e.*, contain too much soluble N). Next, gather the materials and blend them into the topsoil. Strive for a blend of 10 percent aged manures, 30 percent composted vegetable matter and 60 percent soil. Finally, test the *pH*. If the blend tests too acidic, add a corrective amendment such as lime; if it tests too basic, add more organic matter.

FIG 8.8 A large holding, like this seven-acre plot at Heyn's Country Gardens near Billings, Montana, may be managed with neat tractor-formed rows.

FIG 8.9 A small holding, like this one-fifth acre plot at the Paul Bowers residence in Grants Pass, Oregon, may be managed with deep, French-Intensive style beds.

Remember, good organic soils are forgiving. Follow the basic directions listed above and allow nature to take care of the rest.

Select the Right Containment: The form an organic soil medium takes is defined, to a great extent, by which crops you grow and how much land you have to work. The general rule is, the smaller the farm, the more intensive the system of soil containment.

By way of illustration, consider Charles Heyn, who grows vegetables, melons and berries on seven acres near Billings, Montana. His eight-inch deep soil medium is organized into neat tractor-formed rows and maintained by an annual application of composted manure and straw.

Paul Bowers grows garlic and shallots on one-fifth acre in Grants Pass, Oregon. His eighteen-inch deep soil medium is organized into a series of wood-framed beds and maintained by frequent applications of animal manures, composts and refined organic fertilizers. The beds are often called "French Intensive," after the "troque" farmers of Paris who developed the technology during the reign of Louis XIV.

Let the size of your farm determine how to contain the soil. If you have a large holding, like Heyn's seven acres, consider building the soil into neat tractor-formed rows. If you have a small farm, like Bower's one-fifth acre, consider establishing French Intensive beds. The smaller your farm, the more intensive should be your control of the soil.

Incidentally, Bowers' one-fifth acre generates $7,000 per year. How much would Heyn's seven acres yield if the acres were operated with the intensity of Bowers' one-fifth acre?

Because of their reliance on the technologies of mass-production, the Goliaths of agriculture find it extremely difficult to manage organic soils. This explains why the technology is often dismissed as being inappropriate. It is, for them. But if you cut their size down to the size of a metrofarm, the technologies of mass production become equally inappropriate.

In fact, a well-designed organic soil provides you with the foundation of a system which can produce more crops per acre, of higher quality, at less expense than can the technologies of mass-production. For working examples of how organic soils are used to achieve this objective, read Heyn, Larkey and The Flower Ladies in Part IV, Conversations.

Consider a Synthetic Soil

Some metrofarms do not require a large parcel of land to yield a large volume of products. In fact, nurseries and greenhouses produce many thousands of high-value plants in containers on fractions of an acre.

Native soil does not provide an ideal medium for container farming. Native soil is subject to compaction which restricts aeration and drainage. It may contain weed seeds, insect and disease pests and chemical contaminants. It is heavy and expensive to handle. It is unlikely to have the uniformity required for standardized watering, feeding and lighting routines.

Deficiencies may be avoided by using a soil consisting of biologically-inert ingredients—a synthetic soil. Ingredients commonly used in synthetic soils include sand, peat moss, perlite, vermiculite and pea-sized gravel.

Cultural practices which use synthetic soils are often called "hydroponics." This term was first used in the 1930's to describe the culture of plants in an aerated, dilute solution of fertilizer salts and water. Today the term is used very loosely to describe the many techniques for delivering fertilizer salts.

Recent technological developments have inspired dramatic advances in nursery and greenhouse farming. Many crops are now grown in synthetic soils charged with water-soluble fertilizers. And since the production systems do not require farmland, they open a universe of opportunity to metrofarmers. You can now grow a large crop on a few square feet in the middle of a city.

BENEFITS OF SYNTHETIC SOILS: The benefits of synthetic soils are ones of utility. Consider how the following benefits can help you increase production and productivity.

Land Costs: Synthetic soils can be used to establish production almost anywhere. In fact, they have been used in production of million dollar-sized crops of culinary herbs and cut flowers in suburban neighborhoods of many large cities. Synthetic soils enable you to grow high-value products in low rent neighborhoods, thereby allowing for an increase in net productivity.

Management: Synthetic soils are easier to manage than mediums built with native soil. The component materials, which are readily available from commercial retailers and wholesalers, are often packaged and ready for instant use. This ease of use can reduce the costs of managing soil for greenhouse and nursery operators, especially where labor would otherwise have to be employed in hauling, mixing and sterilizing native soils.

Product Quality: As mentioned, synthetic soils are uniform and easy to control. Uniformity facilitates standardized watering, feeding and lighting routines required in greenhouse and nursery production. This standardized production yields a consistent product of uniform quality, which is a prerequisite for successfully entering wholesale markets.

Pest Control: Greenhouse and nursery operations are extremely vulnerable to infestations of insects and diseases. Restricted environments must be kept clean and sterile to prevent infestations of pests from raging out of control. Since synthetic soils are, or should be, sterile in the first place, they are more conducive to maintaining the hospital-like cleanliness required.

Tax Benefits: Elaborate hardware used in greenhouse and nursery production systems can provide healthy tax benefits. The forty-acre solar greenhouse mentioned in the introduction to Part I of this book was funded by wealthy individuals in search of three-for-one tax benefits as well as operating profits. Tax benefits, however, come and go at the whim of government. If tax breaks are here today, they will likely be gone tomorrow.

LIMITATIONS OF SYNTHETIC SOILS: The limitations of synthetic soils are ones of cost. Consider how the following costs can reduce productivity.

Infrastructure Costs: Production systems based on synthetic soils often require use of containers, greenhouse structures, automatic irrigation systems, fertilizer injectors and so on. This infrastructure is expensive to build and operate. While capital costs may be amortized over the life of the system, they can prove difficult for a beginning farmer in need of immediate cash flow. To overcome this limitation, both Gerd Schneider and Jan LaJoie (Part IV, Conversations) started with small systems and built up with cash generated from sales.

Materials Costs: Most synthetic soils are blended from materials purchased from commercial sources. Lacking advantages of economies of scale, small-scale growers often have to pay premium prices. This limitation may be eased somewhat by joining a co-op buying organization.

FIG 8.10 Container grown crops, like the semi-tropical fruit trees from the Seaside Banana Gardens near Ventura, California, are made possible by the flexibility of synthetic soil ingredients.

Management Costs: In production systems based on synthetic soils, one must keep in balance all chemical, physical and biological soil processes which regulate growth. In other words, you must provide the guidance Mother Nature provides in systems based on organic soils. This is not an easy job.

First, it is difficult, if not impossible, to blend a nutrient solution which will provide all of the essential nutrients required by a growing crop. There is simply no way to know which nutrient is required, in what amount and at what time. While it is relatively easy to provide nitrogen, phosphorus and potassium, it becomes more difficult to provide the secondary macronutrients and almost impossible to provide all micronutrients. Since synthetic soils, by definition, do not provide nutrients, most crops grown in synthetic soils are deficient in one or more essential nutrients. And deficient crops are vulnerable crops.

Second, synthetic soils lack colloidal properties of organic soil. This means you must feed the crop frequently while carefully monitoring the quality of the nutrient solution for possible concentrations of toxic salts. Feeding and watering are technically difficult jobs. Many greenhouse and nursery operators have experienced loss as a result of a badly miscalculated application of fertilizer.

Third, synthetic soils lack the natural *pH* controls of organic decomposition. Without this natural buffering action, plants become subject to the kind of violent swings in acids and alkalines which result in adverse growth, yield and quality. Synthetic soils are unforgiving. One small error can result in a major loss. Consequently, the more complicated the production system, the more critical becomes the role of manager. Management is a job for the technologically proficient, and management costs are therefore an important limitation with production systems using synthetic soils.

Water and Fertilizer Costs: Since synthetic soils lack the colloidal and buffering action of organic soils and are subject to concentrations of toxic salts, they must be watered and fertilized frequently. Water and fertilizer, together with application costs, are therefore a major expense in production systems based on synthetic soils.

Maintenance Costs: Production infrastructures with features like automatic watering systems, fans, pumps, heaters and other mechanisms must be carefully maintained. The more complicated the systems become, the more vulnerable they become to breaking down. There are literally hundreds of things which can go wrong, ranging from a sudden and overwhelming infestation of insects to a heater malfunction. It takes a skilled technician to maintain equipment-intensive systems and this know-how costs time and money.

ESTABLISH SYNTHETIC SOIL: The job of building a synthetic soil begins with selecting the right content and containment. Content is the materials of which the soil will be composed. Containment is the vessel or container in which the soil will be kept and managed.

Select the Right Content: Good soil provides a sufficient and available supply of nutrients, a well-textured structure and a neutral or near-neutral *pH*.

In systems using synthetic soils, nutrients are supplied by water-soluble synthetic fertilizers which are mixed into the soil or into the water. *pH* is controlled by carefully monitoring the water and soil and then adding corrective amendments such as lime when required.

Soil structure is provided by a variety of materials which contribute texture, support and drainage. Periodic applications of water and fertilizer then supply required moisture and nutrients. Following are descriptions of popular materials used in building synthetic soils.

Peat Moss is partially decomposed sphagnum moss. It is extremely light, can hold up to thirty times its dry weight in water, permits good aeration, is sterile and can hold fertilizers. Peat moss has an acidic *pH* (4.5) which can be corrected with lime. Because of its popularity, peat moss is becoming increasingly more expensive. As a result, many commercial soil blends contain substitutes such as partially decomposed fir bark. Commercial blends may also contain lime to neutralize *pH* and a wetting agent to help dampen the dried moss.

Perlite is made by heating particles of volcanic sand until they explode into little white popcorn-like puff balls. Though perlite does not hold moisture or nutrients or aid in the capillary movement of water, it is highly valued as a soil conditioner because it is light and extremely resistant to compaction.

Vermiculite is made by heating particles of mica until they expand into accordion-like granules. Vermiculite contains small amounts of potassium, magnesium and calcium and is capable of holding other nutrients. It also is valued for its light weight and its capacity to hold water and provide for aeration.

Other popular inert materials include pea gravel, sand, wheat straw, sawdust and fir bark. Each material has a set of unique properties with respect to its capacity to hold and drain water and store nutrients.

Many commercial nurseries blend synthetic soil mediums with sterilized topsoil. One popular recipe combines one-third sterilized topsoil to a two-thirds mix of equal parts peat moss, perlite and vermiculite. Synthetic soil blends can either be mixed on site from bulk materials or purchased pre-mixed and packaged from commercial sources.

Select the Right Containment: As mentioned earlier, the benefits of synthetic soils are ones of utility. This utility becomes evident when you consider the number of different systems used for containing and managing synthetic soils.

Be aware of a tendency to overspend in elaborate soil systems. Because of recent developments in hydroponics and the media's propensity for featuring fancy gadgets and devices, new production systems based on synthetic soils tend to receive a great deal of attention. Though the highly technical systems make good copy, many may not be cost-efficient.

A couple of years ago an advertisement appeared in *The Wall Street Journal* for a portable hydroponic growing system. This system, the ad claimed, had potential to generate about $150,000 of profits per acre/year. A demonstration featuring the system was located in San Diego, California. The demonstration consisted of an acre's worth of pipes, tanks, timers, pumps, filters and other equipment. It would cost more than $60,000 per acre to establish production with this system. A reasonably experienced organic grower could accomplish roughly the same production objectives for about $600 per acre. Months later the demonstration was deserted and became covered with weeds.

Let the crop, logistical considerations of the farm, and need to produce the most for the least define what kind of system to build. If the crop is a propagation product, like Gerd Schneider's and Jan LaJoie's, the right form is likely to be heated beds, flats and containers. If the crop is greenhouse tomatoes, the right form is likely to be long, trough-like containers or even simple plastic bags. Each system of containment must provide for drainage, be readily managed and be cost effective.

Synthetic soils are effective in enclosed or restrictive environments, where diseases are difficult to control or where an inexpensive source of power is available. Conditions like those mentioned are often found in inner-city neighborhoods, nurseries and greenhouses. In fact, synthetic soil mediums have made

FIG 8.11 Though complicated hydroponic production systems make interesting stories in local newspapers, few are commercially viable. This system near San Diego cost many thousands of dollars to build but failed in a few months.

possible million dollar-sized crops of culinary herbs and cut flowers in New York City's rubble-strewn South Bronx, where good soil is very hard to find. For working examples of how synthetic soils are used, read Schneider and LaJoie in Part IV, Conversations.

APPLY FERTILIZER

Take raw elements from the environment and, through the medium of crops, convert the elements into products for which consumers pay money.

For every ounce of crops, one ounce of raw elements will have been processed from your environment. Three of the elements— carbon, hydrogen and oxygen— are free. You must provide the rest through a program of fertilization.

The objective of a fertilizer program is to provide for maximum growth and yield with minimum resources. If your production system has been designed properly, you will be able to convert elemental nutrients into marketable products with the intensity of F.H. King's "farmer of forty centuries," who sustained the lives of 20 from the proceeds of a one-half acre farm.

This section describes essential nutrient elements, tells how to analyze a production system's nutrient requirements, and how to establish the organic and synthetic fertilizer programs commonly used in metrofarming.

Know Nutrients

There are 16 elements known, at this writing, that are essential to plant growth and development. Three of the elements— carbon, hydrogen and oxygen— are processed from air and water through photosynthesis.

The remaining 13 elements are often called "soil nutrients" because, in nature, they are made available through decomposition of organic and mineral matter in the soil.

Soil nutrients are categorized by amounts required for growth and development. Categories are primary macronutrients, secondary macronutrients and micronutrients. Soil nutrients are building blocks of marketable products but, unlike carbon, hydrogen and oxygen, they are not free. It pays to become a soil nutrient authority because if one nutrient is missing or not available, yield and productivity will suffer accordingly.

PRIMARY MACRONUTRIENTS: The nutrients required most by growing crops are nitrogen (N), phosphorus (P) and potassium (K).

Each is listed in the three-digit code on commercial fertilizer labels. A fertilizer labeled "0-10-10," for example, contains 0 percent available N, 10 percent available P and 10 percent available K. All three macronutrients are rapidly depleted by vigorously-growing crops and must therefore be frequently replenished.

Nitrogen is used in synthesis of amino acids, which in turn form proteins, N is the principal building material for all living cells. N is also used in synthesis of chlorophyll, nucleic acids and enzymes. It ranks fourth after carbon, hydrogen and oxygen in percentage of total dry matter of plants and is the nutrient most often deficient in cultivated soils.

Nitrogen is mobile within the soil environment. It can escape back into atmosphere or leach away into subsoil. N is also mobile within plants. When the supply of N runs low, it migrates from older growth to new. Symptoms of N deficiency include a slowing of growth rate, yellowing of older leaves (chlorosis) and dying-back of older growth.

Nitrogen fertilizers are typically used early in the season to promote vegetative growth and, in special circumstances such as the propagation of cuttings, are used to force new growth and development.

Phosphorus is used in formation of nucleic acids, storing and transferring of energy, cell division, rooting, maturation and development of fruits and seeds.

Phosphorus is not mobile within plants or soil. Symptoms of P deficiency include slow or stunted growth, purplish coloration on foliage, delayed maturation and poor flower, fruit or seed production.

Phosphorus fertilizers are typically used to promote root development, increase maturation rates for plants in cold weather and to increase production of flowers, fruits and seeds.

Potassium is used in formation and transfer of energy and for opening and closing of stomata. K is also used in formation of starches, sugars and oils— the building blocks of cell membranes.

Potassium is mobile within plants but is not subject to leaching in the soil. Deficiency symptoms appear first in older leaves and include dying tips on older leaves, weak stem and branches, poor flower and fruit production and slow or stunted growth.

Potassium fertilizers are used to improve resistance to inclement weather, insects and diseases. Fertilizer blends of P and K (*e.g.*, 0-10-10) are often used to promote flower, fruit and seed production.

SECONDARY MACRONUTRIENTS: The macronutrients required to a lesser degree than the primary ones are calcium (Ca), magnesium (Mg) and sulphur (S).

Secondary macronutrients are listed in the "Guaranteed Analysis" section on fertilizer packages. However, since Ca, Mg and S are not depleted as rapidly as primary macronutrients, they are often not included in commercial fertilizers.

Secondary macronutrient deficiencies may occur in intensively cultivated soils, sandy soils, heavily irrigated soils or as regional geographical phenomena. Oregon, Washington and Idaho, for example, have large tracts of soil deficient in sulphur.

Calcium promotes root growth, strengthens cell walls, influences uptake of other nutrients, neutralizes toxins and encourages flowering, fruiting and seed production.

Calcium is not mobile within plants, so deficiency symptoms appear first in new growth and include dying back of growth tips, abnormally dark appearance of foliage, premature shedding of flowers and weakened stems. Ca deficiencies may exist in highly acidic soils.

Magnesium is a constituent of chlorophyll and therefore an essential element in photosynthesis. Mg also serves as an activator of enzymes, regulates the uptake of other nutrients, promotes formation of oils and fats and aids in movement of starches through the plant.

Magnesium is mobile, so deficiency symptoms appear first in older growth and include chlorosis within the veins of leaves. Mg deficiencies may exist in sandy soils and in soils which have been fertilized with high concentrations of soluble potassium.

Sulphur is used in synthesis of proteins and is also a factor in the formation of oils which give a plant its characteristic aroma and protection from adverse

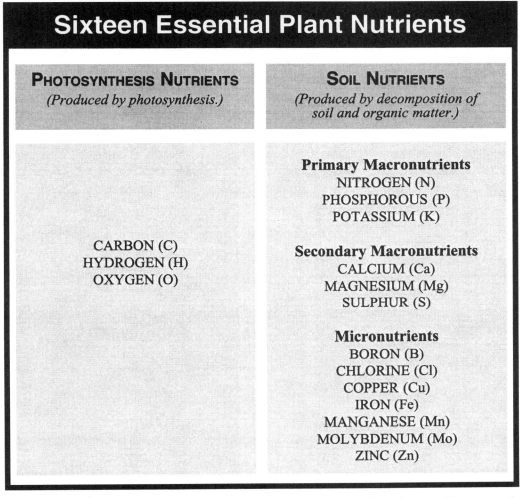

Sixteen Essential Plant Nutrients

PHOTOSYNTHESIS NUTRIENTS
(Produced by photosynthesis.)

CARBON (C)
HYDROGEN (H)
OXYGEN (O)

SOIL NUTRIENTS
*(Produced by decomposition of
soil and organic matter.)*

Primary Macronutrients
NITROGEN (N)
PHOSPHOROUS (P)
POTASSIUM (K)

Secondary Macronutrients
CALCIUM (Ca)
MAGNESIUM (Mg)
SULPHUR (S)

Micronutrients
BORON (B)
CHLORINE (Cl)
COPPER (Cu)
IRON (Fe)
MANGANESE (Mn)
MOLYBDENUM (Mo)
ZINC (Zn)

FIG 8.12 There are 16 elements known to be essential for plant growth and development. Three are made available through photosynthesis and 13 through the decomposition of soil materials or through fertilization.

weather.

Sulphur is subject to being leached from soil but is not mobile within plants, so deficiencies appear first in new growth and include an anemic green color, spindly growth and retarded maturation. S fertilizers are often used to protect plants from cold, damp weather.

MICRONUTRIENTS: The seven micronutrients are Zinc (Zn), Iron (Fe), Manganese (Mn), Copper (Cu), Boron (B), Molybdenum (Mo) and Chlorine (Cl).

Though micronutrients are used by plants in extremely small amounts, they are essential. If one is missing or not available, growth and development will suffer accordingly.

Fertilizing with micronutrients is a sensitive business because the line between deficiency and toxicity is thin and ill-defined. A boron deficiency, for example, might be corrected by adding ounces over acres. Micronutrient deficiencies may exist in intensively cultivated soils, sandy soils, soils with an acidic or basic *pH*, frequently irrigated soils and, obviously, synthetic soils.

Zinc is used in synthesis of enzymes. Zn deficiencies are first noticed on new growth tips and include rosetting of terminal leaves and interveinal yellowing.

Iron is a constituent of chlorophyll and therefore an essential element in photosynthesis. Fe deficiencies may exist in poorly-drained soils with a high *pH*. Symptoms include interveinal yellowing of young leaves and dying back of branch ends.

Manganese stimulates growth processes and assists Fe in formation of chlorophyll. Manganese is mobile so deficiency symptoms begin with interveinal yellowing of older leaves.

Copper is an activator of enzymes. Cu is highly toxic in small amounts and deficiencies are extremely rare.

Boron functions in differentiating cell growth and also regulates metabolism of carbohydrates. B is as often found in toxic levels as in deficient levels. Deficiency symptoms include dying back of stem tips and wilted or yellowed leaves.

Molybdenum is required for utilization of N. Mo deficiencies are extremely rare.

Chlorine is required for photosynthesis. Cl deficiencies are extremely rare.

In addition to 13 soil nutrients, many other elements have been identified in examinations of plant tissue. What role, if any, they perform in growth and development of plants is unknown at this time. For this reason, they are not considered essential plant nutrients.

Know Nutrient Requirements

Each crop requires a unique combination of soil nutrients. Each environment makes a unique combination of soil nutrients available. The difference between what is available and what is required defines what you must do to establish a successful fertilizer program.

EVALUATE REQUIREMENTS: There are three steps to determining which soil nutrients your crop requires for optimum growth and yield. The steps are the first toward earning the "green thumb" sensitivity it takes to win the competition for consumer dollars.

General Growth Characteristics: First, study basic characteristics of the crop's growth and development. Characteristics define, in general terms, which nutrients the crop uses most. For example, a quick look at a stand of sweet corn tells us its basic characteristic is one of green vegetative growth. This green growth tells us corn requires a large amount of nitrogen. A root crop like garlic, on the other hand, puts most of its growth and development into roots and bulbs. From this characteristic we may presume garlic will require relatively large amounts of phosphorus and potassium.

Seasonal Growth Characteristics: Next, analyze seasonal growth characteristics because they define, in general terms, the crop's short-term nutrient requirements. Though the stand of sweet corn requires a large amount of nitrogen through its vegetative cycle, it will also require phosphorus and potassium to bear its seed. And though garlic requires a large amount of phosphorus and potassium to build its roots and bulb, it also requires nitrogen to sustain its energy-producing vegetative growth.

Published Research: Finally, evaluate the crop research published by seed companies, land grant colleges, extension services and others. The studies may provide, in exacting detail, information about the crop's need for secondary and micronutrients as well as correct timing of fertilizer applications.

Evaluate Availability: There are three ways to determine availability of nutrients: a soil test, a tissue test and yield observations.

Soil Test: For a good idea of which nutrients are available, take a sample of your soil and subject it to chemical analysis. Samples tested should be representative; if there are several soil types within the production plot, test a sample of each.

There are several kinds of soil test kits available for on-site use through farm and garden supply centers. If a more accurate accounting is required, as may be the case where a micronutrient deficiency is suspected, hire the services of a commercial test laboratory.

County extension agents can provide information about general nutrient characteristics of local soils as well as directions to reputable local test labs.

Yield Observation: For a good idea of overall nutrient availability, keep accurate records of your crop's growth performance.

Many nutrient requirements can be detected by watching how the crop performs. For example, if a fast-growing crop like corn displays lackluster growth with yellowing tips on older leaves, one can conclude that the soil is deficient in nitrogen because the N is mobile and travels to young growth.

An educated eye and good records are needed for yield observations. After you have watched a crop for a period of time and kept accurate account of its growth and development, you will have earned the ability to detect deficiencies in primary macronutrients and soil *pH* levels. Until then, you may have to rely on an educated eye and accurate records of others.

Tissue Test: For a precise picture of nutrient availability, take a sample of plant tissue and subject it to chemical analysis.

Tissue tests are most often used for solving difficult problems, such as detecting micronutrient deficiencies. Tests tell exactly which nutrients are getting to the plant and in what quantity. By seeing which nutrients are not present, we can determine which are needed. This procedure requires the services of a professional test laboratory.

Know which nutrients are required by your crop and which are made available by your environment. The difference defines what you must provide with a program of fertilization.

There are two programs to consider, organic and synthetic. In organic fertilizer programs, nutrients are made available through decomposition of biotic materials in the soil. In synthetic fertilizer programs, nutrients are suspended in solution and fed directly to plants.

Following are guides to establishing organic and synthetic fertilizer programs. Which program will support your high-density, low-cost system of production?

Consider Organic Fertilizer

Organic fertilizer programs provide a sufficient and available supply of essential nutrients by amplifying the natural processes of soil.

Start with a layout which takes maximum advantage of environmental elements which govern growth, including air and soil temperatures, wind direction and speed and so on. Then build a soil with good structure and *pH*. Finally, add naturally-occurring biotic materials rich in essential nutrients. Nature's decomposers then break down organic compounds into available inorganic nutrient elements.

Establishing the kind of organic fertilizer program it takes to support high-density, low-cost production requires thought and preparation. Following are two steps which will help you succeed.

KNOW ORGANIC MATERIALS: Begin by developing a good understanding of the nutrient personality of organic fertilizer materials. There are three materials to consider: green manures, composts and concentrates.

Green Manures: Herbaceous plants which are plowed under while still green to enrich soil with nutrients, build texture, reduce erosion and prevent nutrient loss through leaching are called "green manures."

Though any green plant can be used as a manure, legumes are most frequently used because they fix atmospheric nitrogen into the soil. Leguminous plants provide up to six times more nitrogen then nonleguminous plants. You can, for example, add up to 400 pounds of N per acre by using a legume, such as clover, in combination with a bacterial inoculant, such as *Rhizobium trifolii*.

Green Manures

Crop	Seeding Rate (lbs/acre)	Region	Comments
Legumes			
Alfalfa	20	all	A deep-rooted perennial, excellent for building sands, clays. Acidic and poorly-drained soils.
Beans, Snap	270	all	Can be harvested before being turned over. High N-fixer.
Beans, Soy	90	all	Inoculate first sowing. Will stand dry conditions. High N-fixer.
Clover, Alsike	10	North	Good for wet and acidic soils. Low N-fixer.
Clover, Bur	30	South	Moderate N-fixer.
Clover, Crimson	30	South & Central	Annual; moderate N-fixer.
Clover, Red	15	North (cool)	Inoculate first sowing. Needs phosphorous and well-drained soil. Low N-fixer. Good mulch.
Clover, White (Ladino)	15	all	Needs neutral pH. Low N-fixer.
Cowpeas	100	all	Very fast-growing annual which thrives in many soils and climates. Excellent soil builder; moderate N-fixer.
Hairy Indigo	10	deep South	Needs warm tempratures. Highly resistant to certain nematodes. Inoculate first sowing; needs P; excellent soil-builder; moderate N-fixer.
Lespedeza, Chinese	25	all	Inoculate first sowing; needs P; excellent soil-builder; moderate N-fixer.
Lupine	80-150	all	Always inoculate; extremely high N-fixer; spring crop in the North, fall crop in the South.
Pea, Field	100	all	Inoculate first sowing; high N-fixer.
Sourclover	20	South & West	Inoculate seeds; sow in fall.
Sweetclover	25	all	Good soil builder; needs a neutral pH and a sufficient supply of P; moderate N-fixer.
Velvetbean	100	South	Excellent soil-builder, N-fixer, for sandy soils. Needs warm temperatures.
Vetch	60	all	Annual and biennial varieties for all areas. Excellent N-fixer.
Nonlegumes			
Barley	100	North	Prefers slightly basic loams.
Buckwheat	75	all	Extensive root systems makes this an excellent soil-builder.
Bromegrass	30	North	Extensive root system; cold-hardy; excellent winter cover crop.
Cow-Horn Turnip	2	all	Extensive root system; excellent soil-builder.
Millet, Pearl	4	South & Central	Fast-growing crop good for controlling weeds.
Oats	100	all	Not good with heavy clays but tolerates a wide range of pH.
Rye	100	all	Extensive root system; excellent soil-builder.
Sudan Grass	100	all	Grows extremely fast in hottest temperatures and provides excellent source of organic matter.

FIG 8.13   Green manures are the green plants plowed under to enrich soil with nutrients, build texture, reduce erosion and limit nutrient loss.

Nonleguminous plants, such as rye, are often used to improve soil texture. A scientist once grew a single rye plant in a box for four months, then measured its roots. He found 14 million roots with a combined length of 387 miles. This kind of root system will loosen a great deal of soil, bring up many minerals and add a tremendous amount of decaying organic material to the soil.

There are three ways in which a green manure can be worked into your fertilizer program: One way is to schedule it into the crop rotation. A common practice is to divide the production unit into three sections, with one being devoted to green manure and the other two to cash crops. Every third year each section is then treated to a green manure nutrient fix. Another way is to schedule green manure as a winter cover crop. Often there is growing time left after the last cash crop of the season has been harvested. This time may be put to profitable use by sowing a quick-growing green manure.

And finally, green manures can be scheduled to fill gaps in the rotation. Often there is a lag time between harvest of one cash crop and sowing of another. If this lag time is of sufficient duration, it can be put to use by sowing a quick-growing green manure.

Use common sense when planning the use of green manures. One rule is to never let green manure go to seed, otherwise it may come back as a weed. Another rule is to plant a cash crop soon after turning a green manure under, otherwise the accrued nitrogen may escape back into the atmosphere or leach down into the subsoil.

Keep the objective in focus while making plans. If you need soil nutrients, turn green manure under while still young, green and juicy; its acids will speed decomposition and the soil will quickly be ready for use again. If you need soil texture, allow the green manure to mature; older woody material and additional roots will take more time to decompose and thus provide better soil texture.

Composts: Composting is reduction of bulk organic materials into a concentrated mass of nutrient-rich humus. This concentrated mass can then be blended into soil, where it will improve texture and add nutrients without causing a delay in production.

There are two ways in which bulk organic materials are composted: One way is in the stomachs of various animals like cattle, chickens and earthworms. This form of composting yields manure. The other way is through the metabolic activity of yeasts, molds and bacteria in compost-pile fermentation.

Though pile composting is an ancient art, its technology has become somewhat lost in the modern rush to synthetic fertilizers. Consequently many people now throw whatever bulk materials they have lying around into a pile and wait for compost to appear.

Composting should be approached as a business, with the objective being to obtain a well-balanced mass of essential soil nutrients and humus for the least cost.

FIG 8.14 Composting is the reduction of bulk organic materials into a concentrated mass of nutrient-rich humus, as demonstrated by Heru Hall, apprentice at the University of California, Santa Cruz Agroecology program.

Begin with a simple design of three layers. On the bottom layer place dried bulk materials rich in carbon like dried leaves, grasses and stems. On the middle layers place fresh materials rich in nitrogen like green grasses and leaves, animal manures and other fresh wastes. On top place topsoil.

Amplify decomposition by applying water and additional nitrogen-rich materials and by periodically turning the pile to increase its supply of oxygen. Within a relatively short period of time, heat of fermentation will reduce the bulk material into a mass of well-balanced nutrient-rich humus. This mass will have a consistent dark, earthy color and will smell clean and fertile.

It is very important to master skills of handling organic materials. If handled improperly, the materials can create all kinds of unpleasant problems like bad odors and flies. While the problems might be tolerated deep in the countryside, they can become significant obstacles in more densely populated environs.

To understand how well organic materials can be handled, consider another of F.H. King's observations in *Farmers of Forty Centuries*: "The International Concession of the city of Shanghai, in 1908, sold to a Chinese contractor the privilege of entering residences and public places early in the morning of each day in the year and removing the night soil, receiving therefore more than $31,000, gold for 78,000 tons of waste." Yes! An individual actually paid the city of Shanghai over one-half million in today's dollars for the privilege of removing human waste!

One would expect an enormous fly problem with so much nightsoil being transported around but King maintained such was not the case. "One fact which we do not fully understand is that wherever we went, house flies were very few. If the scrupulous husbanding of waste refuse so universally practiced in these countries reduces the fly nuisance and this menace to health to the extent which our experience suggests, here is one great gain."

A well-designed composting system produces a clean, fresh-smelling mass rich in nutrients, humus and texture. A poorly-designed system, on the other hand, can produce big trouble in the neighborhood.

Concentrates: Many bulk plant and animal materials are refined into concentrated meals and powders. The objective of refining a bulk material, like animal bones, into a concentrated product, like steam-treated bonemeal, is to make the nutrients more readily available and the material easier to handle.

Refined organic products are available through many commercial farm and garden supply sources. However, the materials are expensive and usage is consequently limited to specific cultural objectives. One objective is to supplement and balance green manures and composts. If, for example, an otherwise complete soil blend is lacking nitrogen, the deficiency can be corrected by adding blood and fishmeal. Another objective is to stimulate specific cycles of growth. Flowering and fruiting, for example, can be encouraged by an addition of a phosphorus and potassium product such as a bone and kelpmeal blend.

Finally, concentrated organics can be blended into a complete organic fertilizer for use on very high value crops. One organic blend registered as a fertilizer in California, for example, contains five concentrates blended into a 6.1 - 7.2 - 1.8 formulation. The product has a near-neutral *pH*, 0.25 percent of sodium chloride and a humus content of more than 50 percent. So well-balanced and complete is this fertilizer, it is often used to grow organic produce in synthetic soils.

BENEFITS OF ORGANIC FERTILIZERS: Following is a summation of the major benefits of organic fertilizers. Evaluate how each benefit will affect design of your production system.

Long-Term Costs: The more organic fertilizers are used, the less expensive they become to use. Why? Organic fertilizers are rich in humus and humus is the colloid— or glue— which binds soil particles together in two important ways:

First, humus is a negatively-charged mass which attracts and binds positively-charged soil nutrients like phosphorus. Second, humus is a sticky substance which traps and holds partially decomposed nutrient-rich material. Negatively-charged nutrients like nitrogen can then be made available through further

Selected Organic Sources of Essential Nutrients

MATERIAL	%N	%P	%K	MATERIAL	%N	%P	%K
Refined and Concentrated Materials				**Animal Manures (cont.)**			
Ash, Incinerator	0	5	2	Duck	.5	1.5	.5
Ash, Wood	0	1	8	Goat	3	2	3
Blood Meal	15	1.5	0	Horse	.5	.2	.5
Bone Meal (burned)	0	30	0	Pig	.5	.3	.5
Bone Meal (steamed)	4	28	0	Rabbit	3	2	1
Brewer's Grains	1	.5	0	Sheep	2	1	5
Castor Pomace	6	3	1	Worm Castings	3	1	1
Cocoa Shell Dust	2.7	1.5	3	**Bulk Materials**			
Coffee Grounds (dried)	2	.5	.5	Alfalfa Hay	3	.5	2
Cottonseed Meal	8	3	1	Alfalfa Straw	1.5	.2	1.5
Crab Meal	10	0	0	Apple Leaves	1	0	.5
Fish Meal	8	6	4	Bean Straw	1	.2	1
Kelp Meal	1	0	5	Cocoa Bean Hulls	1	0	2
Hoof and Horn Meal	12.5	1	0	Corn Silage	1	.2	1
Lobster Shell Meal	5	4	0	Cotton Gin Trash	1	1	1
Milorganite	6	3	1	Clover	.5	.5	.5
Molasses Residue	1	0	7	Grain Straw	.5	0	1
Sewage Sludge (digested)	2	3	0	Granite Dust	0	.1	5
Tankage (animal)	11	15	1	Grass (Immature)	1	.2	1.5
Winery Pomace	1.5	1.5	1	Greensand	0	0	7
Animal Manures				Oak Leaves	1	.5	0
Bat (Guano)	12	5	2	Peach Leaves	1	1	.5
Chicken (Aged)	2	1	1	Raspberry Leaves	1.5	.5	.5
Chicken (Fresh)	4	3	2	Rock Phosphate	0	18	0
Cow (Dairy)	1	.5	.5	Seaweed	.2	.1	.6
Cow (Feedlot)	2	.5	2	Tobacco Straw	2	0	7

N=Nitrogen   P=Phosphorous   K=Potassium

*The nutrient percentages listed in this chart are averages. Actual percentages will vary from source to source.
** Analysis lists only primary macronutrients. Most organic materials also contains secondary and trace nutrients.

FIG 8.15 Many plant and animal materials are reduced into concentrated meals and powders that are easy-to-handle and rich in available nutrients.

Each application of an organic fertilizer adds essential nutrients to the soil and builds the soil's capacity to hold even more nutrients until they are needed by growing plants. And so the more organic fertilizers are used, the less frequently they need to be used.

Operating Costs: A well-designed program of organic fertilizers can reduce or eliminate the need to buy expensive commercial fertilizers. Organic farmers simply do not need synthetically-derived commercial fertilizers because everything their crops need is in the soil. (For an amusing example of this independence, read Charles Heyn's discussion of "Plant Two" in Part IV, Conversations.)

Available Nutrients: Though organic materials may exhibit a preponderance to one or two macronutrients, they also contain smaller amounts of secondary and micronutrients.

Consider kelp, which is often used as a source of potassium. According to various chemical analysis, kelp may contain up to sixty-two other mineral elements. Since organic materials like kelp contain so many nutrient elements, their decomposition in soil provides plants with a virtual feast of available nutrients. Plants can then select whichever nutrients they need, when they are needed.

This abundant supply of available nutrients promotes the kind of slow, healthy growth required to resist pests and maximize quality and yield.

Soil Texture and Condition: In addition to providing nutrients, organic fertilizers also improve soil structure and tilth by adding fiber, humus and humic acids which increase permeability of soil. This improved texture and conditioning facilitates root growth and improves uptake of nutrients and water.

Water and Irrigation Costs: Because of roughage contained in organic fertilizers and colloidal properties of its humus, production may require only one-third to one-thirtieth the amount of water required by systems based on large tracts of saline soils. Reducing water requirements by such an extent will result in significant savings in water and irrigation labor, especially in areas where water is an expensive commodity.

Soil Micronutrients: The humic acids in organic fertilizers dissolve soil particles and free minerals which would otherwise remain locked-up in the particles.

Soil pH: The decomposition of organic fertilizer helps maintain a neutral soil *pH*. In addition, it helps buffer against abrupt *pH* changes when acid or alkaline substances are added to the soil. A well-balanced soil is a forgiving soil and forgiveness can result in big savings, especially where unskilled labor is involved in important cultural practices such as fertilization.

Product Quality: Crops grown with organic fertilizers may qualify as "organic produce" and earn a premium price. One reason for this premium price is the many accounts of pesticide contaminations seen in popular media. Another

Application Rates for Selected Organic Fertilizers

SOURCES OF NITROGEN
Application Rates for Each 2% of Nitrogen Needed

MATERIAL	% NITROGEN	PER 100 SQ. FEET	PER ACRE (IN LBS.)
Alfalfa Hay	3	3.6 lbs.	1440
Bat Guano	12	14 ozs.	335
Blood Meal	15	10 ozs.	265
Bone Meal	4	2.5 lbs.	1000
Cocoa Shell Dust	2.7	3.75 lbs.	1500
Cottonseed Meal	8	1.25 lbs.	500
Fish Meal	8	1.25 lbs.	500
Hoof and Horn Meal	12.5	13 ozs.	320
Milorganite	6	1.6 lbs.	665

SOURCES OF PHOSPHOROUS
Application Rates for Each 4% of Phosphorous Needed

MATERIAL	% PHOSPHOROUS	PER 100 SQ. FEET	PER ACRE (IN LBS.)
Bone Meal (Steamed)	2.8	12 ozs.	280
Cottonseed Meal	3	7 lbs.	2500
Fish Meal	6	3 lbs.	1350
Miloganite	3	7 lbs.	2500
Rock Phosphate	30	12 ozs.	270

SOURCES OF POTASSIUM
Application Rates for Each 2% of Potassium Needed

MATERIAL	% POTASSIUM	PER 100 SQ. FEET	PER ACRE (IN LBS.)
Ash, Wood	8	20 ozs.	500
Fish Meal	4	2.5 lbs.	1000
Granite Dust	5	2 lbs.	800
Greensand	7	24 ozs.	570
Kelp	5	2 lbs.	800

FIG 8.16 The application rates listed assume an average nutrient content for each material. Actual contents and rates vary from source to source.

reason stems from the perception that organically-grown produce is more nutritious than produce grown with synthetic fertilizers on large-scale farms.

LIMITATIONS OF ORGANIC FERTILIZERS: Following are significant limitations of organic fertilizers. Consider how each limitation will affect the design of your production system.

Start-up Costs: A program of organic fertilizers is relatively expensive to establish. It costs time to locate and evaluate all necessary ingredients; money to purchase them; time to gather and compost them into a suitable form; and more time to move them into the production unit and mix them into the soil. It is always easier, in the short term, to walk into a store and buy a bag of synthetic fertilizer.

Purchase Price: Many of the best organic materials, like manures, are difficult to locate and procure, especially in urban and suburban environs of metrofarms. And concentrates, such as bone meal, are expensive and likely to become even more so with an increase in demand by an increased number of organic gardeners and farmers.

Processing: Organic fertilizers, if not well-handled, may create odor and insect problems. Enough resources must be invested to build a proper system for processing and handling materials.

Product Quality: Organic fertilizing materials do not yield a consistent amount of specific nutrients. This lack of consistency makes it difficult to plan the kind of mass-production regimes required in nursery and greenhouse operations. As a consequence, organic products may lack the uniformity required to satisfy mass-market standards.

ESTABLISH AN ORGANIC FERTILIZER PROGRAM: Fundamental design of an organic fertilizer program is, naturally, quite simple:

Begin with a well-balanced blend of materials rich in nitrogen, phosphorus and potassium (*e.g.*, 3-3-3). This well-balanced blend will contain many additional nutrient elements and will consequently almost certainly provide secondary macronutrients and micronutrients required as well. Next, mix the blend into the topsoil so decomposition can provide plants with the right nutrients, in the right quantity and at the right time. Following are steps which will help you achieve the design objective for the least possible cost.

Calculate Material Cost: First, determine which materials are the least expensive in your area. The cost of any given organic material can vary greatly from one location to the next. Where chicken manure may be cheap in one location, fishmeal may be least expensive in another.

Consider how farmers work with this fact of life. Charles Heyn operates across the road from a dairy farm. Heyn worked out a trade in which he provides straw for the dairy's barns in exchange for manure. Paul Bowers, a garlic grower in Grant's Pass, Oregon, parks an empty trailer in a convenient location and the city's Park Department fills it with grass clippings and leaves, providing a nearly-free source of good compostable materials.

Evaluate Material Personality: Determine the general personality of each organic material. Consider salt content, *pH* level and humus content; all affect the material's value. Chicken manure, for example, tends to have high concentrations of nitrogen, phosphorus and sodium chloride (salt). While nitrogen and phosphorus are valuable nutrients, sodium chloride will contribute to toxic buildups of salt in the soil.

Evaluate Material Nutrient Content: Next, consider the specific nutrient personality of each material, which will vary from source to source. For example, I recently subjected two samples of bonemeal to a chemical analysis. The analysis showed bonemeal from China contained three percent more available phosphorus than did bonemeal from the United States. (Was the difference caused by ways in which the cattle were fed or by how materials were processed?)

Combine and Apply: Finally, combine organic materials you have selected into a well-balanced blend of nitrogen, phosphorus and potassium. Use a soil test kit to help maintain this nutrient balance and a neutral *pH*. Then mix the blend into the topsoil and let nature do the rest.

Though an organic fertilizer program is relatively expensive to establish, it will become less expensive with use and may provide you with means to win the competition for consumer dollars. To see how organics are used in this competition, read Heyn, Larkey and The Flower Ladies in Part IV, Conversations.

Consider Synthetic Fertilizer

There are many situations in which organic fertilizers simply do not make sense. An inner-city nursery specializing in container-grown ornamentals, for example, would find it extremely difficult to obtain and manage organic fertilizing materials at a cost effective price. Synthetic fertilizers provide many metrofarmers with a cost-effective means to produce.

Synthetic fertilizers are water-soluble inorganic nutrients which have been synthesized from natural substances. Unlike organics, synthetics bypass natural soil processes to feed plants directly.

The technology was born in 1840 when German chemist Justus von Liebig subjected animal bones to a bath of sulfuric acid. This bath "unlocked" the bones' phosphorus, making the nutrient immediately available to plants.

Synthetic fertilizing materials can be spread on topsoil, mixed into soil or mixed into irrigation water. Once dissolved, nutrients are taken up by plant roots in whatever combination they were added.

Establishing a synthetic fertilizer program which can support high-density, low-cost production requires thought and planning. First know the materials; then design the program.

KNOW SYNTHETIC MATERIALS: Immediate availability of synthetics does not provide for a large margin of error. If you apply too much, your crop will burn; too little, the crop will be deficient. If you apply the wrong combination, other nutrients will become locked-up in a chemical imbalance. Designing a synthetic fertilizer program therefore requires a considerable degree of technological expertise. Begin by developing a good understanding of basic fertilizing materials.

Bulk Synthetic Fertilizers: Bulks are single-nutrient fertilizers which can be applied to satisfy specific culture requirements.

Following are several examples of synthetics and their use: Urea is a dry nitrogen product, manufactured from natural gas, which is applied for a quick boost to a crop's vegetative cycle. Superphosphate is a dry phosphorus product which is used to boost flowering and fruiting cycles. Muriate of potash is a dry potassium product which is applied to improve resistance to inclement weather and to increase flowering and fruiting.

Blended Synthetic Fertilizers: Blends are made by combining selected bulk materials to obtain a more complete fertilizer. "Complete," in this case, means the fertilizer has all three primary nutrients (*e.g.*, 20-20-20). "Incomplete" means the fertilizer has only two primary nutrients (*e.g.*, 0-20-20).

There are innumerable N-P-K combinations which can be made with bulk materials (*e.g.*, 16-16-16, 0-25-25, 6-20-20). Blended fertilizers are made by mixing bulk elements together or by chemically binding different elements into one material. Binding prevents segregation of constituent materials which differ in size, shape and weight. Chemically-bound fertilizers also can be coated with slow-release resins, making this form a very popular one with container farmers because one application may last four months and more.

Blends can be purchased from a manufacturer or custom blended on-site from bulk supplies.

Liquid Synthetic Fertilizers: Liquids hold nutrients in solution with water. There are two methods of making liquid fertilizers, batching and neutralization.

The simpler method, batching, consists of dissolving solid materials such as ammonium phosphate, potassium chloride and urea in water to give the desired grade of fertilizer. The second, neutralization, consists of combining phosphoric acid with ammonia, followed by addition of other nitrogen and potassium-bearing materials.

Liquid fertilizers are popular for several reasons: They can easily be combined with liquid pesticides for a one-shot-does-all application; they can be applied with automated irrigation systems for a significant savings in application labor costs; they are more adaptable to foliar spray applications; and they provide for a more uniform application.

BENEFITS OF SYNTHETIC FERTILIZERS: Following is a summation of the

Nutrient Content of Select Bulk Synthetic Fertilizers

NITROGEN CARRIERS

Fertilizer	%N	%P P₂O₅	%K K₂O
Ammonia			
Anhydrous	82	-	-
Aqua	24	-	-
Nitrate	33	-	-
Nitrate-Sulfate	28	-	-
Sulphate	20	-	-
Chloride	26	-	-
Urea	45	-	-
Ureaform	38	-	-
Calurea	34	-	-
Sodium Nitrate	16	-	-

PHOSPHOROUS CARRIERS

Fertilizer	%N	%P P₂O₅	%K K₂O
Mono-ammonium Phosphate	11	48	-
Diammonium Phosphate	16-21	48-53	-
Ammonium Phosphate Sulfate	16	20	-
Ammonium Polyphosphate	15	60	-
Nitric Phosphate	10-22	10-22	0-16
Nitrate Phosphate	20-27	12-14	-
Phosphoric Acid (normal)	-	54	-
(super)	-	76-80	-
Potassium Metaphosphate	-	26	33
Superphosphates (Single)	-	20	-
(Concentrate)	-	45	-

POTASSIUM CARRIERS

Fertilizer	%N	%P P₂O₅	%K K₂O
Potassium Chloride (Muriate of Potash)	-	-	60-62
Potassium Sulfate	-	-	50
Potassium Nitrate	13	-	44
Potassium Metaphosphate	-	26	33

FIG 8.17 Synthetic fertilizers are inorganic nutrients which bypass the natural decomposition processes of soil to feed plants directly.

major benefits of using synthetic fertilizers. Factor the benefits into the design of your fertilizer program.

Purchase Price: Synthetic fertilizers are relatively inexpensive to mass produce. Many companies manufacture them and many stores sell them. Competitive factors tend to keep the price of synthetic fertilizers relatively low.

Handling Costs: Synthetic fertilizers may contain very high concentrations of nutrients (*e.g.*, 36-30-30). Consequently, a large supply of nutrients can be purchased in a small package. All one need do is drive down to a store and pick up the package or make a call and have it delivered. It is easy to keep handling costs to a minimum when a large amount of nutrients is purchased in a small package.

Product Quality: As synthetics contain a consistent quantity of available nutrients, growing crops respond to an application quickly and with uniform growth characteristics. This enables a grower to establish standardized watering and feeding routines which yield a uniform grade of product. And in the mass-market, uniformity is often the sole criterion for product quality.

Growing Conditions: As synthetic fertilizers bypass normal soil processes, plants can be provided with essential nutrients in much lower soil temperatures than would otherwise be possible, thus allowing for an earlier start and later finish.

Forcing Control: As synthetic fertilizers provide readily-available nutrients, they can be used to force growth into more desirable directions. For example, if a nursery grower wants to take cuttings from mother plants, the plants may be induced to shoot tender new growth ends out with an application of soluble nitrogen.

LIMITATIONS OF SYNTHETIC FERTILIZERS: Following is a summation of major economic limitations of synthetic fertilizers. Be certain to factor the limitations into the design of your fertilizer program.

Purchase Price: The price of synthetic fertilizers is determined by forces outside of one's control. An energy cartel in the Middle East or a teamster's union in the Middle West can have a profound impact on cost and availability of synthetic fertilizers.

Start-Up Costs: Some synthetic fertilizer programs require an expensive infrastructure of pipes, metering devices, timers, valves mixers and so on. While this equipment can save labor costs, it can also add considerably to a farm's start-up costs.

Product Quality: There are two reasons why synthetic fertilizers simply cannot provide all thirteen essential soil nutrients in a formula which can satisfy the exact requirements of a growing crop.

First, the line between micronutrient deficiency and toxicity is thin and ill-defined. Most commercial fertilizers therefore avoid this problem by blending only primary and secondary macronutrients. Second, it is virtually impossible to know the exact nutrient requirements of a crop because growth and development change from

Common Sense Guidelines for Applying Synthetic Fertilizers

DO NOT	feed young seedlings a diet of fertilizer designed for mature plants. Young seedlings can generally grow for an extended period of time without fertilizer; a hot dose of soluble nutrients may easily destroy their roots.
DO NOT	feed a dry plant. Always water first, apply the fertilizer, and then water again.
DO NOT	apply fertilizers when transplanting. Allow the roots some time to adjust before making demands upon them. Use rooting hormones to help overcome transplant shock.
DO	follow label directions. Fertilizer manufacturers must conform to strict label laws. Try it their way first; observe the results, and then try it your way.
DO	apply fertilizer to where it will do the most good. Build a basin around the base of your plants so the water and fertilizer will not flow away to feed weeds.

FIG 8.18 The immediate availability of synthetic nutrients demands that common sense application guidelines be followed.

hour to hour, day to day and so on. Though knowledge of the role nutrients play in growth and development has improved dramatically, there is still no way to measure the exact day-to-day needs of a crop.

Consequently, when a synthetic fertilizer is used in conjunction with an inert soil, as is common in nursery and large-scale field operations, plants will likely be deficient in some essential nutrients. And deficiencies limit yields, increase costs and reduce product quality.

Water Costs: Synthetics contain high concentrations of soluble salts. After a number of applications, salts can build up to toxic levels in the soil, thereby limiting nutrient uptake and increasing the need for additional water to dissolve the salts, carry them away and clean the soil.

Management Costs: Since synthetic fertilizers provide immediately available nutrients, there exists the possibility of burning plants by providing too many nutrients. The design and management of a synthetic program therefore requires a considerable degree of technological expertise. This technological expertise is an expensive commodity.

Disposal Costs: In addition to providing essential nutrients to growing crops, synthetic fertilizers can contaminate the environment. Highly mobile nitrogen salts which have seeped down into groundwater aquifers, for example, have been associated with a disorder called "blue-baby syndrome," in which newborn babies are deprived of oxygen. Disposing of waste fertilizers can therefore mean expensive problems, especially for farmers who use large volumes of synthetics while operating in or near a city.

ESTABLISH A SYNTHETIC FERTILIZER PROGRAM: The objective is to satisfy nutrient requirements by providing a balanced formula of water-soluble nutrients. Following are simple steps which can help you achieve this design objective for the least possible cost:

Evaluate Forms: Determine which form of fertilizer will work best in your production system— dry bulk, dry blend, liquid bulk or liquid blend.

Allow logistics of production to determine which form is best. If you are working open ground, for example, consider a dry form which can be spread onto the topsoil and watered into the root zone. If you are working containers, consider time-released granules which can be mixed into the soil or liquid blends which can be fed through drip irrigation. Many nursery and greenhouse growers combine liquid fertilizers with slow-release blends of dry, ensuring each plant a small dose of essential nutrients on a regular basis.

Evaluate the Material's Personality: Synthetics, like organics, have a nutrient personality. This personality must be acknowledged before you can determine how best to satisfy the needs of your crop.

Potassium chloride (KCl), for example, may be the least expensive source of soluble potassium for your vineyard of winegrapes. However, too much KCl may adversely affect the taste of the grapes. The best fertilizer may therefore be the more expensive potassium oxide (K_2O).

Match the nutrient requirements of your crop with the personality which can best satisfy requirements. Then determine whether you can buy this personality in a commercially-packaged fertilizer or whether you will have to blend it from bulk ingredients.

Shop Prices: Synthetics are made by many manufacturers and sold by many distributors. Force manufacturers, distributors and retailers to compete for your business. Shop prices and save money.

Synthetic fertilizers can enable you to establish right in the middle of a hungry market. For working examples of metrofarmers who use synthetic fertilizers, read Schneider and LaJoie in Part IV, Conversations.

Combine Fertilizer Programs

Organic and synthetic fertilizer programs discussed in preceding pages represent extremes. There is a middle ground in which one kind of fertilizer is used to supplement the other. This middle-ground approach can help you achieve many cultural objectives.

For example, container crops grown with synthetic soils and fertilizers may display deficiency symptoms like lusterless leaves and weak blossoms. This deficiency may be corrected with a foliar-feeding of water-soluble organic kelp, which will provide all required secondary macronutrients and micronutrients. Or a stand of organically-grown sweet corn may display yellowing leaves of a nitrogen deficiency. This deficiency may be easily corrected with a low-cost synthetic urea.

The principal disadvantage of combining organic and synthetic fertilizers is loss of the market designation, "organic produce."

Take raw elements from the environment and convert the elements into marketable products. Become a master of the elements by establishing a fertilizer program which can satisfy the needs of your high-density, low-cost system of production. Keep the "farmer of forty centuries" in mind. Perhaps you, too, will be able to sustain the lives of twenty from proceeds of one-half acre.

MANAGE WATER

Water is the lifeblood of production. It provides the crop with turgidity or support. It is a source of photosynthetic nutrients hydrogen and oxygen. It is a carrier of soil nutrients. It removes waste particulates through flow of the transpiration stream. If there is too much or too little water in a farm production system, crop growth and development will suffer accordingly.

The management of water is therefore a prime factor in determining yield and productivity. As the intensity of production is increased to metrofarm levels, more and more attention must be paid to water management.

This section provides simple steps to managing water. The first step is to measure the difference between water required and water available. The next step is to establish the irrigation and or drainage systems necessary to provide the difference.

Evaluate Water Requirements

Knowing when and how much to water is a critical skill of production. To master this skill you must establish a system for determining whether water available will be equal to water required.

WATER REQUIRED: Water required for a crop may be expressed as the amount required to sustain optimum growth and development. This amount is determined by the nature of crop, its stage of growth, fertility of soil, humidity of atmosphere, air and soil temperatures, wind speed and amount and intensity of light. Given so many variables, how can you determine how much water is required?

One way to know how much water is required is to watch flow of the transpiration stream. This cyclical flow, as defined in Chapter One, is the path water takes as it flows from atmosphere into soil, from soil into plant and from plant back to atmosphere. If a crop does not get enough water to replace water being transpired, it wilts. If it gets too much water, it drowns. Both conditions are "telltales" for flow of the transpiration stream through a crop.

By using telltales to visualize flow of the transpiration stream, you will be able to see exactly how much water a crop requires at each stage of growth and development. You will see a ten-foot tall stand of sweetcorn in mid-July transpiring a tremendous volume (250 to 300 pounds of water for every pound of dry matter produced). You will see a flat of freshly made floribunda cuttings on a rooting bench transpiring very little.

Visualizing the flow of water from atmosphere to soil, from soil to plant and from plant back into atmosphere requires a considerable amount of sensitivity. It is one of the green thumb technologies you need to succeed. When you have developed the ability to see the flow of water, record it for future use. Records will help you manage the water requirements of your crop through its various stages of growth and development.

WATER AVAILABLE: There are two ways to determine water availability, feel and projection.

Feel Test: A feel test reveals soil moisture available now. It consists of taking a handful of soil and feeling for moisture content. Accuracy of this test depends upon experience and sensitivity of the tester.

There have been many mechanical innovations on the feel test. For example, electronic metering devices can now feel for soil moisture levels and then communicate the levels directly to automated sprinkler systems. Accuracy of mechanized feel tests depends on the technology of metering devices and maintenance.

Projection: Published data from a variety of sources often tell what water is likely to become available in the future. This information is especially significant in managing seasonal water needs.

For example, a stand of sweet corn in mid-July will transpire a huge volume of water. Meteorological records will give a good indication as to whether or not it will rain enough to satisfy the needs. On the other hand, a vineyard of winegrapes at harvest may be destroyed by too much rain. Meteorological records will also

Moisture "Feel" Chart

Degree of Moisture	Feel	Percent of Holding Capacity
Dry	Powder dry	0
Low (Critical)	Crumbly, will not form a ball	Less than 25
Fair (Usual time to irrigate)	Forms a ball, but will crumble after being tossed up and down several times	25-50
Good	Forms a ball that will remain intact after being tossed five times, will stick slightly with pressure	50-75
Excellent	Forms a durable ball and is pliable; sticks readily; a sizable chunk will stick to the thumb after soil is squeezed firmly	75-100
Too Wet	With firm pressure can squeeze some water from ball	More than capacity

FIG 8.19 The "feel test" consists of taking a handful of topsoil and feeling it to determine the amount of moisture available.

provide a good indication of the presence of this danger.

Data are likely to be available for whichever kind of water your system relies on, including irrigation district ditches, aquifer levels and even seasonal adjustments for the price of city water.

Know the difference between water needed and water available. This difference must be managed for optimum production and productivity.

Establish Irrigation and Drainage

Optimum water content of a well-textured soil is about two-thirds saturation. This is the level at which soil microbes and plants work at their highest efficiency. If there is too little water available, irrigate. If there is too much water available, drain.

IRRIGATE: Irrigation is the centuries-old technology of providing water to growing crops. Since irrigation has a major impact on both costs and yields, you must install an irrigation system which will deliver a sufficient amount of water for the least cost.

There are three types of irrigation systems typically used in metrofarming: surface flow, sprinkler and drip. Following are descriptions of each. Which irrigation system will enable you to satisfy your water requirements for the least cost?

Surface Flow Irrigation: The application of water by surface irrigation is used where water resources and topography of the land permit. Water is conveyed to fields in open ditches or pipes and then allowed to flow into the production plot by various control devices such as small dams or siphon tubes.

Distribution of water over the plot may be accomplished by either flooding the entire field ("flood irrigation") or by restricting the flow to a narrow channel ("furrow irrigation"). Provisions such as drainage canals and ponds must be provided to remove waste water and reduce pooling of water.

Surface flow irrigation must be carefully controlled for two reasons. First, water is an expensive resource and much can be wasted in surface flow. Second, a rapid flow of water may leach away valuable water-soluble nutrients and topsoil.

As long as there is an abundant source of inexpensive water uphill from the plot, surface flow irrigation is the least expensive method of irrigating crops. The disadvantage is labor costs. Constant supervision is required to prevent furrow streams from uniting and forming large channels.

For a working example of surface flow irrigation, read Heyn in Part IV, Conversations.

Sprinkler Irrigation: Many innovations in sprinkler irrigation have been made possible by technological advances in metal and plastic. Consequently there are now many types of sprinklers available, including impulse, oscillators, revolving arms,

FIG 8.20 Water in a surface flow irrigation system flows from a high point through pipes or ditches to a low point.

FIG 8.21 Water in a sprinkler irrigation system is pumped under pressure through an enclosed pipe or hose to its destination, where it is broadcast through a sprinkler.

micro-jet and sieve.

There are several advantages to using sprinklers. They provide an even and controlled application rate, which results in a more efficient use of water (except in hot, dry and windy areas where evaporation may be higher than with surface flow). They reduce runoff and erosion through a slower rate of application. And they increase the productive value of land which is too steep to permit proper use of other irrigation methods.

The primary limitation of sprinklers lie in the high initial purchase cost, which may be quickly offset by reduced labor costs. Another limitation is the cost of power to maintain suitable water pressure for sprinklers. Finally, areas with high winds may lose too much water to evaporation to be efficient.

For a working example of the use of sprinkler irrigation, read Larkey in Part IV, Conversations.

Drip Irrigation: Drip irrigation consists of applying water through mechanical emitters located at selected points along water-delivery lines. Some types of emitters release a periodic drip of water at an average rate of one gallon per hour. Other types sprinkle water at varying rates. Drip systems deliver water directly to where it's needed, when it's needed.

Drip irrigation provides four significant benefits: First, by allowing for application of water directly to the crop's root zone through a mechanized system of timers, injectors and emitters, drip irrigation can reduce costs of water and irrigation labor. Second, by delivering water to the crop, instead of to the field, drip irrigation can reduce costs of controlling weeds. Third, by requiring less space than conventional irrigation systems, drip irrigation allows more space to be devoted to production. And finally, by preventing soil from becoming too wet or dry and by keeping salts more dilute, drip irrigation allows for use of marginal waters.

Drip irrigation also provides four significant limitations: First, by requiring a complex maze of emitters, water lines, meters, filters, screens, injectors, pressure regulators and timers, it is costly to establish. Second, because it is complex and subject to occasional mechanical breakdowns, drip irrigation is costly to manage, especially if the know-how must be purchased on the open market. Third, by feeding fertilizes salts directly to the crop's root zone, drip irrigation can allow dangerous concentrations of mineral salts that will adversely affect growth and development. And finally, pressure differences on steep slopes may cause plants on the bottom to receive more water than plants at the top, thus creating different rates of growth and development.

The benefits of drip irrigation outweigh limitations in many kinds of farm operations. For example, I recently visited a 16,000 acre pineapple farm on Lanai, Hawaii, which was irrigated with drips. Then, upon returning to the mainland, I helped install drips on a vineyard of *Cabernet* wine grapes in Southern Oregon.

There are many irrigation technologies available to metrofarmers. In fact, irrigation systems are now designed to satisfy specific needs of each metrofarm. A safe way to establish an irrigation system is to begin with a simple, low-cost design and then experiment, on a limited basis, with more complicated and expensive systems.

DRAIN: Too much water in the soil reduces growth and development several ways. It prevents plant roots from exhaling carbon dioxide, prevents aerobic soil bacteria from functioning, allows for accumulation of toxic minerals and compounds and maintains low soil temperatures. Excess water, in other words, will drown, poison, freeze and rot your crop.

Excess water, therefore, must be removed by a system of drainage. There are surface and subsurface drainage systems. The kind of drainage system suitable for your system will likely be determined by the lay of your land.

In many areas, however, economics of drainage must now include waste removal costs of agricultural chemicals. If you have a chemical-intensive production system, costs can prove substantial, as a neighboring California rose farmer recently discovered when he was forced to build a series of catch ponds by local government.

Surface Drainage: When the slope of land permits, excess water can readily be removed by a series of ditches. Large ditches are dredged out of the mud to take excess water downslope. Smaller ditches and channels are then cut into the contour of the slope to allow water to drain into large ditches.

FIG 8.22 Water in a drip irrigation system is pumped under pressure through an enclosed pipe or hose to its destination, where it is narrowcast through emitters. Top: Watsonville, CA; strawberries. Bottom: Lanai, Hawaii; Pineapples

Subsurface Drainage: Much of the land under intensive cultivation now has lines of drainage tile which run through the relatively wet areas. In fact, even agricultural lands in the arid west contain vast fields of drainage tile to remove excess salts.

The tile, which is perforated cement or plastic pipe placed below the root zone, collects subsurface water and carries it down a carefully graded path out of the production zone. Main lines of tile are placed to follow natural lines of drainage. Lateral tiles are placed at regular intervals to feed the main line.

The extent to which tile drainage is necessary is determined by the topography of land, type of soil, amount of precipitation and the crop. Systems of drainage tile are commonly installed in areas where water accumulates near the surface of the soil. In areas where water cannot be drained without an exorbitant expense, it would be wise to find either a water loving crop like cranberries or move on to greener pastures.

CONTROL PESTS

Your crop is like a savings bond. After it matures you can harvest it and profit from its accrued growth. Pests, however, do not have to wait for your crop to mature before they profit.

Insect, bird, animal and disease pests feed on crop growth and development. Weed pests steal sunshine, soil, nutrients and moisture from your environment. People pests steal into the field at night and spirit away your valuable products. If pests are not controlled, they will reduce yields, increase costs and diminish quality. The objective, therefore, is to establish a pest control system which will afford maximum control for minimum cost.

There are two steps to establishing an effective pest control system: First, establish a common sense line of defense. Next, establish means to regain control should pests breach your defense and flourish.

Establish a Line of Defense

Common sense precautions, when deployed in a systematic way, can help you build a fortress-like line of defense around your crop. Following is a brief discussion of the more important precautions. Consider how each will enable you to protect the economic viability of your crop.

HEALTHY CROPS: Nature provides each plant and animal with defenses, such as phytotoxins, which can ward-off harmful pests. However, the organism must be healthy before its immune system can be effective. If the organism is stressed by deficiencies of nutrients, moisture or sunshine, it will become weak and vulnerable to attacking insects and diseases. A healthy crop is a pest-resistant crop. Common sense precaution number one, therefore, is to reduce the threat of pests by growing healthy crops.

GENETIC VARIETY: Some genetic varieties are more vulnerable to pests than others and there is no way of knowing, for certain, which variety will be most vulnerable. Consider the AXR #1 grape root stock.

In the early 1900's, an epidemic of root louse, called phylloxera, decimated the wine-grape industry. New plantings were made with a root stock found to be resistant to the phylloxera, the AXR #1. An estimated 70 percent of all grape vines in California's Sonoma and Napa Counties are now planted to the AXR #1 root stock. However, a new strain of phylloxera, called "Type B," has recently emerged to launch a full-scale invasion on AXR #1 acreage.

Lacking genetic diversity, grape growers who planted exclusively to AXR #1 root stock are now fully exposed to the dangerous Type B phylloxera. Consequently, thousands of acres of premium wine grape vines are now being taken out of production. Common sense precaution number two, therefore, is to reduce your exposure to pests by providing genetic variety.

ROTATION AND DIVERSITY: Monoculture is the production of the same crop in the same soil year after year. This practice provides a good environment for soilborne diseases and pests. Continuing with the phylloxera example mentioned above, imagine the Type B's joy at finding their favorite meal, AXR #1, planted to thousands of acres without an obstacle in sight!

This vulnerability can be reduced by rotating crops from section to section, by growing a variety of crops and by providing a genetic variety. Certain control crops can also be planted to discourage pests. Common sense precaution number three is to confuse pests by diversifying your cropping schedule.

CLEANLINESS: They say cleanliness is next to godliness. This is certainly true for controlling insect and disease pests. Staying clean includes removing weeds and trash, maintaining restricted environments such as greenhouses in a near-antiseptic state and cleaning and storing tools. Staying clean can have a direct impact on pest control. Consider, again, the business of growing wine grapes.

One of the toughest pests to overcome for growers of the grape has been bunch rot, which is simply a fungus which attacks nearly ripe bunches of grapes preceding harvest. The solution to bunch rot has been liberal applications of fungicides. However, fungicides have been under attack by consumer groups and some have even been taken out of use by government.

Growers experimenting with various cultural practices, called "canopy management," discovered bunch rot could be controlled by pruning away the covering leaves and allowing air to circulate around bunches. By cleaning the environment around the grape bunches, instead of spraying, growers controlled the pest, reduced costs and increased the value of the product. Common sense precaution number four, therefore, is to stay clean.

TIMELINESS: Costs of controlling many pests can be reduced by proper timing of cultural practices. Weeds, for example, are much easier to remove in the spring, when the soil is soft and weeds small, than in the heat of an August afternoon, when soil is rock-hard and weeds are tall. Common sense precaution number five is to use good timing.

BARRIERS: Many pests can be discouraged by erecting a barrier to their movement. Insect pests can be hindered by barriers of marigolds, birds can be dissuaded with scarecrows, neighborhood thieves can be discouraged with a good fence and "No Trespassing" signs. The effectiveness of barriers, however, needs to be maintained with frequent applications of personal attention, otherwise the scarecrow will become just another roost. Common sense precaution number six is to discourage movement of pests by erecting barriers.

ATTITUDE: A healthful attitude is a valuable asset in the struggle to control pests. This attitude begins with the realization that most environments are well-populated with pests, and each pest has its season. Some panic at the first sign of an insect or disease, and panic will never aid in control of anything. Consider a neighbor, for example.

This person became known around the neighborhood as "The White Terror." Whenever she found an insect, she would dress up in white clothing, don a gas mask and spray pesticides. Her plants, thoroughly stressed from so much pesticides, were forever infested with pests. Amused neighbors seldom sprayed and rarely had damaging pest problems.

A healthful attitude begins with recognizing pests cannot be eliminated from the face of the earth and a few pests will always get through your line of common sense defenses. Maintain a healthful attitude.

Most pest control problems can be controlled by establishing the common-sense line of defense mentioned above. Russell Wolter, an organic grower of vegetables near Carmel, California, described how effective precautions can be in a conversation with a visitor to his farm. "I sell my crops through a conventional produce broker. One day the broker came out to look at the maturing crops in the field. When I mentioned that I do not use pesticides, he became very nervous and insisted that I protect the crop with at least one application of insecticide. After the crop had been harvested and sold he came back for another chat and said, 'Aren't you glad that you sprayed?'

"'Spray?' I replied. 'I thought you said pray! I came out here every day and prayed that the bugs wouldn't get my crop. And they didn't.'"

Common-sense precautions notwithstanding, pests can break through your line of defense and threaten the economic viability of your crop. Should it happen, there are two course of action you can take to regain control: organic and chemical. Following is a brief discussion of each.

Consider Organic Pest Controls

The term "organic" has come to mean many things to many people. California has even established a legal definition for the term. The law, section 26569.11 of the California Health and Safety Code, says, in essence, any product labeled "organic" must be produced without benefit of synthetically-derived fertilizers or pesticides.

Legal definitions notwithstanding, there is still a good deal of disparity when it comes to organic produce. There is the organic of neglect, as in apples filled with coddling moth worms from a neglected orchard. And there is the organic of high-density metrofarming, as in the polished blue ribbon winners which take consumers' top dollar.

The objective of establishing organic pest controls is to produce maximum quality for minimum cost and still win "organic" designation. When employed with diligence and persistence, organic techniques listed below can help you regain control of pests which threaten the economic viability of your crop.

LAYING ON OF HANDS: Though seemingly out of place in a world of high technology equipment and chemicals, the laying on of hands still provides metrofarmers, operating on small parcels of land, with a cost-effective means to control insect and disease pests.

The technology consists of simply removing trouble by hand. For example, if you find a diseased branch on your stand of dwarf apple trees, prune the trouble out and burn it immediately. The key to exercising effective hand controls is observation, and it is relatively easy to keep a small parcel of land under close observation. Stay vigilant while doing other chores. When you see a pest problem, attack it immediately.

PEST PREDATORS: Many centuries ago, Chinese farmers used the predatory ant *Oecophylla smaragdina* to control caterpillars and beetles in their citrus orchards. Specialists brought ant colonies in from the countryside and sold them to farmers.

The predator business is alive and well today, with specialists selling aphid midges, lacewings, lady bugs, praying mantids, predatory snails and others through retail nurseries and mail-order catalogs.

Some predators, like praying mantids and lacewings, feed on the body of a pest; others, like the *Trichogramma* wasp, lay eggs inside the pest's body. One company, Bunting Biological Control Ltd. of Great Horkesley, England, advertises their thrip-eating anthocorids feed by "inserting the proboscis-like mouthparts into the prey and sucking out the body juices."

There must be a sufficiently large population of pests for an application of predators to be effective, otherwise predators will simply go away in search of food.

TRAPS: Many pests can be controlled with traps which appeal to the need for food, shelter or sex.

Food traps lure a pest to food, where it can then be captured or killed. An effective way of controlling snails and slugs, for example, is to fill a shallow dish with stale beer. Snails will crawl into the dish for beer and drown.

Shelter traps lure a pest to shelter. Snails and slugs which escape a food trap, for example, might be induced to crawl under some strategically placed boards, from which they can be collected and disposed of during the day.

Pheromone traps lure a pest to a sex bait, where it can be trapped in a sticky substance or diseased with a biological agent. Pheromone traps not only provide an excellent means for control but also help one keep track of insect activity in the production plot.

Color traps lure a pest to a specific color, where it can be trapped in a sticky substance. Schoolbus-yellow boards covered with 90 weight oil provide an excellent means for controlling infestations of white flies.

BARRIERS: Pests can also be captured in barriers placed in strategic locations. Barriers are simply traps which lack the power of attraction. To control coddling moth larvae, a band of sticky tanglefoot can be placed around the apple tree's trunk. Similarly, a strip of salt can be placed across the migration path of snails.

BIOLOGICALS: Many pests can be controlled by interfering with their biological processes. There are bacteria, virus and fungi controls which can infect pests with diseases; and there are hormone controls which limit the pests' ability to reproduce.

The bacteria *Bacillus thuringiensis*, or Bt, is one example. When sprayed on leaves of a plant or tree, Bt infects the leaf-eating larvae of many butterflies and moths. The larvae immediately quit eating and die shortly thereafter. Bt can be purchased as a wettable powder or spray under a variety of commercial labels. It does not affect people, pets or birds; it is fast-acting and closely approximates the kind of control afforded by chemical pesticides.

ORGANIC SPRAYS: Organic sprays are ones made of naturally-occurring substances. They control pests by either repelling or killing them.

Repellents are used on chewing insects like cutworms, grasshoppers and caterpillars. They are made by mixing powerful substances like garlic and red pepper or by employing sticky substances like water-soluble kelp. (The application of kelp also provides a good supply of essential micronutrients.) Repellents are also used to control large pests. One popular deer repellent, for example, is a concentrated wildcat (predator) urine.

Killer sprays are used on juice-sucking insects like aphids, flies and scale. Insecticidal soap and water, for example, coat the pest's body and prevent it from breathing. Other examples of killer sprays include solutions of water and nicotine and of water, oil and sulphur.

RECOMBINANT GENE TECHNOLOGIES: Recent developments in the science of genetics have given us the ability to take genetic characteristics of two different organisms and recombine them into one.

At this writing, for example, scientists are splicing the gene for the toxic protein in Bt directly into plants. When a pest takes a bite of the plant, it gets a mouthful of the Bt toxin and dies.

There are limitations to this approach. One is a crop with Bt toxin spliced into its genes will likely encourage evolution of a Bt resistant pest. Another danger is commercial gene-splicing enterprises are likely to develop crops more resistant to pesticides, thereby allowing for increased usage of pesticides. Take advantage of recombinant genetic technologies but as always, be alert and ask questions.

For working examples of organic pest controls, read Heyn, Larkey and The Flower Ladies in Part IV, Conversations.

Plant Pest Barriers

PESTS	PLANT BARRIERS
Ants	Pennyroyal, spearmint, tansy
Aphids	Anise, *Alliums* (garlic, chives, onions, etc.), nasturtium, petunia, pennyroyal, southernwood, spearmint
Borer	Garlic, tansy, onion
Cabbage Maggot	Celery, catnip, hyssop, mint, rosemary, tomato
Chinch Bug	Soy beans
Cutworm	Tansy
Flea Beetle	Catnip, mint, wormwood
Fruit Tree Moths	Southernwood
Gopher	Daffodils, castor beans, gopher plant (*Euphorbia lathyrus*)
Leafhopper	Geranium, petunia
Mice	Mint
Mites	*Alliums* (chives, garlic, onion)
Mole	Castor beans
Nematodes	Tagetes (Mexican, French, and African Marigolds, dahlia)
Rabbit	*Alliums* (chives, garlic, onions)
Snail	Rosemary, wormwood
Tomato Hornworm	Borage, marigold, basil
White Fly	Marigold, nasturtium

FIG 8.23 The destructiveness of pests can be limited by erecting barriers to their movement.

Consider Chemical Pest Controls

Some pest situations require drastic action. Pests breaching the defenses of an unbalanced environment like a greenhouse or nursery, for instance, could quickly destroy many thousands of high-value plants growing on fractions of an acre. To maintain control and save the crop, many growers rely on quick and reliable controls afforded by chemical pesticides.

Pesticides are chemical compounds designed to kill. As there are different kinds of pests, there are different kinds of pesticides, including herbicides for weeds; insecticides for insects; fungicides for bacteria; fungi and viruses; nematocides for soilborne pests like nematodes; rodenticides for rodents; miticides for mites and defoliants for unwanted plant growth.

Though pesticides provide a quick and predictable means for controlling pests, they are poisons and must be used with extreme caution, especially in populated environs of metrofarms. A sudden shift of wind, for example, could spread a great deal of ill will around the neighborhood. Following are simple steps to designing a pest control program using chemical pesticides.

PEST IDENTIFICATION: First, identify exactly which pest is causing your problem. Is it an insect, fungus, mite or mammal? There are many good information sources which can help you identify common pests. They include resource books published by agricultural chemical companies, fellow farmers, the local nursery, the county extension agent or cooperative-extension service. Be certain of the cause of your problem before trying to solve it.

PESTICIDE SELECTION: The best pesticide is the one which provides the control desired; no more, no less. Many pesticides have been formulated to control a broad spectrum of pests. One popular rose product, for example, contains carbaryl (insecticide), malathion (insecticide), folpet (fungicide) and kelthane (insecticide). This shotgun approach to pest control is expensive, dangerous and guaranteed to destroy whatever kind of natural balance may exist in your production system. Be selective. Use one pesticide to control one problem.

FORMULATION: The best form of pesticide can be determined by accounting for such factors as size of the pest problem and available equipment. Pesticides are available in aerosol cans, dusts, wettable powders and sprays.

Aerosols are convenient because they require no mixing or cleanup; they are also very expensive and are therefore used only for very small problems. Dusts can be purchased in hand-applicators, making the form nearly as convenient as aerosol cans. However, hand applications of pesticide dusts are uneven; some plants get covered and others do not. Wettable powders and emulsified concentrate sprays are liquid formulations most frequently used in commercial operations. Liquids are inexpensive and, when used in conjunction with a good compressed-air applicator, provide the best coverage. However, liquids also require mixing,

FIG 8.24 Traps appeal to the need for food, shelter or sex. This pheromone trap, for example, lures pests to a scent and then traps them in a sticky substance.

handling, cleaning and storing.

LABEL DIRECTIONS: Pesticides are chemicals designed to kill. They are dangerous. Read label directions carefully and resist the temptation to overkill by doubling recommended dosage.

EQUIPMENT: Dedicate equipment specifically for mixing and applying pesticides. Keep this equipment out of general-use circulation.

MIXING: Mix only the amount of pesticide required to achieve control. Effectiveness of pesticides left in containers diminishes rapidly, creating a waste of money and a danger to children and the uninformed.

PROTECTIVE GEAR: Protect lungs, eyes and skin with protective clothing and masks. Wash thoroughly after each application.

THE UNINFORMED: Post signs warning others of pesticide dangers after you leave the area. The signs will help protect neighbors from your pesticides and help protect you from their lawyers.

STORAGE SAFETY: Store pesticides in a safe place. A safe place is dry, well-ventilated, clearly-marked, secure and locked. (Note: In farm country, pesticides have become a favorite target of thieves.)

DISPOSAL: The temptation to simply dump unwanted pesticides is great. After all, disposal costs of toxics in metropolitan environments is great. However, what may seem to be the least expensive means of disposal, such as dumping them into vacant lots down the road, may prove the most expensive if you are discovered. Check with local authorities for proper disposal procedures for pesticides and pesticide containers. Though proper disposal of pesticides may be an expensive proposition, dumping them into the neighborhood will be the most expensive proposition of all.

Consider Integrated Pest Management (IPM)

The organic and chemical pest controls mentioned above represent extremes of pest management. Large areas between are called "integrated pest management" or IPM. Practitioners of IPM use common sense and organic controls to maintain pest populations below economically damaging levels; then, if a pest population grows out of control, they use chemical pesticides to regain control.

FORCE GROWTH AND DEVELOPMENT

Good production systems have environments which allow crops to achieve their maximum level of natural growth and development. The best systems have controls which allow for forcing of natural growth and development to much higher levels of productivity.

Forcing is an ages-old practice of bringing plants or their desired parts such as flowers and fruits to maturity out of their normal season or out of their normal environment. Forcing is an important technology because it allows for maximum utilization of minimum resources.

It can even be argued that development of forcing controls gave rise to the industry of farming small parcels of land in metropolitan areas, at least here in the West. Consider the following: When Louis XIV was King of France some 300 years ago, he had an appetite to match his position. According to W.H. Lewis, in *The Splendid Century*, the king's typical supper would consist of "four plates of soup, a whole pheasant, a whole partridge, two slices of ham and a salad, some mutton with garlic, followed by pastry, and finished off with fruit and hard boiled eggs." From his seat on the throne, Louis was in the position to demand fresh food, year round.

Louis' royal edict strained the ability of an agriculture based on Thirteenth Century technology. So, to please his king and save his neck, La Quintime, the royal gardener, set about developing new technology. He used glass cold frames, heated soil beds, crop rotations and other space intensive techniques to force year round production of fresh fruits and vegetables for the king.

La Quintime's techniques were quickly adopted by the market gardeners who farmed small parcels of land on the outskirts of Paris. Word of this supply of fresh food spread quickly to the capitals of Europe and commerce— or "troque"— developed between Paris and other population centers like St. Petersburg.

In time troque farmers— or "truck farmers" as Americans call them— became prosperous and established a guild to protect their valuable technology. And so, by the hunger of a demanding king, emerged the technology of French-Intensive agriculture and the industry of truck farming.

Forcing growth and development can enable you to increase production, extend the season, reduce costs and increase the value of your product. There are two kinds of forcing controls. One manipulates the environment; the other, the crop itself. This section will help you develop both controls.

Force the Environment

Plants grow and develop in direct response to their environment. Elements of environment which regulate growth and development include period and intensity of light, air and soil temperatures and relative humidity. If you develop the means to control the elements, you can force growth and development into new, more desirable directions. Following are methods for controlling the environment. Use controls to force your crop into new, more desirable directions.

SOIL MULCHES: Mulches are protective coverings used to control soil temperature and moisture levels. There are organic mulches like hay, straw, grass clippings, leaves and sawdust; and there are synthetic mulches like plastic film, fiberglass, aluminum foil and stones.

Soil temperatures can be raised or lowered with proper mulches and, since temperature regulates many of the metabolic processes of growth and development, mulches provide an effective forcing control. Some mulches increase soil temperature, which allows soil to be used earlier in the season. Some mulches lower soil temperature, which allows soil to be used in the heat of midsummer. Growing a crop out of its normal season is a very effective way to increase its value.

Mulches can also be used to reduce costs. A good system of mulches can reduce evaporation of water by 10 to 50 percent, reduce weed control costs, diminish soil erosion, reduce loss of transplants, diminish soil compaction, conserve essential soil nutrients and protect ripening crops like strawberries from dirt and soilborne pests.

Mulches do have their costs. Time and money are needed to obtain materials and manage them. Mulches can provide a habitat for pests, prevent rainfall from reaching soil and may cause damping-off.

Many farmers disagree about the economic benefits of mulches. Some, typically those who operate on large parcels of land, argue mulches are too expensive; others, ones who operate on small parcels of land, say the savings more than justify the costs. The true value of mulches can only be discovered through a cost/benefit analysis for each production system.

PORTABLE SHELTERS: There are many kinds of small portable shelters which can be placed over young plants as protection from adverse air temperature.

Frost caps are preformed caps made of translucent paper or plastic. Tunnels are custom-built mini-greenhouses, constructed with plastic sheeting and wire frames. Cold frames are shallow rectangular soil beds covered with glass or plastic film. These devices provide a protected environment in which you can germinate seeds, nurture seedlings and force maturation through frosts of early spring and late autumn. By forcing production during off-seasons, you can gain a competitive edge in the competition for consumer dollars. For a working example of frost caps, read Heyn in Part IV, Conversations.

HEATED BEDS: Heated beds are soil beds which contain the means to control temperatures. There are many ways to control temperature. In the days of Louis XIV, beds were heated by the decomposition of horse manure. Today, beds are heated with electric cables, steam pipes or lights. Heated beds are often used in propagation work because warm soil temperatures encourage rooting. For working examples of how heated soil beds are used, read Schneider and Heyn in Part IV, Conversations.

LIGHTS: Life cycles of many temperate-zone crops are governed by the photoperiod of sunlight. In Spring, as the hours of sunlight increase, plants go into a vegetative cycle. In Autumn, as the hours of sunlight wane, plants go into a flowering and fruiting cycle. Growth cycles can be forced by using supplemental lighting to control the photoperiod. If vegetative growth is desired, turn the lights to Spring; if flowering and fruiting are desired, turn the lights to Autumn.

Agricultural lighting has improved dramatically over the past decade. Much of this progress is due to high-wattage metal halide lamp originally developed to illuminate large sports stadiums and provide a sun-like light for television cameras. Many commercial greenhouse and nursery operators now use halide and sodium vapor lamps to force out-of-season growth from high-value crops such as cut flowers.

GREENHOUSES: The ultimate control is the one earned when you build your own environment within a greenhouse. A well-designed and constructed greenhouse can give you the ability to control the period and intensity of light, air and soil temperatures and relative humidity.

A greenhouse is an enclosed structure devoted to propagation of young plants and to production of crops out of their normal season or out of their normal environs.

The essential structure consists of a framework of wood, metal or plastic and a transparent or translucent covering made of glass, fiberglass or plastic. More elaborate structures also include environmental controls like heaters, lights, ventilators, irrigation systems and drains.

Need: There are, perhaps, millions of different kinds of greenhouses. Your greenhouse should reflect the economic reason for which the greenhouse is needed.

Heyn, Larkey and The Flower Ladies use greenhouses to get an early start on the season. Since their greenhouses supplement outdoor production systems, they tend to be simple, low-cost designs, consisting of basic frameworks with

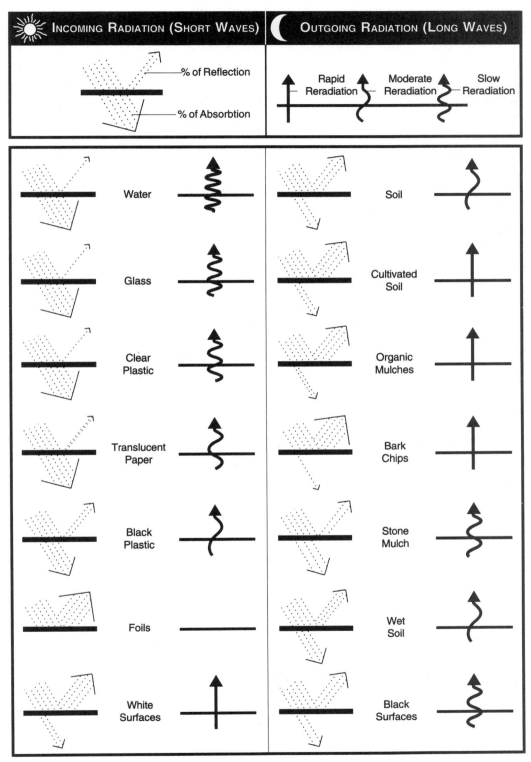

INCOMING RADIATION (SHORT WAVES)	OUTGOING RADIATION (LONG WAVES)

% of Reflection

% of Absorbtion

Rapid Reradiation — Moderate Reradiation — Slow Reradiation

Water — Soil

Glass — Cultivated Soil

Clear Plastic — Organic Mulches

Translucent Paper — Bark Chips

Black Plastic — Stone Mulch

Foils — Wet Soil

White Surfaces — Black Surfaces

FIG 8.25 Mulches provide a way to control soil temperature, and by extension, crop growth and development.

FIG 8.26 This frost cap protected the young melon plant from the last frosts of a Montana winter and allowed for an early start on the growing season.

plastic film coverings.

Because of the Heyns' location in Central Montana, electric cables were added to heat the soil beds. Heated beds enable the Heyns to force growth very early in the spring and consequently they now market vegetable starts to home gardeners. However unremarkable the supplemental greenhouses may appear, they do provide growers with a good headstart on the growing season.

Schneider and LaJoie use greenhouses to propagate nursery products. Since the greenhouses are principal units of production, the design tends to be more complete. Each features an enclosed structure with heaters to control air and soil temperatures, ventilation fans and automated irrigation systems.

In order to generate income as quickly as possible, Schneider and LaJoie based their designs on the least expensive materials. And though unremarkable in appearance, the designs enable Schneider and LaJoie to grow many thousands of high-value nursery products on very small parcels of land.

Finally, greenhouses are used as complete production systems for high-value products such as vegetables and cut flowers. A commercial tomato green-house near Redding, California, for example, produces up to 50 tons of salad tomatoes per year in a half-acre of hydroponic greenhouses. This greenhouse can even produce through winter because environmental controls in the greenhouse maintain year-round temperatures of 75 degrees.

FIG 8.27 Frost tunnels are custom-built miniature greenhouses. The tunnels in the top photo were designed for an experimental unit in Aptos, California, while the ones in the bottom photo were built for a commercial unit near Grants Pass, Oregon. Photo of Oregon unit by Scott Murray.

Design: Designing a greenhouse requires research and planning. Following are three steps you can take to develop a greenhouse design.

First, identify the economic reason for building a greenhouse. If you simply want a headstart on the growing season, consider an inexpensive wood frame and plastic skin. If you want a year-round production unit, consider an elaborate structure with the means to control the environmental elements which affect crop growth and development.

Next, research standard greenhouse designs. Many improvements have been forced on greenhouse design by the advent of energy shortages. Today's passive-solar designs, when properly constructed and managed, can reduce energy costs dramatically.

Finally, visit operating greenhouses and examine their strengths and weaknesses. One possible source of inspiration is the local college, which may have working examples of the latest greenhouses designs. Another source of inspiration is

FIG 8.28 A well-designed greenhouse, like this tomato unit in Redding, California, allows for control of the period and intensity of light, air and soil temperatures and relative humidity.

professional nursery growers. Jan LaJoie took a job in a local nursery before building her greenhouses.

Force the Crop

There are three kinds of controls which can be used to force the crop itself. When properly integrated into the production system, each control can also provide you with the means to force growth and development into more favorable and profitable directions.

PHYSICAL CONTROLS: One way to improve a crop's productivity is to improve its shape, size and orientation to its environment.

There are thousands of physical controls used in the production of crops: grape vines are trained, apple trees pruned, cattle castrated, Christmas trees shaped and so on. Though physical controls are typically associated with the culture of longer-living crops such as trees, many annuals like tomatoes and cucumbers can also be forced into new, more favorable directions.

If you have 10,000 acres in tomatoes, the controls may not make much sense. However, if you have one half-acre in solar greenhouses, you can, like King's farmer of forty centuries, train each plant to a "strong, durable, light and very readily removable trellis." You can also prune inefficient growth near the ground in order to force more productive growth on the trellis.

BIOLOGICAL CONTROLS: Natural growth can be forced into more favorable directions by manipulating a crop's biological processes.

There are many biological controls used in the production of crops: normal-sized fruit trees are grafted to dwarfing root stock; seedlings are hardened-off before being transplanted into a field; the flowering cycle of cut flower crops is initiated with supplemental lighting. Biological controls provide an effective means for redirecting growth and development into more profitable directions.

CHEMICAL CONTROLS: Chemical regulators can also be used to force more profitable growth. The regulators are chemical compounds synthesized from hormones normally produced by a plant or animal. When applied in very small amounts, regulators can promote, stop or modify many physiological processes.

There are two kinds of plant regulators commonly used. One forces growth, the other, development. There are many specific controls within each kind. Two which have been studied the most are auxins and gibberellins. Both regulators have been used for many centuries. Kelp, for example, has long been recognized for its ability to promote germination and rooting. Recent scientific analysis confirms kelp is rich in auxins and gibberellins which promote germination and rooting processes.

FIG 8.29 Workers near Smith River, California remove blossoms from lily plants to force growth back into bulbs, which are then harvested and sent to market.

Scientists are continually finding and synthesizing new plant regulators, many of which go into widespread use, especially in the nursery industry. One such regulator is a growth retardant.

Nursery growers often have trouble keeping greenhouse-grown products such as flowering bedding plants from growing too tall in their containers. This stretched growth reduces product value. Stretching could be decreased by reducing air and soil temperatures; however, reducing temperatures would also delay maturation and flowering. Chemical growth retardant enables the production of short, healthy plants in full bloom.

Some crop forcing controls were developed millenniums ago; others were developed only recently. To find the best way to force your crop, research long-standing and current cultural techniques. Though this research will cost time, your investment will be returned in increased production and productivity.

MANAGE EQUIPMENT AND LABOR

It requires time, money and know-how to establish production. The typical low-on-cash start-up begins with much time and little money: You turn the soil with a shovel, plant the seeds by hand, weed the plot with a hoe and take out the trash.

Later, money replaces time: You hire a tractor to turn the soil, plant the seeds, weed the plot and so on.

Substituting money for time enables you to focus on more productive aspects of your business or to devote time to another, more profitable endeavor. How you substitute money for time is very important. Though equipment and labor are investments which can increase the capacity of your metrofarm, each must be made with a dollars and cents objectivity.

Select Equipment

One aspect of the ancient Chinese production system King described stands out above all others: Physical work! Few residents of metropolitan areas would consider an occupation requiring so much physical labor! Few have to because modern times have made many labor-saving devices available at little cost.

When it comes to making equipment decisions, however, farmers often become afflicted with "new-paint" disease and lose their ability to think in terms of costs and returns. Consider a California neighbor who had one acre planted to kiwi and raspberry.

One summer day this fellow became frustrated with the five-foot tall weeds which had taken over his acre and went shopping for a solution. He returned with a new rototiller. It was silver, red and black; it even had fender skirts. Though the new tiller was able to knock down the weeds, a sharp hoe wielded in spring would have been far more effective in controlling weeds and generating profits.

This fellow is not alone in suffering from new-paint disease. There are very good reasons why the malady is common to farmers everywhere. First and foremost is taxes. If you have a choice between giving money to government for taxes or giving money to an equipment dealer for a new piece of equipment, you take home the equipment. Occasional government efforts to stimulate the economy with allowances for "accelerated depreciation" and "investment tax credits" merely amplify the tendency.

New equipment also provides a source of prestige. The neighbor's colorful new rototiller attracted far more attention than a sharp hoe raised at the right time. Though the new rototiller gave the neighbor the pride of ownership, it did not help him generate profit on his kiwi and raspberry plantings and consequently weeds eventually won the battle for his acre.

Equipment decisions must be governed by economic objectivity. Place all gains on one side of the scale, including increased yields, tax benefits and prestige. Place all costs on the other side of the scale, including purchase price, interest charges, depreciation, upkeep, operating expenses and storage charges. Then allow the scales to make the decision. The following steps will help you make objective equipment decisions.

START-UP REQUIREMENTS: Determine which equipment will be essential for establishing production. Many establish production with little more than ordinary gardening tools. The Flower Ladies, established production with shovels, hoes, cultivators and an old rototiller. More complicated systems, like the automated greenhouses at Jan LaJoie's Plant Carrousel, require more elaborate equipment like thermostatically- controlled heaters. Determine which equipment you need to establish production and then factor the equipment into start-up costs of your business.

EXPANSION REQUIREMENTS: Determine which equipment you will need to expand production, reduce costs and make more efficient use of your time.

During their start-up period, The Flower Ladies stored finished wedding floral arrangements in the cool air of a creek bottom. After they became established, the Ladies purchased a walk-in cooler. This cooler enabled the Ladies to store more arrangements for less cost. (No raccoons!) And, as the cooler was next to their workshop, it enabled them to make more efficient use of their time.

Though not essential, optional equipment like The Flower Ladies' walk-in cooler can increase profits significantly and should therefore be planned as follow-up investments.

ACQUISITION: Determine whether to purchase, lease, rent or barter your equipment. In general terms, equipment used frequently, as on a daily or weekly basis, should be purchased or leased; whereas equipment used infrequently, as on an annual basis, should be rented or bartered. New-paint disease notwithstanding, there is little economic justification for buying a new tractor if it will be used only once or twice per season. By renting a tractor or the services of a custom-operator, you can conserve precious operating capital.

QUALITY COMPARISON: Make certain the equipment can do the job. Metrofarming is a business which requires industrial-grade strength and durability. Consider the hoe: A quality hoe will hold its edge far longer than a cheap one. Though this quality will cost a few more dollars at the sales counter, it will save more time and money out in the field. This is true for tractors as well.

Make certain the equipment can be serviced. Use equipment which can be serviced this season, next season and many seasons into the future. As stressed, you do not want to purchase equipment unless it is needed. If it takes three months to repair a tractor needed today, the tractor is going to be one very expensive investment.

Finally, ask for a demonstration and some customer recommendations. Verify the manufacturer's claims before investing your money. If the equipment is guaranteed, make certain it is a written guarantee.

PRICE COMPARISON: Determine which vendor has the best price for the

FIG 8.30 This "Down to Earth" farm equipment belongs to Russell Wolters of Carmel, California, who uses it to plant, fertilize and harvest the many kinds of crops growing in his fields.

best equipment. Use an objective criterion, such as horsepower per dollar. Examine this ratio on several different models. Check fuel consumption to price, energy output to price and so on. Comparisons will reveal which equipment represents the best value. Shop around for the best price. You will save money by forcing vendors to compete for your business.

MAINTENANCE: Take care of your equipment. Design a good maintenance program for cleaning, sharpening, lubricating and storing. Maintenance reduces equipment costs by extending the productive life of the investment and by maintaining its resale value.

Though many metrofarms are not equipment-intensive businesses, a well-selected piece of equipment can increase capacity significantly. Good equipment does not need to be a complicated product of high-technology covered in new paint. Russell Wolter, a successful California vegetable grower, uses a machine which looks like it was jury-rigged from parts salvaged from 30 years of family cars. The contraption helps Wolter plant, fertilize and harvest the many different vegetable crops growing in his fields. Though it does not turn many of Wolter's neighboring farmers green with envy, few pieces of new equipment can do so much work for so little cost.

Hire Labor

Most metrofarm's labor is supplied by the owner and family. Their labor is, as demonstrated by the individuals featured in Part IV, Conversations, one of the principal strengths of metrofarming. There comes a time, however, when additional hands can be used.

Labor is one of the most expensive production inputs in farming, which explains why many large-scale farmers came to rely on the inexpensive labor of "illegal" immigrants. Illegals, as many big operators freely admit, worked extremely hard without complaint for subminimum wages and could be housed in subpar facilities.

Metrofarmers, on the other hand, are blessed by their proximity to the vast pool of labor in the city. There is always a surplus of idle hands within the city and its environs, and a surplus of anything means reasonable prices. The problem is finding people who are dependable, hard-working and honest. Though it is not an easy problem to overcome, there are some common-sense guidelines you can follow to find good legal labor.

PART-TIME HELP: Much of your need for additional labor will come from seasonal production requirements like planting, weeding, harvesting and processing. Though requiring comparatively little technological know-how, the tasks do require hard work and perseverance. There are a couple of good ways to satisfy this need for part-time help.

One way is to advertise. Try the local paper, the local school or college, the senior citizens center or even the unemployment office. Recently the Orange County, California, unemployment office failed to post job openings for the local strawberry harvest because officials thought nobody would be interested. But when a raid on illegals picking the berries by the U.S. Border Patrol was reported in the local press, the unemployment office was flooded with angry requests for the jobs.

Another way to secure temporary labor is to induce family, friends and neighbors to participate in an old-fashioned "labor bee." The bee is half work and half social. People gather to do something together, such as build a barn, and then to socialize and enjoy each other's companionship.

Nancy Tappan, of Rogue River, Oregon, organized a bee to build soil beds for her commercial crop of garlic and shallots. She contacted a local private school and suggested a little work party. Saturday morning a busload of healthy kids pulled in the drive and, within a few hours, built the soil into production-ready beds. The kids were then treated to a barbecue and party.

FULL-TIME HELP: When your metrofarm has become an established income producer, you may want to hire full-time help. This full-time employee will be like a partner because you will rely on him or her to conduct the day-to-day operation of your business. And, like all partnerships, there must be giving and

FIG 8.31 This labor bee was organized by Nancy Tappan of Rogue River, Oregon, who called a school and suggested a work party in exchange for a how-to-grow crops seminar and a poolside barbecue.

taking on both sides. The employee must be hard-working, honest, reliable and intelligent; the employer must develop the potential of an employee through training and motivation. Following are steps to hiring and developing a good full-time employee.

Employee Requirements: Before you hire someone for a full-time position, determine if he or she can produce more than he or she will cost. This will tell you if you really need the employee or if you can do better by rearranging production and hiring a part-timer. If your evaluation is positive, prepare a job description which defines specific duties; list employee specifications (*e.g.*, physical, educational, experience) based on the job description; and then determine how much you are willing to pay.

Applications and Resumes: Have prospective employees fill out written applications and provide resumes. Read each carefully and eliminate undesirable applicants. Hold personal interviews with likely applicants and encourage them to talk about themselves and their work experience. Ask specific questions: What did you do on your last job? Why did you leave? Complete your "due diligence" with a complete check of all references and verify information listed on the written application. Ask previous employers the question: "Would you hire this person back?" Hire the best applicant.

Employee Introduction: Make certain your new employee is properly introduced to the job. Take time to welcome him or her and describe the business and where he or she will fit in. Explain the rules of dependability, safety and quality and tell how he or she is going to be trained.

Employee Instruction: Tell your employee what to do. Many people cannot fully understand a verbal description of a complicated procedure, so break the job down into specific, easy-to-understand steps. Show the process. Help him or her understand by giving a hands-on demonstration. Next, have your employee do the job and explain it as he or she goes along. Finally, follow up by checking with him or her frequently in the first few days and then taper off as he or she works into the job.

EMPLOYEE MOTIVATION: As mentioned earlier, labor is one of the greatest expenses in farming. It makes sense, therefore, to help your employee perform at or near his or her maximum level of ability by providing sufficient motivation.

Employee surveys point out that if a fair wage is paid, other objectives become more important than money. The objectives include recognition, good working conditions, good administration, job status and security. In addition, an employee will not strive for an employer's goal unless he or she accepts the goal as being worthwhile and important.

The key to motivating an employee, therefore, is to find a way to convince him or her that achieving your goal will help to achieve his or her goal. Following are some common-sense guidelines which, if followed with a little sensitivity, will help you motivate an employee to perform at the maximum level of his or her ability.

Be Personable: Though there must always be a barrier between an employer and employee, you can learn to treat your employee well. Get to know his or her family. Invest a moment or two from time to time and talk about his or her hobbies and interests. Treat him or her to a dinner once in a while. Offer to help whenever possible.

Keep Informed: Let your employee know about his or her personal progress and, where possible, your plans for the business.

Listen: A good employee will have good suggestions and complaints. Listen and be willing to adopt suggestions and improve working conditions wherever possible.

Confer a Title: Give your important employee something to tell friends and family (*e.g.*, "I am the foreman.").

Critique Performance: Recognize good work by praising it in public.

FIG 8.32 An employee is your partner in sustaining the economic viability of your business.

Recognize bad work by reprimanding it in private.

Provide Benefits: Provide your important employee with benefits like holidays, a vacation, overtime pay and rest periods.

The key to effective management of equipment and labor is to recognize that each investment must produce more than it costs. As a business person, you must resist the temptation to buy new equipment because it has new paint or to hire the neighbor's kid because you want to be a good neighbor.

Make your equipment and labor investments on the basis of a thorough cost/benefit analysis. By doing so, you can ensure your investments will increase the capacity of your metrofarm to produce more for less.

PROVIDE SHELTER

There was a time when the big red barn served as the focal point of the farm infrastructure. Though the red barn has gone the way of the 160-acre homestead, the need for shelter still exists.

Shelters can increase the productive capacity of your metrofarm by providing a comfortable place to work, store equipment and crops and sell products. In addition, a shelter can increase the value of real property and provide tax benefits.

Your shelter may be as simple as a small roadside stand or as complex as a multipurpose facility for work, storage and sales. Ideally, each shelter you provide will conform to existing building codes and serve to reinforce your image as a producer of quality farm products. (Shoddy structures badly in need of paint weaken your position in the mind of consumers and may draw the attention of a local building inspector.)

Following are general guidelines for providing shelter for work, storage and marketing.

Develop Work Space

A well-designed work space can make it easier and less costly, to maintain equipment, transplant seedlings, mix fertilizers, process harvested crops and so on. To realize how valuable a good work space can be, consider processing crop commodities into consumer-ready products.

Many farmers transport their harvested crops directly from field to market. This is a commodity business in which the farmer must take whatever the market will give for raw commodities. The big money, as every good business person knows, is made by processing raw commodities into consumer-ready products. Buy a handful of corn for two cents, run it through a couple of machines, into a colorful box, and then sell it as breakfast cereal for two dollars.

Many metrofarm crops can be processed into consumer-ready products. The Flower Ladies (Part IV, Conversations) provide a very colorful example: When the Ladies began selling their crop of cut flowers, they received one-half dollar or so per bouquet. The Ladies then processed their flowers into wedding floral arrangements which sold for up to one hundred dollars and more.

In fact, value can sometimes be added to a crop commodity simply by packaging it as a product. Scott Murray, a grower of specialty vegetables near San Diego, California, found he could gain two weeks of shelf-life by packaging his sweet-basil in plastic bags. The bags also enabled Murray to rapidly expand his wholesale market because wholesalers did not lose as much money on spoiled basil.

The minimum requirement for processing a raw crop commodity into a consumer-ready product is a good work space. Both The Flower Ladies and Murray used a good cool work space to keep their products fresh and to maintain worker efficiency. Think ahead to the various tasks of your production cycle. Allow the tasks to define what kind of work space to provide. Keep the design cheap, simple and presentable, like the red barns of yesterday.

Develop Storage Space

A good storage space can increase profits by protecting equipment from the environment and by reducing the rate of crop spoilage.

Consider how a good storage facility can be used to reduce equipment costs: While a shovel left out in the rain may last two seasons, one cleaned after use and hung in a dry protected space may last 15 seasons. A shovel amortized over 15 years is less expensive than one amortized over two years. Furthermore, storage helps maintain resale value, and the 15 year-old shovel might be valued at two or three times the two year-old shovel. Multiply the savings by every piece of equipment on the farm.

Consider how a good storage facility can be used to reduce the rate of crop spoilage. Crops mature in their respective seasons. When crops are harvested their value falls with the flood of supply in the market. By storing the crop and preventing its spoilage until demand goes up, you can take advantage of high off-season prices.

FACILITY REQUIREMENTS: There are two kinds of crop storage facilities, "common" and "cold." Common storage provides space, insulation from temperature extremes and ventilation. Cold storage adds hardware for artificially cooling. All storage facilities should be well-drained, clean and vermin-proof. Complete facilities also include provisions for controlling elements which diminish product value, including the following:

Temperature: In general, the higher the temperature, the more quickly the crop will decompose. The ideal temperature for storing many crops is just above freezing.

Humidity: Evaporation, or shrinkage, is a function of temperature and relative humidity. The shrinkage loss of many crops can be reduced by maintaining a high level of relative humidity. Exceptions include such crops as onions and sweet potatoes.

Air: Oxygen fuels respiration and therefore increases the shrinkage loss of many crops. This loss may be reduced by keeping the storage facility airtight or by adding an inert gas.

Quality Control: Shrinkage losses with stored crops are always heavy enough. Storing bruised, damaged, diseased or pest-infested produce is a sure way of inviting disaster. As the saying goes, "It takes only one rotten apple to spoil the bushel."

FACILITY MANAGEMENT: Following are means for managing environmental elements which affect shrinkage rates within a storage facility:

Insulate: The simplest way to control air temperature and oxygen content is to insulate the storage space with a fibrous or cellular material. Good insulation is a poor conductor of heat, moisture proof, easy to handle, durable, odorless and cheap. Popular insulating materials include styrofoam, fiberglass and soil.

Circulate: The objective of circulating is not to replace inside air with outside air but to move the air around in order to maintain uniform temperatures. The air in tall storage spaces may be stirred with a small fan or two.

Humidify or Dry: There are many electrical humidifiers and driers available for controlling relative humidity. This kind of equipment may be especially important in areas of climatic extremes such as deserts or rain forests.

Refrigerate: The most complete way to control temperatures in a storage facility is to install a refrigeration unit. The Flower Ladies' business bloomed, literally, when they purchased a second hand refrigerated storage unit.

FIG 8.33 Storage space helps increase productivity by extending the useful life of equipment and by maintaining the quality of harvested crops.

Storage facilities protect equipment and crops from time. Allow the needs of your equipment and crops to tell you what kind of storage facility to provide.

Develop Marketing Space

One advantage to metrofarming lies in proximity to the market. A direct marketing facility will provide you with the means to exploit this advantage.

Many metrofarmers begin direct marketing with existing facilities such as a barn or packing shed and then develop a permanent facility when business permits. A new direct marketing facility should be designed with a thorough cost/ benefit analysis in mind.

Direct marketing costs include principal, interest and depreciation costs for the facility, labor to sell products and sales supplies like bags, cash registers and sign materials. Returns include the elimination of middlemen, transportation, standardized containers, grading, inspection and packaging costs.

Following are simple guidelines for establishing a direct marketing facility:

DESIGN: Your direct marketing facility should be designed to handle the volume of business you expect. It should be built of inexpensive materials and the design should include provisions for expansion.

EQUIPMENT AND MATERIAL: Include all lights, drains, display racks and refrigeration units required for processing and selling your crops. If possible, include separate rooms for preparation and sales.

IMAGE: Your marketing facility is often the public's first look at your business. And, as another old saying goes, "You never get a second chance to make a good first impression." Design your direct marketing facility to look good. Shoddy construction and a rundown appearance will diminish your image as a producer of quality products.

INCLEMENT WEATHER: Your direct marketing facility should be constructed to allow continued sales in adverse weather. This need was well-illustrated by the rain which recently flooded Half Moon Bay, California's Pumpkin Palace one weekend before Halloween. Half Moon Bay bills itself as the "Pumpkin Capital of the World," so their Halloween rain was a serious business setback.

ADEQUATE PARKING: Adequate provisions should be maintained for cus-

FIG 8.34 A direct marketing facility, like the one at the Gizdich Ranch near Watsonville, California, provides an effective way to exploit a nearby population center.

tomer parking. Parking spots should be well-defined and the lot should be surfaced to prevent dust from becoming an overwhelming problem.

To find the right design for your direct marketing facility, visit as many of the direct facilities as possible. Carefully list positive and negative aspects of each place you visit, then draft a design which incorporates the best for your direct facility.

SCHEDULE CROPS

This chapter has presented steps to establishing an efficient system of production. Now it is time to put the system to work with a crop schedule. The crop schedule is a written plan which includes goals, policies and procedures necessary to generate maximum return on investments of time, money and know-how.

Begin with a simple schedule and improve upon it as you grow. Experience will provide you with the know-how to design a schedule as efficient as the one used by the Chinese farmer King observed in *Farmers of Forty Centuries:* "... so not a minute had been lost in the change of crops. On the contrary, this man had added a month to his growing season...."

Schedule Primary Crops

Start with crops which will pay bills and generate profits. They are your money crops. The first scheduling objective is to ensure the maximum production of money crops.

PLANTING DATE: Farming is a cyclical business in which there is no clearly defined beginning or end. The first step in scheduling production is consequently a bit like jumping onto a merry-go-round in motion: You pick a horse and jump on when the horse comes around. A good horse to jump on is the planting date.

Weather: In conventional farming, the first safe planting date is the day after the last frost. This conventional planting date is a good place to begin scheduling. Research weather and determine an average date of the last frost for your area.

Forcing Controls: Next, determine which forcing controls you need to get an early start on the season. Frost caps, for example, will enable you to plant seeds before the last frost (*e.g.*, Heyn, Part IV, Conversations). Or a greenhouse will enable you to transplant mature, hardened starts into the field on the day after the last frost. Use forcing controls to gain a competitive edge on conventional planters. When the last frost arrives, be in the ground with a crop which is already well on its way to winning consumer dollars.

Succession Plantings: Many fresh market crops require a succession of planting dates. The objective of staggering planting dates is to ensure that you do not enter the marketplace with your entire production on the same day. If you plan to direct-market sweet corn, for example, establish a succession of planting dates so part of your crop is yielding a fresh product week after week.

PRE-PLANTING PRODUCTION: Have everything ready to go on time so you can successfully meet your planting date and keep costs at a minimum.

Plan backwards from the planting date and schedule tasks which must be accomplished in order to get your crops into the ground. Make a list: When must seeds be ordered? When must they be germinated? Which forcing controls must be used to meet the planting date? What must be done to make forcing controls operational? What materials must be purchased? By thoroughly planning all of your pre-planting tasks, you can reduce costs by taking advantage of off-season sales, bartering with your neighbor for use of his tractor and working during your spare time.

GROWTH CYCLE PRODUCTION: Plan ahead from your planting date and schedule tasks which must be accomplished to grow the crop.

Make another list: How often will you need to irrigate? Fertilize? When is the optimum time to remove weeds? Which pests will most likely attack the crop? When will the crop be most vulnerable to pests? What are the chances for inclement weather? What materials and equipment will be required? How can materials and equipment be obtained for the least cost? Will additional labor be required?

Think your way through the growth-cycle and eliminate, as best you can, any expensive surprises which can reduce growth and development, increase costs and diminish product quality.

HARVEST AND POST-HARVEST PRODUCTION: After a crop has been harvested, its quality can only deteriorate. Ensure maximum product quality by timing the harvest and by planning post-harvest handling.

Begin by estimating your crop's maturation date. Though seed companies publish estimated maturation rates, the real rate will vary from location to location. Experience and accurate records will help you come close, but even experienced estimates are vulnerable to vagaries of climate.

Make your estimation by adding the published growing time to your planting date and then begin a list of post-harvest tasks: Will additional labor and equipment be required for harvest? What facilities will be required to process the crops for market? How will the crop be transported to market? Will advertising and promotion be required?

After a crop has been harvested, natural forces which act upon it can only diminish its market value. By planning your way through post-harvest production, you can ensure an orderly and profitable transition from farm to market.

Schedule Secondary Crops

The Chinese farmer King observed in *Farmers of Forty Centuries* grew two crops of greens by the time his primary crop of cucumbers had reached its productive stage. Secondary crops, in other words, enabled the farmer to increase profits without increasing the size of his farm. If your primary crop will not utilize 100 percent of your system's capacity, consider generating more income by scheduling secondary crops to fill voids in the rotation.

There are three kinds of secondary crops. Consider each carefully and determine which is best for your business. After you have selected a secondary crop, schedule it by using the procedures outlined for scheduling primary crops.

COMPANION CROPS: As the name implies, companion crops are ones which make good companions for primary crops. There are two techniques for scheduling good companions.

Similar Maturation Rates: Two or more crops of similar maturation rates can be grown in the same soil at the same time. They must be physically compatible; each must avoid competing with its companion. Compatibility traits include a short crop and a tall one, a deep-rooted crop with a shallow rooted one, a sun-loving crop with a shade-loving one and so on. A ground-hugging winter squash, for example, would provide a good companion for a sky-reaching sweet corn. Scheduling companion crops with similar maturation rates will allow you to harvest two or more crops at the same time.

Different Maturation Rates: Two or more crops with different maturation dates can be grown in the same soil at the same time. King's Chinese farmer, as mentioned, produced two crops of quick-growing greens while his primary crop of cucumbers was maturing. This companion crop technique can enable you to generate cash-flow from a short-season crop to finance the harvest and marketing of your primary crop.

SUCCESSION CROPS: After your primary crops have been harvested, consider putting the soil back into production with a second, third or even a fourth crop.

Succession cropping is a technique which is frequently employed in production of fresh-market produce. The basic technique employs forcing controls like hot beds and greenhouses to grow out large and relatively mature starts. When a crop is harvested, large starts are immediately rotated into the field, thus reducing the amount of time required in the ground.

One example of a three-crop succession is an early spring crop of radishes followed by a summer crop of melons and an autumn crop of spinach.

Succession crops should also be good companions. They should not deplete the soil of nutrients required by the next crop, nor should they attract and harbor pests which might destroy the following crops.

EMERGENCY CROPS: Production does not always work according to

FIG 8.35 "This year we have a better schedule. We know exactly what to plant in each row, how far apart each plant must be and when to rotate those plants out and others in. This year we are going to triple our production." From Scott Murray, a second-year grower of specialty vegetables near San Diego, California.

schedule. After you have scheduled primary and secondary cash crops, plan for the unknown by scheduling emergency crops. The best emergency crops are those with a short growing season, are simple and inexpensive to produce and have a ready, if not ideal, market.

Schedule Other Economic Crops

In addition to money crops, there are others which can be scheduled into the production rotation for economic gain. Remember, the objective of scheduling is to ensure the maximum utilization of resources. If your soil is fallow, you lose money.

GREEN MANURES: Green manures are crops which are plowed under while still green or which are composted and then returned to the soil to increase the soil's nutrient content, structure and tilth. By scheduling a quick-growing green manure to fill voids in the production rotation, you may be able to reduce costs of fertilizer, water, pest control and labor. In addition, the resulting healthy soil may increase the quality and value of your money crops.

COVER CROPS: Cover crops are those which protect soil from the erosive effects of wind and rain. Metrofarms, by definition, operate within tightly limited spaces, and it pays to protect every square inch. By scheduling a cover crop like rye or a rye-based blend, you may be able to reduce erosion control costs and add valuable moisture-holding tilth to your soil.

PEST CONTROL CROPS: Pest control crops are ones grown for their ability to help control insect pests. There are two important points to remember when scheduling pest control crops. First, young plants are less effective than mature plants. Mature plants have enough growth and development to produce essential oils and aromas which repel insects. Second, a variety of pest control crops is more effective than a single one.

REVIEW

To survive and prosper, you must win the competition for consumer dollars. To win, you must overcome the obstacles of a small parcel size and a shortage of capital. Overcome the obstacles by incorporating elements of high-density, low-cost production into one smoothly functioning system. Elements to incorporate include the metrofarm's layout, its soil and how the soil is fertilized, management of its water, control of its pests, forcing of its crops, utility of its shelters and efficiency of its schedules.

Start with a simple system. Keep accurate records and research production technologies with the intensity of an industrial spy. It may cost a season or two to establish a successful system. The Chinese farmer King observed, after all, built his system upon forty centuries of collective experience.

Recently, a second-year grower of specialty vegetables near San Diego, discussed his metrofarm in a letter. "I have the beds scheduled right down to the week," he said. "Last year we only made ten thousand per acre, but this year we have a better schedule. We know exactly what to plant in each bed, how far apart each plant must be, and when to rotate those plants out and others in. This year we are going to triple our production."

Though farming in the midst of the ultramodern San Diego freeway society, this fellow is happily following the footsteps of the "farmer of forty centuries" King observed somewhere in ancient China.

CHAPTER EIGHT EXERCISES

i   List six contemporary facilities designed to produce more from less (*e.g.*, electronic microprocessors, steel minimills, neighborhood microbreweries). Describe, with one sentence, why each facility can produce more from less.

ii   Begin a new subsection entitled "The System" in the "Production" section of your metrofarm binder. Use the land selected in exercises 5*i* to 5*v*, and the crops selected in exercises 6*i* to 6*v*. Sketch a layout which will take maximum advantage of the farm's microclimate.

iii   Select a soil medium. List materials required to build your medium. Describe how you will arrange the medium to grow the maximum amount of products in the minimum amount of space (*e.g.*, tractor-formed rows, French-Intensive beds, plastic vessels, etc.). Finally, integrate the soil medium into the map of your production system (Exercise *ii*).

iv   List physical, biological and chemical controls which can be used to force your crop's growth and development. Select controls which will increase yields, improve quality and reduce costs. Where appropriate, as would be the case with a greenhouse, integrate these controls into the map of your production system.

v   List the pests which will compete with you for growth and development of your crops. List pest barriers and other common-sense ways to prevent pests from growing out of control. List organic, chemical or IPM pest controls you will use should pests invade your production system and grow out of control.

PART III

MARKETING

INTRODUCTION

The fortunate few go to market protected by patents, trademarks, copyrights and other proprietary restrictions. You, on the other hand, will likely have to fight it out in the no-holds-barred my-potato-against-yours world of the free market.

If demand for your product is high and the supply low, you can sit back and wait for consumers to beat a path to your door. However, when supply and demand have achieved an equilibrium, as they tend to do, you need to implement a well-conceived and coordinated marketing plan. Consider, by way of example, the marketing efforts of two San Francisco area metrofarmers:

After a considerable amount of study and hard work, a salesman of printing services succeeded in producing a bumper crop of organically-grown carrots on the large lot next to his suburban home. When he trucked the harvested crop to a produce broker, however, he was offered only the going price for commercial, field-grade carrots. "Makes no difference to me whether your carrots are organic or not," said the broker. "I sell them all for the same price." Rather than accept the broker's offer, which was far below his expectations, the salesman trucked his carrots back home and sat on them until it was too late.

Jeff Larkey (Part IV, Conversations) also produced a crop of organically-grown carrots. Through a careful survey of the local market, Larkey found a retail outlet which catered to vegetable juicers. Juicers prefer fresh, locally-grown carrots to those produced in distant growing regions, kept for months in cold storage and then shipped into the area for redistribution. Larkey began selling his carrots through the store and, to advertise just how fresh they were, did not trim away the green foliage tops. Local juicers bought Larkey's sweet, juicy carrots and spread the news. Other consumers bought. Larkey now does a brisk trade in many fresh products through this store. And he sells for a handsome profit.

You develop advantage by producing the maximum amount of products of highest possible quality with a minimum amount of resources. Your advantage can be protected and your competitive position improved by investing the same kind of dedication into marketing. Part III of *MetroFarm* will help you make this investment. Chapter 9 will help you prepare for the market and Chapter 10 will help you generate the maximum return on your sales efforts.

Preparing for Market

*People forgot that the most important event on earth
each year is the harvest.*

Thomas Malthus

You will not forget. Harvest is when your investments of time, money and know-how are taken from the ground and sent into competition for consumer dollars. This Chapter will help you prepare for the competition. It includes steps to planning quality, harvesting, sorting, storing, processing and packaging. The proper planning of each step will help you win the consumer dollars.

PLAN FOR QUALITY

Quality and perception of quality are what distinguish metrofarm products from mass-produced ones of the competition. Your crop products will reach the peak of their quality at harvest; thereafter, forces which act upon the products only serve to deteriorate their quality. The first step in preparing for market, therefore, is to plan ways to maintain the quality of your products. This step begins with recognizing elements of quality and forces which affect quality.

Recognize Elements of Quality

Quality refers to essential characteristics of a product, including appearance, texture, flavor and food value. Following is a brief discussion of each characteristic:

APPEARANCE: "Eyes are the first to feast," is a popular expression among French chefs. It means a customer's first look tells whether your product has a uniform shape and size, is free of bruises and blemishes, has good color and is free of extraneous matter like twigs and dirt. Many customers use the first look as their sole criterion for judging quality. Therefore, when you take your product to market remember another old saying: "You never get a second chance to make a good first impression."

TEXTURE: A customer can feel a product and judge it by whether it is too hard or soft, too granular or smooth, too crisp or wilted, too fibrous or lacking in fiber and so on. Tough kernels of sweet corn, stringy celery and wilted lettuce are examples of poor texture. Though many properties of texture depend upon cultural practices and inherited characteristics, they can also be influenced by the product's stage of maturity at harvest and how it is handled following harvest.

FLAVOR: Flavor is a combination of taste and odor. Taste depends on soluble compounds recognized in the mouth; odor depends on volatile compounds recognized in the nose. Many judge flavor by simply biting into a product to see if it has a pleasing taste. Others, like judges at a wine-tasting, conduct elaborate tests for both taste and odor. The soluble and volatile compounds which constitute flavor are often present in such small quantities so as to defy analysis. Compounds can be greatly altered in harvest and post-harvest handling.

FOOD VALUE: Though a customer may purchase a farm product because it looks good, feels right or has an excellent flavor, the product may be needed to sustain the lives of the customer's family. It stands to reason, therefore, food value is also an important factor in evaluating quality. Principal elements of food value are carbohydrates, proteins, minerals, vitamins, fiber and oil. The amount of each element in a product is determined by genetic heritage and cultural practices. However, food value can also be significantly altered by harvest and post-harvest handling. Sweet corn, for instance, will lose half its sugar content in twenty-four hours if stored at room temperatures.

Recognize Forces Which Affect Quality

After the crop has been harvested it becomes subject to forces which can only deteriorate its quality, including respiration, moisture loss, decay, diseases and insects, rough handling and temperature extremes. Following is a brief discussion of

FIG 9.1   After harvest a crop becomes subject to forces which can only deteriorate its quality.

each force. Plan to control the relevant ones as you prepare your crops for market.

RESPIRATION: Respiration is the taking up of oxygen, the combining of oxygen with sugars through enzymatic synthesis and the giving off of carbon dioxide and heat.

Respiration does not end after a crop has been harvested. On the contrary, it continues burning up stored energy until the crop has been consumed by man or nature. The process is governed by temperature: Warm temperatures speed respiration, low temperatures retard it.

The most perishable products are those which respire fastest, like berries, peas and sweet corn. The least perishable are those which respire at the slowest rate, like apples, cabbages and potatoes. It follows that products with the fastest rate of respiration present the greatest problems in post-harvest handling.

Plan to limit respiration losses by controlling temperature from harvest to market. In general terms, the lower the temperature, the slower the respiration rate.

MOISTURE LOSS: Many fresh crop products are 80 to 95 percent water. Products can lose water weight, shrivel and deteriorate if exposed to adverse conditions after harvest. Conditions which govern moisture loss after harvest include surface area, relative humidity, wind velocity and temperature. Each condition, as briefly described below, may be controlled by storing products in a cool, humid environment.

Surface Area: Consider the surface area of your crop. If it has a large leaf area, like lettuce, it will likely lose moisture rapidly. If it has a waxy surface, like watermelon, it will likely retain moisture for a much longer period of time.

Humidity: Humidity is the measure of moisture present in the atmosphere. Air with a low relative humidity increases the rate of moisture loss; air with a high relative humidity retards loss. Relative humidity is an important consideration in areas of extremes, such as a desert or a rain forest.

Wind Velocity: Air flowing past the surface of a product will replace the moisture-laden air surrounding the product with fresh, dryer air, thereby increasing the rate of moisture loss.

High Temperature: High temperature increases the capacity of air to hold water and reduces the surface tension of moisture within a product, thereby speeding evaporation.

DECAY: All living organisms are covered with microscopic bacteria and mold spores which, given the right conditions, begin feeding off the host and cause its decay. Organisms must be controlled after harvest. Various chemicals and gases have proven effective in controlling post-harvest decay; however, low temperatures are perhaps the most effective and least dangerous means for retarding microscopic agents of decay.

PESTS: High temperatures, humidity and unsanitary conditions increase potential for disease, insect and rodent damage to harvested crops. Though various chemicals and gases are frequently used in controlling pests in post-harvest processing and storage, a clean, temperature-controlled storage facility can also provide an effective means for reducing pest damage.

HANDLING: Poor handling during and after harvest often causes cuts, bruises and other blemishes. Defects diminish the appearance of a product and provide a place for the microorganisms of decay to begin their work. Plan ways to reduce post harvest handling to a minimum and save quality.

TEMPERATURES: Not all crop products can withstand low temperatures after harvest. Tomatoes and bananas, for example, will not ripen if stored in low temperatures. When a crop is stored at the wrong temperature, its quality may be adversely affected and its value diminished.

Your well-designed metrofarm will deliver the highest possible quality for the lowest possible cost. This quality will not improve after harvest; it will only diminish. The forces which affect quality must therefore be controlled from harvest through the final sale to a customer.

HARVEST THE CROP

City people often think of the harvest as being a time of joyful celebration. They imagine a Thanksgiving-like picture of farmers singing happily while "bringing in the sheaves." Experienced farmers know better.

Harvest, more often than not, takes place in the heat of battle: A hail storm is brewing on the horizon; a forecast for an extreme frost hits the news; an urgent order from a broker sits on the desk. Plan your harvest to satisfy the urgency of the moment and to maintain quality of your crop products. Make provisions for the harvest itself and for handling the crop after it has been harvested.

FIG 9.2 The last grapes of the first harvest are in, to the delight of Nancy Tappan, Marlene and Kelsey Olson and Vernon Hixson at Venture Vineyards in Rogue River, Oregon.

Plan the Harvest

As stated, the objective of planning a harvest is to ensure your crop arrives at market in prime condition. There are a number of factors to consider in your planning. They include time of day, the crop's maturity, its size and weight, schedules and supervision.

TIME OF DAY: Crops with a slow rate of respiration, like apples, cabbages and potatoes, mature slowly and stay ripe for a considerable length of time. Crops with a fast rate of respiration, like berries, peas and sweet corn, mature rapidly and do not allow much leeway.

Since temperature affects the rate of a crop's respiration and consequent maturation and decay, it provides an important key to determining when a harvest should occur. If your crop respires rapidly, plan to harvest during the cool temperatures of early morning. If your crop respires slowly, relax and harvest at your leisure.

If you must harvest a temperature sensitive crop during the heat of day, plan for emergency cooling procedures immediately following harvest. Emergency planning can pay for itself by maintaining product quality.

MATURITY: When will your crop reach its peak of maturity? This is an important consideration for those who strive to gain advantage by producing vine-ripened quality. Their products will taste better and consequently have more value if allowed to ripen on the vine than the products grown by large-scale competitors, which may be ripened in storage with ethylene gas.

In addition, the longer a crop is allowed to grow, the more weight it will produce. This additional weight, however, does not always improve quality or increase income. Consider the zucchini squash: Though zucchini reaches its optimum quality when young, small and tender, it is often allowed to over-ripen to increase its size and weight. The resulting jumbo-sized zucchini is tasteless and almost impossible to give away.

Finally, delaying harvest past maturity may increase your crop's vulnerability to insects, diseases, rain, frost and premature seeding. Capitalize on your proximity to market. Plan the harvest for the peak of your crop's maturity.

SIZE AND WEIGHT: The physical characteristics of some crops at harvest are important for a legal reason. There are marketing orders in effect which restrict the size and weight of certain products to a narrowly defined range.

Generally speaking, marketing orders are legal mechanisms sponsored by industry and sanctioned by government to support commodity prices by restricting supply. If fruits of your nectarine trees are not the right size, for instance, you may not sell them in wholesale markets.

These marketing orders are designed and put into effect to keep dilettantes from flooding the market with excess supply. Plan ahead for these orders. Will they affect the marketing of your crops? If so, how will you satisfy their requirements?

MULTIPLE HARVESTS: Some crops require you to schedule a series of harvests. Apples, for example, ripen at different times on the tree. The repeated harvest of an apple crop can often help to maintain high quality— especially in hot weather— and provide more income. Can you increase yields by scheduling multiple harvests? How will you manage the increased costs of harvest labor?

HARVEST SUPERVISION: The breakage and bruising which might seem innocent enough to hired hands in the field may reduce a crop's quality and value considerably. The only way to limit poor handling by harvest labor and equipment is to personally supervise the harvest.

Plan Post-Harvest Handling

After a crop has been harvested it must be trimmed and cleaned. Though "separating the wheat from the chaff" may seem like unpleasant work, it is work which affords you another opportunity to maintain product quality and value. Following is a brief discussion of post-harvest chores to be considered.

FIG 9.3 Mike Watchorn, of the Hog Island Oyster Company in Tomales Bay, California, washes his oysters to ensure that customers receive a clean, fresh product.

CULTURE PRACTICES: Many post-harvest handling chores can be reduced by good cultivation practices during the growing season.

For example, when California fruit growers irrigated their orchards with overhead sprinklers, mineral-laden water left white splotches on the fruit. Consumers saw the splotches and thought them to be residues of pesticides and consequently did not buy the fruit. It would cost a lot of time and money to remove the mineral splotches from a commercial-sized crop of fruit, consequently many growers switched to drip irrigation and eliminated the problem.

Research the cultural practices used in production of your crop and determine what new techniques been developed to reduce post-harvest processing.

CLEANLINESS: Cleaning is vital to the maintenance of quality, especially in markets where customers are conditioned to antiseptic cleanliness of modern supermarkets. If a retail customer sees your product contaminated with dirt, weeds and other extraneous matter, your reputation as a producer of quality goods will suffer accordingly.

Logistics of cleaning depend on the crop. Root crops, like potatoes, onions and carrots, may require scraping, washing or scrubbing. (As I discovered with my first crop of garlic!) Leaf crops, like lettuce and spinach, may simply require a good rinsing. Other crops, like small fruits and flowers, may not require cleaning at all.

Cleaning a large crop by hand can be time-consuming and expensive. This chore can be lessened to some extent by mechanizing the process with high-pressure water sprays and various other cleaning machines available through commercial sources.

Plan to clean your crop if necessary. Your investments will be repaid many times over when customers see its scrubbed quality.

TRIM: Remove damaged, dead, discolored, inedible or unusable material from your crop product. Amount and degree of trimming depends on the type of crop being harvested: Some crops, like celery, are trimmed quite closely; others, like cabbage, are trimmed more loosely to leave a protective layer of outer leaves. Trimming reduces a bulk crop into a marketable product; saves storage, transportation and handling costs; and improves appearance and value.

Plan the steps necessary to clean and trim your crop down to the essential product. Many crops are now cleaned, trimmed and packaged in the field during harvest. This field work can reduce costs of post-harvest handling and speed crops to market.

SORT FOR GRADE

All farm products are not created equal. Even apples harvested from the same tree vary in size, shape, color and the number and size of blemishes. To facilitate marketing your crops, consider separating each into a uniform grade. This process is called "sorting for grade."

Recognize the Objectives of Grading

Sorting requires an investment of time, money and know-how. Why bother? Why not simply sell everything in one big bunch and call it a day? Because there are important ways in which sorting for grade can improve your competitive position.

INCREASE VALUE: By sorting top grades from bottom ones, you may be able to generate higher prices than if you simply sell everything as one. For example, if you sell all the apples from your tree for a single price, the price will most likely be one paid for a low or medium grade apple. By sorting your apples into different grades and packaging them in separate containers, you may be able to sell a gourmet grade for a top price, a commercial grade for a medium price and the rest for juicers. The total received for your three grades will likely exceed the total earned by lumping all apples together as one.

REDUCE COSTS: Wholesalers and retailers like to order by specific weight and grade. They like to say, for example, "Send me ten cartons of Extra Fancy." A uniform grade is a prerequisite for distribution through intermediaries because it

FIG 9.4 Marlene Olson and Nancy Tappan set aside the top grade of Rogue Valley Gardens' Alliums for shipment to a New York City department store.

eliminates the need for personal inspection of each package and allows for smooth flow of commerce between farmer and consumer.

REDUCE SUPPLY: One way to raise prices is to reduce supply. In fact, much of the periodic oversupply of farm products—and resulting depressed prices— can be reduced by withholding inferior grades. Withholding is often enforced by Marketing Orders sponsored by industry and sanctioned by government. Grades which cannot be sold for a price high enough to justify transportation and marketing costs are then sorted into cull grades.

REDUCE SPOILAGE: "It only takes one rotten apple to spoil the bushel." There is a considerable amount of truth to this old saying. Sorting for grade is the perfect place to cull defective products which may spoil valuable ones.

Establish a Uniform Grade

Uniform grades may be based on taste, price, appearance, texture, nutrient content, uniformity, wholesomeness and even cultural practices employed in production. Grades may be based on all of the characteristics, none or something else entirely. There are typically three ways in which uniform grades are established. Following is a brief discussion of each:

FARMER ESTABLISHES GRADE: The simplest way to sort for grade one used by farmers who sell direct. An apple farmer selling direct at a farmers market, for example, might put the best apples in one container, average ones in another and

United States Department of

Product	Current Grade Name	Grade Definition
MEAT	USDA PRIME*	Very tender, juicy, and flavorful; has abundant marbling.
	USDA CHOICE*	Quite tender and juicy, good flavor; less marbling than prime.
	USDA GOOD	Fairly tender; not as juicy and flavorful; has least amount of marbling.
LAMB	USDA PRIME*	Very tender, juicy, and flavorful; generous marbling.
	USDA CHOICE*	Tender, juicy, flavorful; less marbling than Prime.
VEAL	USDA PRIME*	Juicy and flavorful; little marbling.
	USDA CHOICE*	Quite juicy and flavorful; less marbling.
POULTRY	U.S. GRADE A*	Fully fleshed and meaty; uniform fat covering; well formed; clean appearance.
	U.S. GRADE B	Not quite as meaty as A; may have occasional skin blemishes; not as attractive as A.
	U.S. GRADE C	May have cuts, tears, or bruises, wings may be removed and moderate amounts of trimming of the breast and legs are permitted.
EGGS	U.S. GRADE AA*	Clean, sound shells; clear and firm whites; yolks practically free of defects; egg covers small area when broken out—yolk is firm and high and white is thick and stands high.
	U.S. GRADE A*	The same as AA except egg may cover slightly larger area when broken out and white is not quite as thick.
	U.S. GRADE B	Sound shells, may have some stains or shape may be abnormal; white may be weak and yolk enlarges and flattened; egg spreads when broken out.
DAIRY PRODUCTS	U.S. EXTRA GRADE	Sweet, pleasing flavor; natural color; dissolves readily in water.
FISH	U.S. GRADE A*	Uniform in size, practically free of blemishes and defects, in excellent condition, and having good flavor for the species.

Agriculture Grading Standards

PRODUCT	CURRENT GRADE NAME	GRADE DEFINITION
FISH	U.S. GRADE B	May not be as uniform in size or as free of blemishes or defects as grade A products; general commercial grade.
	U.S. GRADE C	Just as wholesome and nutritious as higher grades; a definite value as a thrifty buy for use where appearance is not an important factor.
FRESH FRUITS & VEGETABLES	U.S. FANCY*	Premium quality; only a few fruits and vegetables are packed this grade.
	U.S. No. 1*	Good quality; chief grade for most fruits and vegetables.
	U.S. No. 2	Intermediate quality between No. 1 and No. 3.
	U.S. No. 3	Lowest marketable quality.
CANNED & FROZEN FRUITS & VEGETABLES	U.S. GRADE A*	Tender vegetables and well-ripened fruits with excellent flavor, uniform color and size, and few defects.
	U.S. GRADE B	Slightly mature vegetables; both fruits and vegetables have good flavor but are slightly less uniform in color and size and may have more defects than A.
	U.S. GRADE C	Mature vegetables; both fruits and vegetables vary more in flavor, color, and size and have more defects than B.
DRIED FRUITS; FRUIT & VEGETABLE JUICES; JAMS, JELLIES, PRESERVES	U.S. GRADE A*	Very good flavor and color and few defects.
	U.S. GRADE B	Good flavor and color but not as uniform as A.
	U.S. GRADE C	Less flavor than B, color not as bright, and more defects.

*Indicates grades most often seen at retail level.

FIG 9.5 USDA grading standards from, "Let the Grade be Your Guide," by Sara Beck, Elizabeth Crosby, and Martha Parris, *Food—From Farm to Table, 1982 Yearbook of Agriculture*, US Department of Agriculture, Washington, DC.

send culls to a juicer. Though products which go directly to consumers are not subjected to a complicated grading structure, they are sorted into uniform grades so products on top are representative of products on the bottom.

FARMER AND INTERMEDIARY ESTABLISH GRADE: A more reliable system for grading is developed when a farmer and intermediaries trade with each other on a regular basis. Their grading system is one based on precedent. When the farmer calls the intermediary and says she has ten cases of "premium" apples, the intermediary thinks back to previous trades and remembers exactly what the adjective "premium" means. The system works as long as the standard is maintained.

INDUSTRY AND GOVERNMENT ESTABLISH GRADE: The most complicated grading system is the one used when farmers sell to the mass market. Since the mass market consists of many sellers and buyers, who may be separated by thousands of miles, it is very important to establish a uniform grade to alleviate the necessity for personal inspection.

Mass markets use grading standards established by the U.S. Department of Agriculture or by industry trade groups. The USDA, for example, has established standard grades for wool, meat, poultry, eggs, dairy products, fruits and vegetables (fresh, frozen, canned and dried), cotton and tobacco.

Grades are based on characteristics unique to each product. For example, beef quality is based on maturity, color, firmness, texture and fat content. Butter quality is based on flavor, body, color and salt content. Various terms are then given to each quality grade, like USDA Prime for the top grade, U.S. Grade A for the medium grade and U.S. Fancy for the lowest grade.

Establish a Sorting System

Buyers are suspicious. Though apples on top of the box may look good, many buyers believe that apples on the bottom will be rotten. Overcome this mistrust by establishing a uniform grade. When properly graded, your package of apples, flowers or ginseng will have the same quality on bottom as on top. Customers will come to trust your packages and consider your product a safe buy. Following are guidelines to consider for establishing a system of sorting for uniform grade.

MARKET STANDARDS: Most established markets use standard grades of one kind or another. If the market is direct, like a farmers market, standards may be quite lax because buyers personally evaluate quality. If the market is indirect, like a mass-market wholesaler on the other side of the country, standards may be quite strict because the buyer cannot see the product and must therefore rely on a standard to judge quality. Look into your market and determine how it grades its products.

Chances are the system used within your market will be the best system to adopt.

GRADING REQUIREMENTS: An elaborate system for grading may not be economically justified if the harvest is small or of uniform quality. The simplest and most profitable system for sorting in situations described may be to separate good, average and cull grades in the field during harvest.

COSTS AND RETURNS: Sorting requires an investment of time, money and know-how. Justify the investments by determining if the price received for the top grade plus the price received for the remaining grades will exceed the price received for a combination grade.

STORE THE CROP

Crops not shipped directly to market after harvest must be protected from the forces of respiration, moisture loss, decay, temperature extremes, diseases and insects. As discussed in the preceding chapter, a well-designed storage facility can provide protection.

There are many kinds of storage facilities. They range in complexity from old-fashioned root cellars to modern cold storage warehouses complete with environmental controls like ice-water spray systems for emergency cooling. The type of facility you select should be determined by the number of products to be stored and the volume of transactions to service. When The Flower Ladies (Part IV, Conversations) began their business, storage consisted of a level spot in a creek bed; later, when the Ladies' volume of business demanded, they purchased a walk-in cooler.

Following is a list of factors to consider before it's time to place your harvested crops in storage.

STORAGE POTENTIAL: Some crops store well; others do not. Storage potential of most crops has been researched and results of the research have been made available to the public. Research public information to determine whether storage is feasible and if special temperature and humidity controls will be required.

PRICE FLUCTUATIONS: In the past, many kinds of crops were stored for later sale at better prices. Today, many of the same crops are produced in favorable climates of distant growing regions and shipped in fresh. Fresh salad tomatoes from Mexico, for example, may now be purchased during the blizzards of a far north winter. Check seasonal price fluctuations for your crops; fluctuations must be great enough to justify storage costs.

CUSTOMER REQUIREMENTS: Most retail and wholesale customers prefer a steady supply with a uniform quality. Storage helps smooth the day-to-day logistics of supplying this demand. For example, storage may enable you to fill an order without having to harvest in the rain, avoid Saturday's oversupply and sell to

Crop Storage Reference Table

Item	Best Temperature (Degrees F.)	Freezing Point (Degree F.)	Preferred Humidity (Percentage)	Sprinkling Desired	Desired Characteristics
Apples	30-32	29.3	90	None	Colorful, uniform, bruise-free
Beans	40-45	30.7	90-95	Light	Crisp, uniform, immature
Beets	32-	31.3	95	Moderate	Small, smooth, firm
Berries	31-32	29.7-30-6	90-95	None	Bright, clean, plump
Cabbage	32	30.4	90-96	Moderate	Hard, heavy, bright color
Cherries	30-32	29	90-95	None	Bright, plump
Corn	32	30-9	90-95	Moderate	Bright, plump, milky kernels
Lettuce	32	31.7	95	Light	Clean, crisp, tender
Onions	32	30.4	90-95	Light	Green, fresh, clean, uniform
Peaches	31-32	30-3	90	None	Bright, fresh, yellow color
Potatoes	45-70	30.9	85-90	None	Smooth, round, firm
Radishes	32	30-7	90-95	Yes	Mild, bright, smooth
Spinach	32	30	90-95	Moderate	Fresh, immature, colorful
Tomatoes	45-50	31.1	85-90	None	Plump, firm, uniform red color

FIG 9.6 Crop storage reference from, *Direct Farm Marketing*, by Jackson and Groder, Extension Circular 945, Oregon State University, Corvalis, 1978, Appendix B, p. 12.

Monday's undersupply or take Sunday off for a family outing and still have products to sell on Monday. Ask customers how they like to be serviced. If storage can help you better those needs, it can help increase your share of the market.

PROCESS THE CROP

Processing consists of preparing a crop for market by subjecting it to physical or chemical change. There are two reasons why crops are processed: One reason is to preserve the crop's value for sales at a later date. Instead of selling all the apricot crop immediately after harvest when prices are low, dry some into a form which will keep until better prices can be obtained. The second reason is to add value. Instead of selling all of the apricot crop as raw apricots, add value by processing some into consumer ready forms like juices, jams, sauces or even pies.

Because of economies of scale available to large, mechanized facilities, processing crop commodities is often considered a battlefield upon which only big operators can survive. Before you give over this valuable turf, consider how processing can be used to generate more income from your small parcel of land.

Process to Preserve Quality

Often it is not possible to sell an entire crop immediately after harvest. One obvious reason is the low prices resulting from post-harvest surplus in supply. Crops not sold immediately after harvest may be stored fresh, as discussed in the previous section, or processed to control organisms which promote decay. Following are six processes for "putting the crop by" to preserve quality for sales at a later date.

DRYING AND DEHYDRATING: Drying is the oldest method of preserving value. Though largely supplanted by other processing technologies, many fruits, vegetables, herbs, meats and flowers are still process dried. Drying and dehydrating consists of removing water by subjecting the crop to fresh air and, in certain cases, supplemental ventilation and heating. Lacking sufficient moisture, organisms of decay will be unable to grow and multiply during storage.

There are two kinds of moisture removal, drying and dehydration. Drying means 80 to 90 percent of the moisture is removed. Dehydration means 96 percent or more of the moisture is removed. The amount of remaining moisture required for safety varies according to whether the product has a low or high acid content, the amount of salt with which it has been treated and, to some degree, upon the type of storage.

Basic equipment required for drying consists of a flat surface upon which products may be laid and cheesecloth or netting to line drying surfaces. Commercial-scale dryers are available with racks, heaters and ventilation fans.

Dry air is the main component for process drying. The lower the relative humidity of air, at any given temperature, the more quickly water can be evaporated. In locations where relative humidity is extremely high, supplemental heat is often added to increase the air's ability to hold moisture. Heated air, however, often causes overdrying in products close to the heat source, so product rotation and air circulation may also be required.

The principal benefit of drying is its relatively low cost. Other benefits include the reduction of the crop's bulk and weight which, in turn, reduces storage and transportation costs. The primary limitation is that many consumers may be inexperienced with dried products and demand, consequently, may be limited.

The Circle I Farm is a five-acre fruit and nut business near Los Molinos, California. How can this small business compete with economies of scale advantages enjoyed by big fruit and nut growers of California's Central Valley? Instead of competing for the fresh fruit customer in distant markets of Los Angeles, San Francisco and Portland, Circle I dries much of its organically-grown fruits and nuts and packages them for sale later. The resulting product, which weighs very little, can be shipped into lucrative off-season markets in big cities for a comparatively low cost and big return. The Circle I makes a little land go a long way by

FIG 9.7 Process drying, as demonstrated by this salmon smoker in Elfin Cove, Alaska, preserves by removing moisture with fresh air and, in some cases, supplemental heating and smoking.

FIG 9.8 Canning destroys harmful organisms with heat and prevents their re-introduction by sealing the product in an airtight can or jar.

selling its organically-grown dried fruits and nuts to specialty markets ignored by big operators.

CANNING: Canning is a process by which harmful organisms in fresh products are destroyed with heat. The product is then sealed in a container— or canned— to prevent reinfection. Farm products which are canned include fruits, vegetables, meats and fish.

There are two canning processes, "boiling water bath" and "pressure cooking." A product's *pH* determines which process must be employed. High-acid products like fruits, berries, tomatoes, pickles and relish may be canned with the relatively simple and inexpensive boiling water bath. Low or no-acid products like meat, fish and many root vegetables must be canned with more complicated and expensive pressure cooking.

Canning costs, therefore, depend upon the type and volume of products to be canned and time and expense of complying with government regulations that have been established to protect consumers against botulism poisoning.

The canning industry has been in decline in recent years. Reasons for this decline include increased year-round availability of fresh foods and proliferation of heavily advertised alternatives. Nevertheless, canning still provides many opportunities to preserve metrofarm products for later sale at a better price.

You can compete with economies of scale advantages enjoyed by large

Examples of High-Acid and Low-Acid MetroFarm Products

HIGH-ACID FOODS

Apples	Apricots	Berries	Cherries
Chili Sauces	Chutney	Citrus Fruits	Conserves
Currants	Fruit Juices	Grapes	Jams
Jellies	Melons	Nectarines	Peaches
Pears	Persimmons	Pickles	Plums
Relish	Sauerkraut	Tomatoes	

LOW-ACID FOODS

Asparagus	Beans	Beets	Broccoli
Cabbage	Carrots	Celery	Cauliflower
Corn	Greens	Meat	Mushrooms
Onions	Parsnips	Peas	Peppers
Pimientos	Potatoes	Poultry	Seafood
Squash	Stews	Stewed Tomatoes	

FIG 9.9 High-acid foods may be canned in a bath of boiling water; low-acid foods must canned in a pressure cooker.

canning operations by limiting your canning efforts to very special products. One way in which metrofarm products can be made special is by identifying them with a specific location— or "appellation." Consider Gilroy garlic, Montana trout and Michigan mushrooms. Location-specific products may be canned and sold through such high-return markets as gourmet food stores, airport gift shops and mail order for a substantial return on your investment.

FREEZING: Harmful organisms can be controlled with extremely low temperatures. Products which may be frozen for later sales include fruits, vegetables, meats, fish and processed convenience foods.

Frozen foods have experienced a tremendous increase in popularity among consumers in recent years. There are several reasons for this popularity: First, frozen foods generally retain more nutrients, color and texture than foods processed through other means and are consequently perceived to be the next best thing to garden fresh. Second, frozen foods are often packaged in user-friendly containers which can be slipped into a microwave oven for an instant meal. Third, it takes a very expensive infrastructure to keep food frozen from processor to consumer, and this infrastructure is now in place in most metropolitan markets.

Freezing costs, which include freezers, special packaging and refrigerated transportation, make the processing option an expensive one for many kinds of metrofarm products. Nevertheless, you can compete. Consider the Gizdich Ranch, a berry and apple U-Pick operation in Watsonville, California. If the Gizdich family had to rely solely on their U-Pick customers, much of the berry crop would not get picked on time and would consequently be lost. To prevent loss, the Gizdich family picks their surplus ripe berries, packages them in plastic bags and then freezes them for sale at a later date.

PICKLING: Pickling controls harmful organisms by immersing the product in a preservative of vinegar or alcohol. Though many of the product's vitamins and minerals are leached into the solution, pickling remains an important method of food preservation, as evidenced by the popularity of tomato catsup. (Currently each citizen of the United States consumes an average of three pounds of tomato catsup each year!) Other popular pickled products include cucumbers, beets, meat parts and eggs.

The pickling process begins with the brine (salt water) soak. There is a short-brine process and a long-brine one. In the short-brine process the product is left overnight in a brine, then packed in jars, covered with pickling solution and processed in a boiling water bath. In the long-brine process the product is soaked in a heavy brine (along with vinegar and spices) and left to ferment for up to one month. The product may then be pickled whole or diced up and combined with other ingredients to make a relish.

You can overcome competition's economies of scale advantages by pickling specialty products. Take an ordinary product— tomato catsup— and use unique ingredients like vine-ripened tomatoes with fresh herbs and spices. Or take a unique ingredient— the cranberry— and make an extraordinary product like cranberry catsup! Your specialty product, if pickled correctly, will taste far superior to mass-produced catsups and sell for a premium price.

CURING: Curing is another moisture removal process for preserving certain vegetables, meat, fish and fowl. This process consists of first removing moisture by immersing the product in dry salt or a concentrated brine. The product is then subjected to the "cold smoke" of a low heat/high smoke fire for several days. Smoke colors and flavors the product, retards growth of organisms and increases moisture loss.

One popular cured meat product is the pork ham. Ham production is, for the most, dominated by large processors who rely on inexpensive feedlot hogs and who benefit from purchasing, facilities and labor economies of scale. However, since a ham's taste is directly influenced by the hog's diet, commercial hams tend to lack a distinctive taste. Sensing opportunity, many small-scale producers have developed market niches for tasty specialty hams such as the German "Westphalian" and Italian "Prosciutto," which are processed from hogs fed on diets of chestnuts and cheese whey.

There are many such opportunities available to metrofarmers. One business which capitalized on this opportunity is Summit Farm, a producer of game birds for restaurant trade in the San Francisco Bay Area. In addition to pheasants and quail, which are sold either fresh or frozen, Summit also offers a smoked Cornish game hen which has a very distinctive and delicious flavor.

A note on curing: Many of the chemical compounds used in curing meats prior to smoking have come under attack by health-conscious consumer groups. Among the compounds are potassium and sodium nitrates (often called "saltpeter") and sodium nitrite. These compounds are used to intensify and preserve color and control growth of harmful microorganisms such as *C. botulinum*. Nitrates and nitrites, in themselves, are not harmful to humans. When metabolized by microorganisms inside the human stomach, however, they may be converted to compounds called nitrosamines, which have been shown in laboratory tests to induce cancer in animals.

Concerns about curing compounds point to additional opportunities for metrofarmers. Cure and package specialty products without potentially harmful chemicals and you will likely attract a clientele of health-conscious consumers.

IRRADIATING: Irradiation kills harmful microbes and insects with gamma radiation. The process works like a medical x-ray. (A chest x-ray requires about one fourth of one rad; sterilizing food requires about four million rads.) Irraditiation was first used in the 1950's and, until recently, was used only sparingly on various spices, potatoes and grains. However, when various toxic fumigants like ethyl dibromide (EDB) were banned, food processors petitioned the government for approval to irradiate. Consequently, irradiation is now approved for fresh fruits, vegetables, herbs, spices and pork.

There exists a considerable amount of controversy regarding the use of radiation to preserve food. Processors say irradiation is harmless and extends the shelf-life of products significantly. An irradiated potato, for example, can be stored for six months without spoiling or sprouting. Many consumer groups, on the other hand, claim radiation also changes the cell structure of food products, causing deleterious or uncharacterized effects on nutrients. Vitamins A, E, K, B1, B2, B3, B6, essential amino acids, cysteine, methionine, histadine, tryptophan and enzymes are said to be either depleted or destroyed by radiation.

Irradiation is clearly a good technology for large-scale packers, processors and food retailers because it reduces losses by extending the shelf-life of perishable commodities. It is a good technology for government agencies like the Department of Energy because it provides a means to generate income from surplus radioactive waste. However, irradiation may be a bad technology for farmers because, by extending the shelf-life of perishable commodities, it will likely serve to increase supply and reduce prices.

Process to Add Value

Processing raw commodities into consumer-ready products is as old as selling loaves of bread instead of yeast and flour. This kind of processing adds value by making products more convenient or more attractive. Following are brief discussions of each way to add value.

CONVENIENCE: "Can she bake a cherry pie, Billy Boy, Billy Boy?" As the song suggests, there was a time when homemakers were judged by their ability to prepare and cook food. However, times have changed and homemakers have moved out into the labor force. This emigration from the home has produced two profound changes in the market:

First, homemakers have less time to buy, prepare and cook food. They now demand products which are ready to eat or easy to prepare. According to processing industry estimates, only about 30 percent of home meals are now prepared from raw materials.

FIG 9.10 Raw farm commodities may be peeled, cored, juiced, cooked or combined into value-added, easy-to-use products like the salad greens featured at this roadside stand.

Second, many consumers simply lack the experience of preparing food and consequently do not know how. Consider the Pillsbury Corporation's recent attempt to remove what seemed an unnecessary direction from its can of corn. ("Put the corn in a pan and heat.") Pillsbury received so many calls from puzzled customers it put the directions back on the can.

Consumers in the city pay premium prices for products which save time and effort. Consider, for example, two commodity prices: Whole fryer chicken: $.79 per pound; Tyson frozen breaded chicken breasts: $4.92 per pound. Raw fruit: $.89 per pound; General Mills' Fruit Roll-Ups: $7.32 per pound.

Processing raw farm commodities to make them more convenient now constitutes one of the largest sectors of the agribusiness industry. Though this sector is dominated by large processors, metrofarmers can also capitalize on the trend.

Joe Carcione, a produce broker from the San Francisco Bay Area, explained how in an interview with the McClatchy News Service: "A produce manager was having a hard time selling the red onions. The dry peel was black and they didn't look very good. He peeled some of them, leaving the shiny purple skin. The sales tripled. So he graduated the peeled onions to the salad department. One day after church... I said to my wife, 'If these red onions sell for seventy-nine cents a pound and clean up off that counter, and these red ones with the skin on don't sell, one of these days you're going to come in here and see a big pile of these peeled red onions and you won't be able to find an onion with the skin on it.' So about a month later we got out of church, went down to the store; there was a huge pile, right in between the lettuce and tomatoes. I said, 'This is the day.' You couldn't find a red onion with the skin on it. I say if I own a store and you're willing to pay me fifty cents a pound more to peel an onion for you, I'll peel them all day."

Metrofarm products can be peeled, cored, juiced, cooked or combined with other commodities into a product which can be taken home in a easy-to-prepare or ready-to-consume form.

The Eva Gates Homemade Preserve Company, a four-person business in the small town of Bigfork, Montana, began with a surplus of berries from Eva's garden. Eva processed berries into jams and jellies using the "open kettle" method. The open kettle method yields preserves which are about 60 percent fruit and 40 percent sugar. Large processors use a pressure-cooking method which yields about 25 percent more juice. The pressure cooker method, however, yields preserves which are only about 45 percent fruit and 55 percent sugar. Consumers could taste the difference. Eva now sells five kinds of fruit preserves and three kinds of syrups through high-return outlets such as gourmet food stores and mail order.

The Gizdich Ranch mentioned previously also makes extensive use of processing to increase returns. In addition to its apple and berry U-Pick crops, customers have an opportunity to purchase freshly squeezed apple juice and farm-baked apple and berry pies. The tasty consumer-ready products attract a large number of customers to the Gizdich Ranch.

APPEARANCE: In addition to essential nutrients, metrofarm products provide a source of romance and fun for consumers who are tired of mass-produced conformity. This trend is clearly evident in the gourmet sector of the food processing industry. The industry's semiannual get-together, called the International Fancy Food and Confection Show, has outgrown all but five convention centers in the United States.

Fundamental characteristics of gourmet markets are uniqueness, premium prices and change. Almost any product sells for a high price, as long as it is not ordinary. However, when a product sells well, it attracts competition which often results in overproduction. Metrofarmers are, by nature, well-positioned to compete in the gourmet market. Indeed, the metrofarm strategy requires unique products and an adaptability to change.

The Flower Ladies, for example, conquer the wedding floral market by growing more than 200 varieties of unique old-fashioned flowers. Variety enables them to sell arrangements with 25 shades of blue flowers and others with the fragrance of a wedding cake. To make their product convenient for time-conscious brides, the Ladies arrange the flowers, deliver them to weddings and install them around the altar. The Flower Ladies' floral arrangements add a great deal of romance and fun to a wedding. Guests admire the arrangements and talk about them in glowing terms. Consequently The Flower Ladies receive a premium price for their convenient, gourmet-grade product and are booked for months in advance.

PACKAGE THE CROP

Packaging is an important step in conversion of crops into products. Good packaging aids movement from farm to market, while poor packaging hinders movement. Plan good packaging for your products because it will help establish and maintain a competitive position with intermediaries and consumers in the market.

Package for Intermediaries

Merchants profit by bringing large units of a product in the back door, separating the units into smaller ones and then selling them out the front door. Products sold to intermediaries should be packaged to aid this movement.

PACKAGE STRENGTH: A package destined for an intermediary's warehouse must be strong enough to withstand his shipping and handling procedures. If packages will be stacked one on top of the other, each should bear its weight without crumbling. Your intermediary loses time and money when a carton breaks open and spills product out onto the floor.

PACKAGE DESIGN: A package should be adapted to the product. It should be the right size and shape and be constructed from material which will afford proper ventilation. It should not have sharp corners or surfaces which can harm the product. It should be designed for convenient and economical packing, handling, transportation and stacking.

PACKAGING STANDARDS: Packages destined for intermediate markets should conform to the legal standards established by industry marketing orders, councils or commissions.

There are three types of containers shipped through intermediate markets. "Nonconsumer" containers are those intended solely for wholesale distribution. "Master" containers are those which contain more than one individual-size package of the same commodity. "Shipping" containers are the ones in which the package or wrapping is used solely for transportation of a product in bulk form.

Standards

FIG 9.11 Standardized containers, even ones with non-standard names, aid the movement of products from farmer to consumer.

for containers destined for intermediate markets can be fairly exacting. Consider, by way of example, the standards mentioned in California's Administrative Code for packaging of apples:

> **1380.19 Standard Containers:** The following containers are authorized for use pursuant to Division 17 of the Food and Agriculture Code. The depth dimension of containers designated below shall include in that measurement any cleats, if used.

CONTAINER NUMBERS AND DESCRIPTIONS

Apples	Depth	Width	Length
AP1	10 1/2"	11 1/2"	18"
(etc., through AP12)			

> All apples, except Lady variety, crab apples, and as otherwise provided in Section 1400.2 when in closed containers, shall be in one of the containers listed. Specific packing instructions are listed in Article 7 of this group.
> In addition, wrapped apples shall be packed only in containers AP1, AP2, AP3, AP4, or AP11. Tray packed apples shall be packed only in container AP11 and placed in layers of 4 to 6 trays, pads, or molded forms. A collar or liner is permitted, provided it does not reduce the internal capacity of the container.

In California alone, about fifty crops are covered by packaging standards. Though total compliance to these standards does not exist, especially in transactions between farmers and small retailers, it pays to plan ahead, especially when shipping to mass-market intermediaries.

PACKAGE MARKINGS: Packages destined for wholesale and retail warehouses should be properly marked with a business logo and an "IRQ" statement. The logo provides a way to increase your name recognition among wholesalers and retailers. Name recognition and the reputation which goes with it is called "goodwill." Goodwill is a tangible asset which can be bought and sold like a tractor or land.

The IRQ statement provides intermediaries with specific information. "I" identifies the package contents (*e.g.*, tomatoes); "R" says who is responsible and lists the address (Sunnyside Farm, Route 1, Any City, State, USA); "Q" is for net quantity in terms of the largest whole unit (10 pounds). The IRQ statement should be placed in a conspicuous position on the package and should help the warehouse person identify it and its contents without having to tear the package apart.

PACKAGE APPEARANCE: Though how a package appears in a warehouse is not vital, it is still important. A package's appearance should enforce the notion that your products are of the highest possible quality. Use attractive packages and display the label and contents to advantage.

Package for Consumers

Whereas a consumer in Two Dot, Montana, buys potatoes in 50-pound sacks, one in New York City buys them in six-ounce shrink-wrapped microwavable trays. From this observation we can generalize the bigger the city, the smaller the package. Packaging is therefore an important element in marketing of metrofarm products.

In an interview in *Food and Drug Packaging* magazine, R. Gordon McGovern, CEO of Campbell Soup Company, talked about the importance of packaging for the consumer: "The whole thing begins with our focus on consumers and how they react to product and package perception. My own observation is that the package is very important to consumers— both from a preservation and a convenience point of view.... Our cost of goods is about 60 percent of the sales price. For example, if a product is sold for one dollar, sixty cents of that is for the cost of goods. Forty-five cents is spent on raw materials and labor and fifteen cents accounts for packaging costs." In other words, 25 percent of the cost of making a can of Campbell's soup is packaging.

Products sold direct through U-Pick, farmers markets or roadside stands are most frequently packaged in simple and inexpensive paper bags or cartons. Value-

FIG 9.12 IRQ declarations tell what is in a package ("I"), who is responsible ("R") and what is the quantity ("Q").

added products sold through high-return markets like gourmet food stores and mail order are packaged attractively. Following are points to consider in designing a consumer ready package.

CONSUMER REQUIREMENTS: First, residents of the metro are in the market for ways to save time and effort. A package should therefore be easy to use. Current packaging successes include squeezable plastic bottles and microwavable trays. (At this writing, nearly 75 percent of all U.S. households have microwave ovens.)

Second, residents of the metro pay a large premium for space. A package should therefore conserve space and be readily stored. Reconsider the bags of potatoes which introduced this section. In the tiny village of Two Dot, Montana, where rent is low and the space required to store 50 pounds of potatoes is readily available, the large packages are popular because they deliver a lower cost potato. On the other hand, monthly rent for square footage necessary to store a 50-pound sack in a New York City apartment would most likely pay for a couple of hundred pounds of potatoes.

Finally, many residents of the metro are extremely design conscious. A package should therefore be attractive enough to give consumers a psychological boost.

BUSINESS REQUIREMENTS: A good consumer package should help build recognition of your name, extend shelf-life of your product and protect against dust, dirt and customer mishandling.

Scott Murray, a grower of specialty vegetables near San Diego, discovered how valuable a good consumer-ready package can be when he packaged his sweet basil crop in plastic bags. Inexpensive bags extended the shelf-life of Murray's basil by two weeks, provided an opportunity to affix a colorful label and made customers happy by reducing spoilage.

PACKAGE TYPES: There are two kinds of packages to consider for the direct market, "consumer" and "gift." Consumer packages are distributed through retail outlets for consumption by individuals. Gift packages are designed to be purchased by one party as a gift for another. Gift packages feature more elaborate designs and higher production costs than do ordinary consumer packages. There are many package designs made from many different materials, including glass, metal, plastic, paper and wood.

CONSUMER PREFERENCES: In a recent survey published in *Food & Drug Packaging* magazine, consumers expressed their packaging preferences as follows:

Package	Percentage Who Favor
Glass	38
Plastic	29
Paper	12
Metal	3

Consumers were also asked to identify packaging features they considered important.

Feature	Important	Unimportant (in %)
Recloseable	75.4	14.4
Tamper-Evident	74.0	16.7
No Refrigeration	53.7	18.0
Microwavable	51.8	21.5
Recycleable	46.0	22.5
Dispenser	44.4	26.4
Shatter Resistant	43.9	28.4
Easy To Open	43.4	33.9

In this survey 64.5 percent of the respondents said it was "extremely important" for a package to be tamper evident, 31.1 said it was "somewhat important" and only 4.3 percent said it was not important at all.

MARKET STANDARDS: There may be legal or professional standards which must be met in packaging for the consumer market. The California Administrative Code, for example, contains the following specifications for consumer packages of apples:

> **1400.33. Apples, Consumer Package Defined:** "Consumer package" means any closed container which will not hold more than 20 pounds net when full. Consumer packages may contain mixed varieties of apples. All apples in consumer packages shall conform to one of the following grades: Extra Fancy, Fancy, or a combination grade of Extra Fancy and Fancy, as defined in this Code.

Package standards may be established by federal or state law and by a federal or state marketing order, commission or council. Check with your County agricultural official for specifics.

PACKAGE DESIGN: Shelves are stacked full of colorful competitors and each is begging to be purchased. Many factors determine winners of this competition for consumer dollars, including size, price, convenience and appearance. A winning design preserves a product's quality, increases its shelf-life and makes it easier to use and more attractive.

Often there may be little observable reason why one package design is selected over another. Why, for example, do hot dogs come packaged in groups of ten, while hot dog buns come packaged in groups of eight or twelve?

Three popular strategies are used in package design: One strategy is to copy a common design, which reduces costly experimentation and presents consumers with a familiar package. A second strategy is to adopt a package which gives more for the money, like two pounds instead of the competition's pound and a half. And a third strategy is to use a package which gives less for the money. The brewers of Bohemia Beer, in a now-classic experiment, decreased the size of their bottles and cans from industry-standard twelve ounces to eleven ounces. Their extra income was then used to buy fancier containers and more advertising. The result: Sales nearly doubled!

The best package is one which presents the buyer with unimpaired quality and an impression of value. At the giant Campbell Soup Company, all packaging decisions are the responsibility of the business unit manager, who is the person responsible for the product itself. As metrofarmer, you are the business unit manager responsible for selecting the right package. This selection is an important one because a good package will sell a product the first time; a good product will sell itself again and again.

Label the Package

A label is a statement associated with a packaged product. This statement may be written, printed or graphic; it may be affixed to, applied to, attached to, blown into, formed into, molded into, embossed on or otherwise appear on the package.

A viable label conveys two kinds of information: First, it gives specific

FIG 9.13 If appearance can make the first sale, quality and value can make the rest.

product information like the IRQ statement discussed in the previous section. Second, it gives general information like the product's brand name. Though a product's name may seem insignificant, it may prove to be the most important marketing decision one makes.

The objective of marketing is to bring customers to a product and to hold their loyalty. This is not an easy chore because, in the "my-potato-against-yours" world of the free market, most farm products are little more than raw commodities. A label and the brand name it conveys help build product loyalty and, in some circumstances, bestow proprietary rights.

Campbell Soup Company's red and white can was made famous by Andy Warhol paintings. When asked if packaging was the medium through which consumers identify Campbell's products, R. Gordon McGovern, the Company's CEO, replied, "Very much so. The red and white can has really been our heritage." Campbell sells a large quantity of mushrooms in its canned soups. Recently the company expanded into the fresh market with a line of mushrooms called "Farm Fresh." Campbell registered the name and consequently now owns it. The Farm Fresh brand name enables Campbell to advertise and build goodwill because the name is a legal property.

Goodwill is a tangible business asset. It is the excess in a purchase price over and above the value given net assets. If the net assets of your farm are valued at $100,000 and someone offers you $175,000, then the goodwill of your farm is $75,000.

To build goodwill, a label, like the package itself, must convey its meaning in an attractive and stylish manner. It must help make consumers want to be associated with the product. Consider the extent to which one vintner goes to make an attractive label: Each year since World War II, world class artists like Pablo Picasso, Andy Warhol and Salvador Dali have been commissioned by Chateau Mouton Rothschild to design art for the company's label. As payment for their services, each artist receives "several cases" of the vintage year for which he or she designed a label. Depending on the year, the grape and the art, several cases of Rothschild wine could equal a small fortune.

Your metrofarm products should be the finest money can buy. Design your packages and labels to let consumers know just how valuable your product is. In a market dominated by heavily advertised red and white cans, you will need all the goodwill you can generate.

REVIEW

Harvest is when your investments of time, money and know-how are taken from the ground and sent into competition for consumer dollars. Take a positive step toward winning the competition by preparing your crop for market.

Begin with quality. Learn to recognize elements of quality and forces which affect the elements. Then look to the harvest. Residents of the city often imagine harvest as being a time of joyous celebration. Experienced farmers know harvest often takes place in the heat of battle: There is a hail storm brewing on the horizon, a forecast for an extreme frost on the news, an urgent order from a broker on the desk. Plan your harvest well because forces which act upon your crop after harvest will only depreciate quality and value. After harvest consider sorting the crop, storing it in a protected environment, processing it into different products and sending it to market in consumer-friendly packages.

Each step can help you increase productivity and gain competitive advantage. Prepare for the market. Deliver a consistent supply of a quality product in an attractive package and win the competition for consumer dollars.

CHAPTER NINE EXERCISES

i   Begin a new subsection entitled "Preparing for Market" and include it in the "Marketing" section of your metrofarm workbook. On a fresh sheet of paper, describe the elements of quality (*e.g.*, size, shape, color) for the crops selected in exercises 6i to 6v. List ways to maintain product quality after your crop has been harvested.

ii   Research seasonal price fluctuations for your crops and determine if crops should be shipped directly to market after harvest or if they should be stored until prices improve. Research the storage history of your crops and list optimum storage temperatures and humidities.

iii   Survey the market to discover how many ways your crops have been processed. Title a fresh sheet of paper with "Processing Options." Divide this page into two vertical columns. In the first column list ways to process your crops to maintain their value. In the second column list ways to process your crops to make them more convenient.

iv   Research logotypes for similar crops and products. On a fresh sheet of paper, sketch three designs for a metrofarm business logotype.

v   Research package standards for your crop products. Design a standard package for shipping to the retail market. On the package's label include an IRQ statement and your business logotype.

Selling MetroFarm Products

*The greatest fine art of the future will be the making
of a comfortable living from a small piece of land.*

Abraham Lincoln

You are the future Lincoln envisioned. You have invested time, money and know-how in the high-density production of high-value crops. Now it is time to take your investments to the battlefield of the marketplace.

Selling consists of an exchange of legal rights: You exchange your right to the crop for another's money or other considerations.

When demand for your crop is high and supply low, selling will be easy and enjoyable. When demand is low and supply high, selling will be difficult and thankless. Most selling, however, will occur in a market which has achieved an equilibrium between demand and supply.

Competition in balanced markets tends to be dominated by giant competitors wielding Goliath-sized marketing budgets. Chapter 10 will help you survive this competition and win consumer dollars. It provides a guide to establishing a price, marketing the product and closing the sale.

ESTABLISH A PRICE

Business began with an exchange of goods: One cave-dweller exchanged his animal skin for another's flint knife. This kind of transaction is called barter.

Barter worked well in caves populated with few people but was too cumbersome to facilitate commerce in more densely populated environs. A standard unit of value was developed, called money, which facilitated the exchange of products, specialization of human resources, expansion of markets and development of civilization.

The value of a product, which is usually expressed in terms of a money price, is the ultimate point of contention in a business transaction. It is the point at which the competition for consumer dollars is waged.

Large-scale sellers enjoy many economies of scale pricing advantages. Big operators can, for example, "average out" prices, selling some products at a loss to lead buyers in and others at whatever the market will bear to deliver the profit. This section of Chapter 10 will help you compete against large-scale sellers. It provides a guide to pricing theory, strategy, tactics and evaluations which will enable you to move with the agility and confidence of a David taking on a Goliath.

Consider Pricing Theory

How valuable is your metrofarm product? Why does this value fluctuate? Answers have been offered by theorists who range the ideological spectrum from Adam Smith to Karl Marx. So many answers being offered tells us establishing a price is not an exact technology like those used in maintaining fertility of soil. Nevertheless, one very simple theory, distilled from observations of many great thinkers, works best in explaining how the market establishes value. It is called the "Equilibrium Theory."

EQUILIBRIUM THEORY OF VALUE: The equilibrium theory maintains that demand and supply bring about a price at which the quantity demanded by buyers equals the quantity sellers are willing to supply.

Should demand exceed supply, the high price discourages buyers, who reduce demand, and attracts sellers, who increase supply. Reduced demand and increased supply then force the price down to its equilibrium point.

Should supply exceed demand, the low price encourages buyers, who increase demand, and discourages sellers, who reduce supply. Increased demand and reduced supply then force the price up to its equilibrium point.

MITIGATING FACTORS OF THE EQUILIBRIUM THEORY: Though the equilibrium theory tells how products are priced, other factors can mitigate its use as a planning tool. Mitigating factors can, temporarily at least, force prices up in a down market and down in an up market.

Product Quality: All farm products are not, as suggested by the equilibrium theory, created equal. Your vine-ripened organically-grown tomato, for example, will taste much better than one picked green in a distant growing region, gassed into

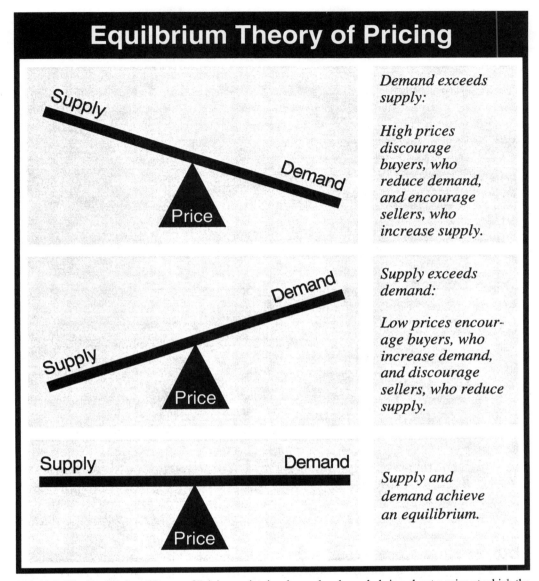

Equilbrium Theory of Pricing

Supply / **Demand** / Price

Demand exceeds supply:

High prices discourage buyers, who reduce demand, and encourage sellers, who increase supply.

Demand / **Supply** / Price

Supply exceeds demand:

Low prices encourage buyers, who increase demand, and discourage sellers, who reduce supply.

Supply / **Demand** / Price

Supply and demand achieve an equilibrium.

FIG 10.1 The Equilibrium Theory of Pricing maintains demand and supply bring about a price at which the quantity demanded by buyers equals the quantity sellers are willing to supply.

redness and then shipped to a local wholesaler. Though both tomatoes may be red and round, they may be placed in different price categories. Increasingly, super-markets establish one price for conventional products and another price for locally-grown organic products.

Importance of Prices: Prices are extremely important in markets where large amounts of raw commodities like wheat and soybeans are traded. In specialty markets, however, prices may not be so important.

For example, a recent price for sweet corn at the Chino family's Vegetable Shop in Rancho Santa Fe, California, was 50 cents per ear. Buyers could have purchased sweet corn at a local supermarket for 15 cents per ear. Judging by the customer traffic in Chino's parking lot, prices did not seem to matter.

Advertising, promotion and publicity are used to compete for the attitude of buyers. Consequently, purchases are often made because of quality, service, prestige, loyalty and whim, rather than price.

Market Conditions: Economic conditions of the times, as well as prospects for future conditions, always affect the price of a product. Pricing is extremely competitive during recessions and depressions because the value of money is relatively high. Pricing is less competitive during times of inflation because the value of money is relatively low. Also important is how buyers see their future economic situation. A person who has a stable source of income, like a government pension, will spend differently than one who relies on personal savings.

Monopolistic Controls: Though the control of a market by one seller or one group of sellers may be restricted by law, situations develop in which this control does occur. Once control has been achieved and consolidated, dominant sellers can collude, if only informally, to control resources, fix prices and trade information. These monopolistic practices can mitigate effects of demand and supply on price.

Pricing Policies: Most large retailers, like integrated supermarket chains, establish prices at the corporate level, post prices on or near their products and then allow the buyer to either accept or reject the prices. This distance between buyer and seller can have a short term impact on effects of demand and supply.

Government Controls: The equilibrium theory is based on a market in which prices are free to move up and down. While dwellers in caves may have enjoyed such a market, it has since gone the way of the dinosaur. Today's market for farm products is often regulated by a multitude of government and industry subsidies, allotments, marketing orders and other controls which can, if only for a short while, completely eliminate effects of demand and supply. In commenting on the U.S. government's control of sugar markets in a recent *The Wall Street Journal* article, Richard Holwill, Deputy Assistant Secretary of State for the Caribbean, said, "It makes us look like damn fools when we go down there (Caribbean) and preach free enterprise."

Develop a Pricing Strategy

As suggested by the equilibrium theory, there are limits to establishing a price: The bottom limit is defined by how much time, money and know-how it takes to produce the product, plus a fair profit. The top limit is defined by the buyers' willingness to spend. Within the limits exist low, medium and high-spectrum approaches for selecting prices.

LOW-SPECTRUM PRICING: Some farmers adopt the "will not be undersold" strategy of selling for the lowest price. This strategy is most often employed by large-scale sellers whose economies of scale allow them to generate a substantial income by taking a small profit on many units.

Metrofarmers use low-spectrum pricing in the following circumstances: (1) They have a large volume of one product, like five acres of carrots. (2) They have below average quality, like frost-damaged fruit. (3) They enjoy a very low cost of production and can maintain a healthy margin by selling low. (4) They want to reduce marketing responsibilities to the minimum. (5) They have an immediate need for cash. (6) They want to capture additional market share.

Paul Bowers is semiretired and grows garlic in Grants Pass, Oregon. His production costs are extremely low and he likes to fish and travel more than he likes to sell garlic. Bowers uses low-spectrum pricing to attract big-city wholesalers, who come to town, pick up Bowers' crop and pay cash. Bowers is left with cash and plenty of time to fish and travel.

Obviously, "will not be undersold" pricing makes it nearly impossible for small-scale sellers to maximize cash return on limited production. However, when fishing time is considered a return, then low-spectrum pricing must be considered as legitimate strategy.

MID-SPECTRUM PRICING: Most farmers adopt the "competitive prices" strategy of selling at or near the middle of a pricing spectrum. This strategy can be seen in action at most farmers markets, where prices are generally set slightly below prices posted at local supermarkets.

Metrofarmers use mid-spectrum pricing in the following circumstances: (1) They have a small or medium-sized volume of products. (2) They have a direct market, like the farmers market, in which to sell. (3) They have a moderate amount of competition with which to contend.

Jeff Larkey (Part IV, Conversations) uses mid-spectrum pricing very effectively. Larkey's farmers market prices are consistently slightly below those of area retailers. Since costs are very low and quality very high, Larkey's pricing strategy helps move products quickly and provides a good profit.

The mid-spectrum strategy of "competitive prices" is the most popular one because it allows buyers a fair price and farmers a fair return on a reasonably quick turnover.

HIGH-SPECTRUM PRICING: Some farmers adopt the "take what the traffic will bear" strategy of charging premium prices. Though taking what traffic will bear has a rather crass ring, it is a widely used indicator for the upper limits at which prices can be set while still maintaining market share.

There are many examples of how high-spectrum pricing has been used to great effect. Consider the Hunt-Wesson Company, which invested $6 million in an introduction of "Orville Redenbacher's Gourmet Popcorn." This popcorn product sold for two and a half times more than the nation's leading brand and even claimed a position as the "World's most expensive popping corn" on the label. Within four years it became the nation's number one popcorn brand.

As demonstrated by this popcorn success, high-spectrum pricing is important for two reasons: First, consumers tend to equate price with quality. High prices therefore lend credibility to a product being positioned in the gourmet grade. Second, high prices afford the best opportunity for maximizing the margin between costs and returns.

Metrofarmers use high-spectrum pricing in the following circumstances: (1) They have a small volume of high-quality products. (2) They have a direct market, like a farmstand or mail-order, in which to sell. (3) They have a good reputation and brand name. (4) They have little or no competition.

The Chino family's Vegetable Shop mentioned earlier charges more than three times the going supermarket rate for sweet corn. Why do people stand in line to pay these high prices? Because the Chino family has developed a good reputation for selling high-quality products through an exciting direct market which has little or no competition.

Clearly, high-spectrum pricing provides the greatest opportunity to realize maximum return from a small parcel of land.

Develop Pricing Tactics

This section describes six commonly-used tactics for setting prices and for making adjustments to prices.

TRIAL AND ERROR PRICING: Most forces which influence prices do not become apparent until after the marketing cycle is over. This is especially true of perishable products, like lettuce, which are sold to markets with pronounced swings in demand and supply. Trial and error is a good pricing tactic under volatile conditions.

Trial and error begins with an educated guess. Survey existing prices and set a reasonable base price. Then determine the validity of your base price by watching the volume of sales: If product sells too quickly, the price is too low; if product sells too slowly, the price is too high.

By way of illustration, consider a recent experience I had while selling a gourmet-grade garlic to a local wholesaler. The base retail price of $3.98 per pound met with stiff resistance from customers and sales were consequently too slow. The second price of $3.25 per pound was very popular and sales were too fast. Most of the garlic sold for a third price of $3.50 per pound. Then, toward the end of the season when demand was high and supply low, the original base price of $3.98 per pound was reestablished.

There is one important caveat to trial and error pricing: Start high because it is easy to lower prices and difficult to raise them.

MULTIPLE GRADE PRICING: Two or more grades of one crop are often priced differently to reflect different quality. Apples harvested from a single tree, for example, will likely yield gourmet, commercial and cull or juice grades. The gourmet grade might be priced high for sales through a local gift store, the commercial grade might be priced moderately for sales to a local retailer and cull grades might be priced low for sale to a juicer.

DISCOUNT PRICING: A discount is a deduction from the list price which is used to manage volume of sales. Following are four commonly used discounts to consider:

Trade Discounts: Wholesalers and retailers which stand between you and the consumers are paid with a trade discount. This trade discount is simply a percentage deducted from the product's retail list price. An example: Instead of selling all of your garlic crop from retailer to retailer, give it to one wholesaler and have the wholesaler do the work. You can pay for this service with a trade discount, which works follows: The retail net price of your garlic is $4.00 per pound and the wholesaler's services cost 40 percent. The net cost for the wholesaler's services will therefore be $4.00 X 40% or $1.60 per pound.

Most intermediaries operate on carefully calculated margins which guarantee a profit on their efforts. You give a list price; they deduct their margin and then pay the discounted price.

Chain Discounts: As the name implies, a chain of discounts consists of discounts in a series, with each discount granted for a specific condition.

A popular chain discount is the "40-10-5." The first 40 percent in the chain is given as a standard wholesale discount, the second 10 percent is given to encourage a volume purchase and the third 5 percent is given to encourage timely payment.

Chain discounts are typically progressive. An invoice for $100 with a 40 - 10 - 5 discount would be figured as follows:

A 40-10-5 Chain Discount

Wholesale Discount:	$100 - $40 (40% of $100)	= $60
Volume Discount:	$60 - $6 (10% of $60)	= $54
Early Payment Discount:	$54 - $2.70 (5% of $54)	= $51.30

Chain discounts allow you to legitimately discriminate between customers by granting or withholding one or more discounts in the chain. And, by adding or subtracting discounts, the chain allows you to change prices without having to reprint a price list.

Quantity Discounts: Some customers want one apple; others want 1,000 apples. Since selling costs may be the same for each transaction, you may want to save money by selling larger quantities to fewer buyers. Quantity discounts are discounts offered to encourage large volume sales.

There are cumulative and noncumulative quantity discounts. Cumulative discounts are those calculated on different purchases over a period of time; noncumulative discounts are calculated on one purchase. For example, a discount given for a weekly purchase of 10 pounds of apples would be noncumulative; the discount given for purchasing 520 pounds over the course of one year would be cumulative.

Quantity discounts are often used to increase sales volume during critical periods. During peak season at a U-Pick berry farm, when berries are ripe and at risk of spoiling, a seller might encourage customers to pick larger quantities by offering progressive quantity discounts like $ .75 for one pint, $2.00 for three pints and $7.00 for twelve pints.

Cash Discounts: It is said "cash talks and all else walks." Cash discounts are often given in recognition of this feat. Cash discounts are inducements for buyers to pay in cash or to pay their bill shortly after receiving the product. Cash discounts are offered in many forms. A common one is the "2, 10 net 30," in which buyers receive a two percent discount for paying in 10 days, with the total being due in 30 days.

TRANSPORTATION PRICING: Someone must pay the freight and prices are often adjusted to account for this cost. There are two ways to account for transportation costs, delivered and F.O.B. ("free on board"). Delivered means the seller assumes all costs of delivering a product to the buyer. F.O.B. means the buyer pays transportation costs from the seller's warehouse. Where F.O.B. sales are the norm, delivered prices may be used to discount the price and increase the volume of sales.

RULE OF 5'S AND 9'S PRICING: Products are often priced immediately below the next level when selling direct to consumers. The reason: Consumers often see a 95 or 99 cent price as being much lower than it actually is. A product priced at $3.99, for example, might actually be seen as a three-dollar item instead of a four-dollar item.

Some sellers, notably those with an established clientele, disdain the rule of 5's and 9's. The Chino family's Vegetable Shop, for example, rounds off each price to the nearest dollar. One dozen ears of the Vegetable Shop's sweet corn may sell for $6.00; one pound of its beans may sell for $5.00.

SALE PRICING: Specials and sales are used to increase the volume of purchases during specific periods of time. Midweek specials are used to increase volume during normally slack middle of the week, evening specials are used to attract working people and closeout specials are used to sell products at the end of the marketing cycle, like the annual Halloween day pumpkin closeout sale.

The loss-leader is another popular sale tactic. A loss-leader is a product which is sold at a loss to attract customers to higher priced goods. Roadside stands near Castroville, California, "The artichoke capital of the world," often advertise "Artichokes, 25 for $1.00!" When motorists stop for the leader, and see the tiny artichokes, they often purchase larger, more aesthetically-pleasing artichokes at the price of two for a dollar.

Evaluate Price Changes

Two things happen when a price is changed: First, customers are forced to reevaluate their desire to buy in light of a new price. If they like the new price, the volume of sales increases; if not, volume decreases. Second, a price change alters the volume of sales required to achieve your business objective. When a price is reduced, volume of sales must increase; when a price is increased, volume may be reduced. The following discussion will help you manage price changes.

ESTABLISH THE SELLING PRICE: The difference between what a product costs and what it can be sold for is called the margin. This margin must be sufficient to pay costs of selling the product, which is called "overhead." There are three kinds of overhead. They are fixed, variable and profit.

Fixed costs are those which must be paid regardless of volume of sales, including rent, insurance, interest, salary, taxes, basic utilities and depreciation. Variable costs are those which change with volume of sales, including hourly wages, supplies, marketing and incremental utilities.

Profit is the return you demand for placing your time, money and know-how at risk in the enterprise of selling. It is important to include profit in a calculation of selling price. Your chances of becoming successful will improve if you plan to profit, rather than to merely cover fixed and variable overhead.

Establish The Selling Price	
PRE-DETERMINED MARGIN	40% MARGIN
$\dfrac{\text{Cost of Item}}{\text{\$1.00 Sales Margin}} = \text{Selling Price}$	$\dfrac{\text{\$.15 Per Pound Cost}}{\text{\$1.00-\$.40 (percent)}} = \text{\$.25 Per Pound}$

FIG 10.2

Many retail sellers calculate prices by adding a standard margin of 40 percent to pay their overhead. The formula for calculating a price by using a predetermined margin is described in Figure 10.2. (Note: Subtracting the margin from $1.00 converts a percentage into a divisible dollars and cents number.)

By way of example, use a standard retail margin of 40 percent to set a retail selling price on apples which cost 15 cents per pound.

Again, establishing a predetermined margin to cover all fixed, variable and profit overhead will help you become a successful seller.

CALCULATE THE BREAK-EVEN POINT: When the total amount of overhead is known and the retail margin established, calculate sales break-even point and what will happen to the break-even point when you decrease or increase prices. The break-even point is the volume of sales required to cover all fixed costs, variable costs and profit. It is the sales objective. To find a break-even point, divide total overhead by the margin per $1.00 in sales.

By way of illustration, sell apples from the previous example at a roadside stand. The total overhead for operating this stand— including fixed costs, variable costs and profit— will be $400 per week.

Apples sold at the standard margin of 40 percent result in a retail price of 25 cents per pound. This means for every dollar taken, 60 cents pays for apples and 40 cents pays for selling apples. What is the volume of sales required to break-even on this roadside sales effort?

Calculate the Break-Even for a 40% Margin

PRE-DETERMINED MARGIN		40% MARGIN	
$\dfrac{\text{Total Sales Overhead}}{\text{Margin Per } \$1.00 \text{ in Sales}} =$	Break-Even Sales Volume	$\dfrac{\$400}{\$.40} =$	$1,000 Sales

FIG 10.3

By dividing an overhead of $400 by a margin of 40 cents per $1.00 worth of apples sold, you can see $1,000 worth of apples must be sold each week to break-even with a 25 cent per pound price.

EVALUATE A PRICE REDUCTION: What happens when the price is changed? In addition to finding the break-even point, the simple calculation used above will help you project the consequences of price changes.

By way of illustration, project what will happen to the break-even point if you have a special weekend sale at your roadside apple stand. The objective of this sale is to reduce the stock of apples prior to arrival of a shipment of freshly picked ones.

In hope of increasing the volume of sales, reduce the price of apples from 25 cents per pound to 20 cents per pound. Since the price paid for the apples is still 15 cents per pound, this weekend sale price will lower your margin on $1.00 worth of apples sold from 40 cents to 25 cents. You need to increase variable costs an extra $40 or so to pay for extra labor and utilities required to manage additional volume of business.

How will this weekend sale affect the break-even point? Plug the new numbers into the formula and see:

Calculate the Break-Even for a Price Reduction

PRE-DETERMINED MARGIN	25% MARGIN
$\dfrac{\text{Total Sales Overhead}}{\text{Margin Per \$1.00 in Sales}} = \begin{array}{l}\text{Break-Even Sales Volume}\end{array}$	$\dfrac{\$440}{\$.25} = \begin{array}{l}\$1,760\\ \text{Sales}\end{array}$

FIG 10.4

This calculation demonstrates that if the price is reduced by 5 cents to 20 cents per pound, $1,760 worth of apples must be sold to break-even. In other words, your price reduction must increase the volume of sales by more than $760 for you to achieve the sales objective of paying for all fixed, variable and profit overhead.

EVALUATE A PRICE INCREASE: What happens to the break-even point when prices are raised? Raise the price of the apples from 25 cents to 30 cents per pound and see what happens.

Since the price paid for apples remains at 15 cents per pound, the new 30 cent price will raise the margin to 50 percent of each $1.00 sold. Furthermore, variable costs can be reduced by $40 or so because the higher prices will likely reduce volume of sales. What will happen to the break-even point with this price increase?

This calculation demonstrates raising prices will increase the margin, reduce overhead and lower sales break-even point. If customers buy $720 worth

Calculate the Break-Even for a Price Increase

PRE-DETERMINED MARGIN	50% MARGIN

$$\frac{\text{Total Sales Overhead}}{\text{Margin Per \$1.00 in Sales}} = \begin{array}{c}\text{Break-Even Sales Volume}\end{array}$$

$$\frac{\$360}{\$.50} = \begin{array}{c}\$720 \\ \text{Sales}\end{array}$$

FIG 10.5

apples at the new 30 cent per pound price, you break-even; if not, you lose.

It is important to understand how selling price affects selling margin. The ability to calculate a margin will give you the agility and confidence required to take on giants of the marketplace and win consumer dollars.

MARKET THE PRODUCT

"He who has a thing to sell and goes and whispers in a well, is not so apt to get the dollars as he who climbs a tree and hollers."

This adage is especially true in markets which have achieved an equilibrium between consumer demand and producer supply. In balanced markets, selling is a competitive business in which the more prospective buyers you reach, the better are your chances of winning dollars.

One way to reach buyers is to dress up, package an attractive sample of your product and knock on doors. This direct approach is the best way to establish a relationship with a relatively small number of wholesalers and retailers.

The other way to reach buyers is to dress up a well-conceived marketing campaign, package the campaign in an advertising, publicity or promotional medium and then allow the medium to knock on the doors for you. This indirect approach is the best way to establish a relationship with a large number of prospective buyers.

This section will help you sell to a balanced market. It provides steps to organizing an effective and efficient marketing campaign and three ways to execute the campaign.

Develop the Marketing Campaign

David beat Goliath with a stone pitched from a well-aimed slingshot. It was a tangible piece of business: David picked up a smooth round stone, placed it into his sling and then hit the giant with enough force to knock him down.

Competition for consumer dollars is not quite so tangible. There is no smooth round stone to pick up, no broad expanse of forehead to target and no slingshot to fire. Nevertheless, you must compete with Goliaths of the marketplace and you must do so with the intangible weapons of a marketing campaign.

Military strategists use "campaign" to describe the period of time in which an army executes a connected series of operations designed to bring about victory. Marketing strategists use the word for the same purpose: It is the period in which a business executes a connected series of operations designed to win consumer dollars. This section will help you campaign with the right weapon, right target and right amount of force.

USE THE RIGHT WEAPON: David's weapon was a smooth round stone pitched from a sling. Though most business persons occasionally feel like pitching a few well-aimed stones at their competitors, such activity is not acceptable in a civilized marketplace. The acceptable weapon is a message which explains why your product should be purchased instead of others available. Successful marketing campaigns consist of strategic and tactical messages.

Strategic Message: Also called "generic" or "institutional" by professionals, the strategic message tells buyers how to think about your product. "Locally owned, locally operated," is an example of an often used strategic message. The objective of this message is to establish a favorable position for your product in the mind of buyers. Then, when buyers walk through the market and are confronted with selection, your product will stand out as a safe and comfortable buy.

Goliaths of the marketplace enjoy economies of scale advantages which allow them to spend hugely to position their products in the mind of buyers. Strategic planning will help you compete with the Goliaths' advantages. Strategic messages convey the result of this planning to buyers.

The Gizdich Ranch, a family owned and operated small farm near Watsonville, California, sells apples in competition with the large-scale growers of Washington State. The Washington State growers invested large sums of money in a strategic effort to position their apples as the market leader. Their effort was seen on television, heard on radio and read in newspapers and magazines across the land. As a consequence, Washington State brand names pushed many small-scale apple brands off the supermarket shelves.

The Gizdich Ranch did not have large sums of money with which to compete in this apple market. Furthermore, the Gizdich Ranch property was located off the beaten path and did not offer the opportunity to sell direct through a roadside stand. How did the Gizdich Ranch compete with the Goliaths of Washington State?

By campaigning with the strategic message, "locally owned, locally operated," the Gizdich family successfully positioned their ranch as the place where residents from surrounding metropolitan areas could enjoy a true family farm adventure. Listen to the variations on this position, as expressed in the Ranch's brochure: "The Gizdich Ranch. It's as American as mother and apple pies, apple juice, apples and pik-yor-sef berries." And, "This third generation, family owned and operated, agricultural business is testimony that what made American great is still viable."

The Gizdich Ranch is much more than a place to buy apples, the strategic message says; it is a place to pick your own apples and berries, watch a juicing operation, shop for antiques and gifts, enjoy a slice of fresh-baked apple pie and show the kids what a real farm is like. The Gizdich family developed a successful metrofarm strategy. This strategy was then reduced into a message which successfully positioned the Gizdich Ranch as a place to go for a wholesome family farm adventure.

Tactical Messages: Strategy tells what to do, tactics tell how to do it. Tactical messages, in other words, are messages which move buyers toward strategic objectives. "On Sale: Buy Now!" is an example of a tactical message. To better understand the relationship between the strategic and the tactical, return to the competition for apple dollars described in the previous discussion of strategic positioning.

As mentioned, Washington State apple growers successfully positioned their brand name as the market leader. This giant cooperative, however, is simply too large to permit effective tactical maneuvering. Consequently, there are still many ways for metrofarmers to compete for apple dollars. One opportunity is the U-Pick.

U-Pick is an attractive opportunity because buyers drive out to your farm, pick fruit, carry it to a checkout stand, pay retail prices and then carry the fruit home. Roughly speaking, U-Pick can deliver twice the profit for half the investment! Buyers, however, must pick when the crop is ready to be picked.

The large apple cooperatives of Washington State cannot rely on U-Pick because there is simply no way to induce enough buyers out of distant cities to pick ripe apples. Consequently, the apples are picked by professional harvesters, packaged in colorful and costly boxes and then sent to brokers, wholesalers and retailers, each of whom takes a fair share of consumer dollars.

The Gizdich family successfully positioned their apple and berry farm as the "Locally Owned, Locally Operated Family Farm Adventure." This investment earned their farm a favorable position with area buyers.

When harvest approaches, the Gizdich Ranch is decorated with colorful signs, scarecrows, boxes filled with apples and well-placed farm implements. When all is ready the tactical message, "Apples now ready for harvest at Gizdich

Nita Gizdich of the Gizdich Ranch in Watsonville checks out some luscious strawberries.

Family Fun On the Farm

As harvest nears, now's the time to pack up the kids and hit the country roads again, for a tour of Santa Cruz County's own rich farmlands

By Anne Baldzikowski

THERE'S nothing like a ride through the country on a beautiful day, so if you've been

description of each farm (not to mention a list of 125 different commodities, tips for an enjoyable harvest and a fresh produce calendar). There are farms where you can buy pigs, dried specialty flowers and antique roses, and farms that let you pick your own produce or even make your own apple butter. The Farm Bureau has also collected information on farms that open their barn doors to visitors, inviting them to sample their fruits of labor.

Here's a sampling of the best of these farms. And remember, before you set off on your tour of farms, you may want to pack a picnic lunch.

tree and pick kiwifruit in November and December, or come up and pick olallieberries from May through July.

John Hudson of Coastways Ranch admits that the scenery is one of the draws to his farm. He enjoys offering people a peaceful getaway and an opportunity to pick and choose the fruits of their choice. While his wife, Katie, cooks up her specialty jams and chutneys, his Aunt Mary is greeting customers, handing out red wagons to hold little ones and harvested goodies, and sharing a bit of local history with inquires.

Real Cows Here

FIG 10.6 The Gizdich Ranch near Watsonville, California is much more than a place to buy apples and berries, the message says; it is the place to go for a wholesome family adventure. Reprinted with permission from the *Good Times*.

Ranch," is sent out and buyers respond. (Without a successful strategic positioning, however, the tactical message would likely have to say, "Special sale, two apples for the price of one at Gizdich Ranch!")

Though Goliaths of the market enjoy economies of scale advantages in building strategic positions, Davids enjoy advantages in tactical maneuvering. Since small-scale sellers enjoy this tactical advantage, many mistakenly focus all marketing efforts on tactical events like the grand opening, spring clearance, back-to-school, going-out-of-business and so on. After all, "On Sale: Two For The Price Of One" will sell more, in the short term, than "Locally Owned, Locally Operated."

To ensure long-term business, a strategic message should be combined with tactical ones. Include the strategic, "Remember, Your Best Buy Is Always At Ben's," when you run the tactical, "Ben's Spring Cleaning Special, Two For The Price Of One." A well-conceived combination of strategic and tactical messages will help you establish and maintain a favorable position.

Effective Messages: Another advantage enjoyed by the Goliaths of the marketplace is their capacity to hire expensive talent. Indeed, money has convinced many great writers of the English language to attempt commercial messages, including Aldus Huxley, Byron, Bernard Shaw, Ernest Hemingway, Sherwood Anderson and William Faulkner.

The writers listed above did not enjoy great success as writers of commercial messages. The reason, one suspects, is simple: Whereas the objective of the greats was to entertain, the objective of the commercial writer is to sell. Entertaining and selling are two different businesses. The four steps listed below will help you create messages which sell.

Capture attention: Capturing attention is more difficult than it might seem. In fact, the more densely populated the market, the more difficult it will be to capture attention.

To fully understand the magnitude of this problem, consider offering a "Good Morning" to two pedestrians, one on the main street of Two Dot, Montana and the other on Fifth Avenue in New York City. The Montanan will likely return your "Good Morning," even if you whisper the greeting from the other end of town on a windy day. Why? Because so few messages are pitched around Two Dot each message pitched is likely to be given careful consideration. The New Yorker, on the other hand, will likely ignore your "Good Morning," even if you shout the greeting from two inches away. Why? Because there are billions of messages pitched around New York City every day, so the pedestrian protects him or herself by ignoring as many as possible. The more crowded the market, the more difficult it will be to capture attention.

Contrast is the technique used for capturing attention in crowded markets. Where others use big bold letters which fill a page, use tiny letters and white space. Where others use a loud voice and fast music, use a soft voice and slow music.

Where others use black and white, add a splash of color. Where others use color, use black and white.

Attention goes to contrast. The way to successfully greet the New York City pedestrian, therefore, would be to politely excuse yourself and then, with a warm and genuine smile, offer a sincere, "Good morning!"

Make the right appeal: David Ogilvy, in *Confessions of an Advertising Man*, said, "We make advertisements that people want to read. You can't save souls in an empty church."

People will want to read if the right appeal is made. Successful appeals are almost always ones aimed at satisfying basic needs and instincts like hunger, sex, health, family, security, beauty, prestige and value. Abraham H. Maslow's "hierarchy of needs" is often used to explain why commercial appeals almost always aim to satisfy basic needs. Maslow maintains human needs are naturally prioritized so a desire to satisfy high-level needs, like an educational accomplishment, is not felt until low-level needs, like hunger, are satisfied. It is therefore easier— and more profitable— to excite interest in a product by appealing to low-level needs.

By way of example, look at how a brewery will hire a young male to read a commercial message about its beer. The odds are overwhelming the brewer's message will feature the appeal of a scantily-clad young female. Does the young male want to get the message?

Identify features and benefits of your product which will appeal to the buyers' basic needs. The Gizdich Ranch message promises a wholesome adventure to families crowded into densely populated metropolitan areas. "Family Fun On the Farm," is the headline below a recent newspaper photograph of Nita Gizdich, the farm's matriarch. Do families living in crowded cities want to listen to the message?

Be direct in making your appeal. Avoid ambiguities, superlatives, generalizations and platitudes. Go straight to the point and tell the truth with an enthusiastic voice and fascinating tone. In *Confessions of an Advertising Man*, David Ogilvy quotes a favorite written by an unsung dairy farmer:

> Carnation Milk is the best in the land,
> Here I sit with a can in my hand.
> No tits to pull, no hay to pitch,
> Just punch a hole in the son-of-a-bitch.

This is an effective message. It captures attention and appeals to the need for convenience. It also demonstrates that what you say is more important than how you say it.

Call to action: A simple way to increase effectiveness of a message, especially a tactical message, is to ask for a specific response. Where appropriate, be insistent and place a condition on the offer, such as "Sale ends Tuesday" or

Steps to Effective Advertising Copy

☑	**SPEAK TO A PERSON**	*Your Halloween pumpkin is ready for picking at Always the Best!* is more effective than *Halloween pumpkins are ready for picking at Always the Best!*
☑	**USE REAL NAMES**	*Always the Best's pumpkin patch* is a more effective way to build a brand name than *Our pumpkin patch.*
☑	**BE DIRECT**	*Always the Best's pumpkin patch offers...* is more effective than *At Always the Best's pumpkin patch you can find....*
☑	**USE ACTION VERBS**	*Always the Best slashes prices!* is more effective than *Always the Best is going to lower prices!*
☑	**CONSERVE WORDS**	*Always the Best, on Smith Road, slashes prices!* is more effective than *Always the Best, which is located on Smith Road, is going to slash prices!*
☑	**STAY POSITIVE**	*Enjoy an Always the Best Halloween pumpkin!* is more effective than *Don't miss out on an Always the Best Halloween pumpkin!*
☑	**USE PRESENT TENSE**	*Choose your favorite pumpkin!* is more effective than *You will be able to choose your favorite pumpkin!*
☑	**AVOID AWKWARD SUPERLATIVES**	*Always the Best's finely shaped pumpkins...* is more effective than *Always the Best's fantastically unbelievably superior shaped pumpkins....*
☑	**LIST FEATURES AND BENEFITS**	*Always the Best's finely shaped pumpkins make perfect Halloween jack-o-lanterns!* will be more effective than *Always the Best's pumpkins are top quality!*
☑	**PAINT A PICTURE**	*Create devilish grins and freakish scowls!* is more effective than *Carve a pumpkin!*
☑	**PROVIDE REAL CHOICES**	*Take your pick from large, medium and small... orange, white and purple!* is more effective than *Hundreds to choose from!*
☑	**BE IMAGINATIVE**	A *Next Week We Plow Em Under Sale!* is more effective than a *Final Pumpkin Clearance Sale!*
☑	**CALL TO ACTION**	*Enjoy Halloween this year... Visit Always the Best's pumpkin patch!* is more effective than *If you want to enjoy Halloween this year, maybe you should think about visiting Always the Best's pumpkin patch!*

FIG 10.7 Thirteen steps to effective advertising copy.

"Hurry for best selection." If you give customers an opportunity to put off buying your product until another day, they will likely put off their purchase.

Repeat names and numbers: Build recognition of names and numbers by repeating them throughout the message. This is especially important when broadcasting telephone numbers and addresses on radio and television.

Strategic and tactical messages are weapons used to establish and hold a favorable position in the market. Metrofarmers who establish and hold the best position will likely win the largest market share and highest prices.

HIT THE RIGHT TARGET: The second element in David's successful campaign against Goliath was his ability to hit the giant's unprotected forehead. This section will help you hit the right target with your message. It provides a guide to identifying the right target audience and to evaluating where the audience can be found.

Target Audience: There is an aphorism about the salesman who is so good "he can sell refrigerators to Eskimos." The real life equivalent is the salesperson who is so good he or she can sell brussels sprouts to children. Regardless of how good the salesperson may be, however, selling brussels sprouts to children will result in much wasted effort.

Truly successful salespersons know how to focus sales efforts on the right audience. The right audience consists of persons with the greatest predisposition to buy the product. Children, generally speaking, do not have the predisposition to buy brussels sprouts. Who does?

There are demographic and qualitative variables which can be used to help identify the right audience for a product. Demographic variables include buyers' age and sex. Market surveys indicate women between the ages of 25 and 49 are the most frequent buyers of brussels sprout. This does not mean some men and senior citizens do not frequently buy brussels sprouts; it simply means most of the frequent buyers are women 25 to 49.

Obviously, not all women 25 to 49 share the same values and preferences. Qualitative variables include psychological and sociological characteristics which affect buying habits. These include income, occupation, education, family life-cycle, social class, race and religion. Qualitative values like those listed provide additional means to refine a target audience. Market surveys indicate *well-educated* women 25 to 49 *with children* are the most frequent buyers of brussels sprouts.

The study of qualitative variables can become very complicated. To keep your study in perspective, remember qualitatives are not very useful for products which enjoy universal appeal. Apples, for example, are enjoyed by everyone and so identifying a narrowly defined target audience for apples may not be worth the time and effort. Qualitative analysis is useful when the product is highly specialized, like the brussels sprout.

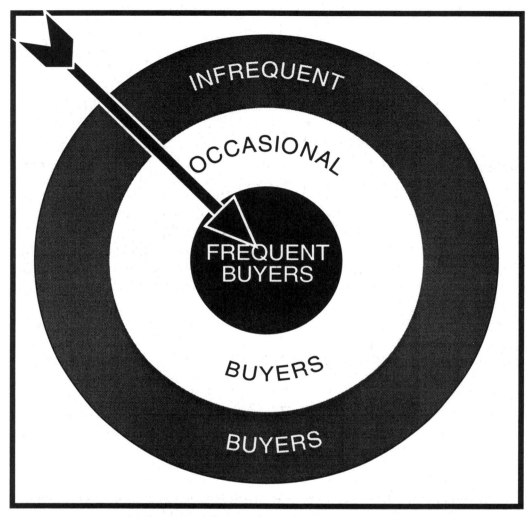

FIG 10.8 Make a target. Place media which serve the greatest concentration of frequent buyers in the bullseye. Hit the bullseye with your advertising dollars.

Target Medium: Media move markets. Media consist of the individual mediums— newspapers, radio stations, television programs— which provide information and entertainment to the public. In providing this service, each medium picks up a group of people who shares the same interests and carries the group down the road of trend and fashion.

In small markets the media might consist of a weekly newspaper and a bulletin board at the local grocery store. In large markets the media is divided into different categories— print, broadcast, mail— and the number of individual mediums within each category is extensive. In a very large market, for example, there might be two daily newspapers, 50 neighborhood weeklies and 200 special-interest tabloids. How do you find which medium will be the most efficient way to deliver your message to the target audience?

First, survey individuals within your target audience. Ask which newspapers they read, which television programs they watch and which radio stations they prefer. Ask if they subscribe to special interest publications, belong to service clubs or attend special events. Individuals within your target audience, if surveyed politely and consistently, will provide a great deal of information about their media habits.

Next, survey likely mediums. Many of the larger ones subscribe to professional services which measure audience. Newspapers subscribe to circulation checking bureaus, television stations to viewer rating services and radio stations to listener rating services. Rating services can provide extremely detailed information. They can, for example, estimate how many women 25 to 49 listen to a specific radio station from 4PM to 5PM on Saturday afternoon.

After the right audience has been identified, make a target. Place mediums which serve the greatest concentration of frequent buyers in the bull's-eye, the ones which serve occasional buyers in the next ring, infrequent buyers next and non-buyers in the last. Now, hit the bullseye with your campaign message.

USE THE RIGHT AMOUNT OF FORCE: How much force did David exert to knock Goliath down? What would have happened to David had he not exerted enough force? Both are questions of great concern to small-scale marketers.

Force is the effort budgeted to compel an audience to receive a message. In small markets, like Two Dot, Montana, the amount of force budgeted may be one handbill posted in front of the local store. In more densely populated markets, like New York City, the force budgeted may be scores of announcements in newspapers, radio and television.

Determining how much force to budget is one of the most critical problems in marketing. If too little force is used, the target audience will not receive the message. If too much force is used, precious time, money and know-how will be wasted. Like David, you must use enough force to complete the job. This section will help. It provides a guide to understanding the elements of force and for using those elements to compel an audience to receive your message.

Elements of Force: The elements of force in a marketing campaign are distribution, reach and frequency.

Distribution is simply the way in which the campaign is scheduled. In a continuous effort, messages are distributed evenly over the entire length of the campaign. In a flighted effort, messages are distributed in concentrated bursts— or flights— at significant times.

Reach is the number of different persons exposed to the campaign's message. A reach of five means five different persons were exposed to the message. A reach of 5,000 means 5,000 different persons were exposed. The more persons you reach, the better are your chances of making the sale.

Frequency is the average number of times each person is exposed to the message. A frequency of one means each person reached was exposed to the message an average of one time. A frequency of five means each person was exposed an average of five times. The more densely populated the market, the greater the need for higher frequency. Where a frequency of one might get the message across in Two Dot, Montana, a frequency of five might be required in New York City.

Finding the effective level of reach and frequency is a "green thumb" skill of marketing. Nobody knows exactly who will receive a message. Nobody knows exactly how many times a message will be received.

In addition, the relationship between reach and frequency is such that only the magic of an unlimited budget would allow for the maximum force of each. In simpler terms: "You can't have maximum reach and maximum frequency, too!" By way of illustration, consider the relationship between reach and frequency in a campaign to sell pumpkins for Halloween. There is 10 newspapers in your market from which to select. To reach the maximum number of people, all 10 newspapers should be employed. However, if you invest your limited budget to buy one ad in all 10 newspapers, the frequency of your message will be reduced to a small, ineffective force. Conversely, if you invest your limited budget to buy 10 ads in one newspaper, the reach of your message will be reduced to a small, ineffective force.

There is a measure of force called the Gross Rating Point— or GRP. This construct is the result of multiplying audience rating (reach) by the number of messages delivered (frequency). Rating services provide the audience estimates used to calculate GRPs. When marketers for Goliath Farms Inc. plan a campaign, they carefully budget what they think is the ideal number of GRPs for each market. The ideal amount of reach, frequency and distribution for your mark campaign depends on your marketing objective.

Following are budgeting guidelines for positioning a product, moving buyers and managing a market. Though only general in nature, guidelines will help you avoid falling victim to overenthusiastic sales representatives, provide a benchmark for measuring progress, give direction to the process of selecting media and provide enough force to compel a target audience to receive your message.

Force a Strategic Message: The first objective of a marketing campaign is to establish a favorable position for a product in the minds of buyers. "Locally owned, locally operated" is an example of a strategic message often used by small business persons. The amount of force required to successfully establish this position depends, to a very large extent, on the product's life cycle.

For a new product or for repositioning an old one, the message must be delivered to the largest possible audience. New product campaigns are therefore given the largest budgets possible to deliver maximum effective reach with minimum effective frequency. In fact, competition for consumer dollars is so

Newspaper Reach and Frequency

	One Ad	Two	Three	Four	Five	Six	Seven
20% net reach:		30%	38%	45%	50%	55%	58%
avg. freq:		1.3	1.6	1.8	2.0	2.2	2.4
25% net reach:		37%	46%	53%	59%	63%	66%
avg. freq:		1.4	1.6	1.9	2.1	2.4	2.6
30% net reach:		43%	53%	61%	67%	71%	74%
avg. freq:		1.4	1.7	2.0	2.2	2.5	2.8
35% net reach:		50%	60%	68%	74%	78%	80%
avg. freq:		1.4	1.8	2.1	2.4	2.7	3.1
40% net reach:		55%	65%	73%	79%	83%	85%
avg. freq:		1.5	1.8	2.2	2.5	2.9	3.3
45% net reach:		60%	70%	78%	84%	87%	89%
avg. freq:		1.5	1.9	2.3	2.7	3.1	3.5
50% net reach:		65%	75%	83%	88%	90%	91%
avg. freq:		1.5	2.0	2.4	2.9	3.3	3.8

FIG 10.9 The force of a marketing effort is often measured in terms of reach and frequency. Reach is the number of different people exposed to a message; frequency is the number of times each person is exposed. Courtesy of the Newspaper Advertising Bureau.

intense in some markets that sellers give new products budgets equal to one and one-half times the annual sales objective. If the sales goal is X, the sellers budget is 1.5X for the first two years in order to establish the position.

It follows that a campaign to introduce a new pumpkin patch, for example, should be given as much time, money and know-how as possible. The maximum number of individual mediums consistent with this budget should be commissioned to run messages with the minimum effective frequency. Finally, messages should be distributed evenly over the course of the entire campaign.

For a well-established product, the objective is simply to remind the audience of favorable features and benefits. Maintaining a favorable position costs much less than establishing one. Consequently, "generic" or "institutional" campaigns are typically given small budgets to deliver minimum effective reach and frequency.

For a well-established pumpkin patch, this guideline suggests the smallest budget possible be dedicated to persistently remind the audience about how much fun they had in previous trips to the patch. This can be done by running a reminder message as many times as the budget will allow in the top one or two mediums for target audience. This message should be distributed evenly over the course of the entire season.

Force Tactical Messages: The second objective of a campaign is to move buyers toward a specific objective. "On Sale, Buy Now!" is a good example of a tactical message. Generally speaking, tactical campaigns are conducted when the desire to buy is greatest or when the need to sell is greatest.

The best time to move buyers is when their desire to buy is greatest. Desire can be measured by evaluating a product's repurchase cycle. Lettuce may be repurchased weekly; wedding floral arrangements once a lifetime.

People desire pumpkins once each year before Halloween. Force must be budgeted to introduce new customers to the pumpkin patch and to remind many established customers of the patch's features and benefits. How can you conduct a new product campaign and a reminder campaign without going broke? One effective way would be to establish a mailing list of satisfied customers and mail each an invitation to return. The largest part of your budget could then be dedicated as a new product introduction, with the objective being to achieve maximum effective reach with minimum effective frequency.

The next best time to move buyers is when the need to sell is greatest. In organizing a campaign for your pumpkin patch, realize that only a day or so separates the propensity to buy and the need to sell. When this transition is made, be prepared with a final Close-Out Super Sale, or be left with a field full of rotting pumpkins.

The need-to-sell campaign is budgeted to achieve maximum effective frequency with messages concentrated before and during the event. In organizing a Close-Out Super Sale, budget to achieve maximum frequency in the top one or two mediums for audience and concentrate distribution before and during your sale. Frequency creates urgency, and urgency moves buyers.

Force the Market: A third campaign objective is to support the market. Two aspects of the market need to be considered when budgeting reach, frequency and distribution. One is the extent of the product's circulation within the market and the other is the extent of the market itself.

The extent to which a product is circulated within an area has a direct impact on the budget for reach and frequency. Obviously, it makes sense to limit reach to the bounds of circulation. Excess budget can then be used to build frequency. (If reach is built too slowly, however, a low awareness level may result in poor sales, and poor sales may result in the seller losing confidence in the product. Consider how a lack of reach would affect the selling of pumpkins from your pumpkin patch. Will your field full of rotting pumpkins affect your confidence in next year's crop?)

The extent of the market also plays an important role in budgeting reach and frequency. Reach is difficult in markets which are geographically and demographically widely dispersed. In dispersed markets, it may take a large number of mediums to achieve a minimum effective reach, thereby limiting your ability to build frequency. (This explains why sellers like densely populated metropolitan areas.) Your pumpkin market consists of families who live within a ten mile radius of the patch. Since the families are relatively easy to reach with a local newspaper, you can afford to build frequency by placing three ads in the newspaper, rather than placing one ad in three newspapers. This additional frequency creates urgency, and urgency moves buyers.

A well-organized marketing campaign uses the right message, to hit the right target, with just the right amount of force. How is this campaign delivered? Look at the organizational flow chart for Goliath Farms Inc. There, under Marketing, you will find Advertising, Public Relations and Promotions.

Advertise

Advertising is a direct way to campaign for consumer dollars: You pay money and receive an audience. However, advertising is a capital-intensive way to compete in which big advertisers enjoy economies of scale advantages over small ones.

To be a successful small-scale advertiser, avoid advantage and compete against disadvantage. One disadvantage of large scale advertisers is distance from the market. When Goliath Farms Inc. campaigns, it does so through an advertising agency whose media buyer views the market through the distant haze of estimates provided by a media rating service.

You live and work in the market. Exploit this position by developing a close relationship with the market's media. Know which medium is hot and which is not. Know which medium is a good buy and which is merely exploiting good will to pay off debt. Use good information to find large audiences at efficient prices.

This section will help you evaluate media in your market. It contains a list of the mediums most frequently used by metrofarmers and lists ways to evaluate which is best suited for delivering your campaign.

SIGNS: One way to advertise is to erect a sign. A sign is a surface upon which a message is placed. The surface is then positioned where passersby can read the message. Signs may be as simple as small placards nailed to posts or as complex as illuminated billboards placed next to busy highways. They have many uses. Charles Heyn posts them along the highway to attract customers to his farmstand, in his farmyard to direct traffic, out in the field to guide the harvest and in his farmstand to list prices (Part IV, Conversations). The following criteria will help you determine the effectiveness of a sign.

Flexibility: Signs can be erected with relative ease, which explains why they are an ubiquitous presence in all markets. Nevertheless, there are two factors to take into consideration when evaluating the flexibility of a sign: presentation and message.

Presentation is how a sign appears. A sign will likely be the first contact a buyer has with a business and the attention a buyer will likely give each sign is minimal. Many small-scale advertisers mistakenly post carelessly scribbled messages on old boards and scraps of paper; the carelessly executed messages, however, tell buyers what to expect from their business. A sign is most effective when it is built from clean, unblemished materials and its message is limited to a few well-chosen words composed in bold, striking letters. "You will never get a second chance to make a good first impression," and a sign may be your only chance to make a good first impression.

Message is what the sign delivers. A strategic message, like "Locally Owned, Locally Operated," requires the sign stand for a long period of time to be effective. Signs which carry strategic messages must therefore be constructed so as to project a good image over a long period of time. Tactical messages, like "On Sale, Buy Now," require the sign stand for a short period of time and convey urgency. Since tactical messages are left in place only a short period of time, it makes sense to deliver them with signs made of temporary materials.

Flexibility is therefore governed by the kind of message being delivered and by the need to present a good impression of your business.

Selectivity: Signs are indiscriminate. They deliver messages to whomever happens to pass by. Location is therefore the means by which a sign selects audience. Though a busy street or crowded highway may deliver a large audience with ongoing frequency, you must determine if this is the right audience or if it is one merely passing through without the slightest intention to stop and shop. Location selects audience; effective advertisers select the right location for their sign.

Cost: Two factors determine the cost of investing in a sign. One factor is production; the other is location. Production costs are determined by adding costs of materials and labor. Location costs, on the other hand, are determined by competition for the audience the location delivers. Locations which deliver a large audience cost more than locations which deliver a small audience.

Legal restrictions placed on signs must also be considered as a location cost. Some locations are very restrictive with respect to size, appearance and placement of signs. Santa Cruz, California, for example, banned billboards within its jurisdiction; billboards "grandfathered" by this legislation became very expensive properties.

To determine total cost of a sign, add costs of production and location. To determine whether the sign will be a cost-effective investment, conduct a Cost Per Thousand (CPM) estimate. A CPM estimate provides an approximate cost for one thousand impressions of the message delivered by the sign. This figure, in turn,

FIG 10.10 A sign may be your only chance to make a good first impression.

provides a way to compare a sign with other signs and with alternative media, like newspaper and radio.

Begin your CPM estimate by counting the audience passing by the sign's proposed location for several representative 15-minute periods on several representative days during a representative week. Extrapolate from this count and estimate the total number of impressions your sign will generate during the campaign. Next, find the CPM by dividing the sign's total costs by its total impressions and multiplying by 1,000.

Try a CPM estimate on a sign for the pumpkin patch worked up in the previous section. Erect an "Always-the-Best Pumpkin-Patch, Coming For Halloween" sign at a great location next to a busy street. This sign will stand for five months, at which time we will harvest the pumpkins and sell them. Will this sign be a cost-effective investment?

Begin by estimating impressions as suggested in the extrapolation technique outlined above. Extrapolating from a count of three representative 15-minute segments during three representative days during a representative week, yields about 2,000 impressions per day. This 2,000 impressions per day, multiplied by 30.416 days in an average month, multiplied by 5 months, gives the sign 304,160 impressions.

Budget a whopping $1,000 to make an attractive sign and another $1,000 to rent its location. Now, divide cost by impressions and multiply by 1,000 to find the CPM.

Cost Per Thousand Estimate for Signage

THE FORMULA	PUMPKIN PATCH
$\dfrac{\text{Cost of Sign}}{\text{Impressions}} \times 1,000 =$	$\dfrac{\$2,000}{304,160} \times 1,000 =$
Cost Per 1,000 Impressions	$6.75 CPM

FIG 10.11

This CPM estimate projects the pumpkin patch sign will deliver 1,000 impressions for $6.57. This estimate provides a means by which the sign can be compared to other signs and to alternative mediums, like newspaper or radio.

NEWSPAPERS AND PERIODICALS: Newspapers and periodicals are regularly scheduled publications directed to audiences which share similar interests. Audiences can be reached with two kinds of advertisements: display and classified.

Display ads, like signs posted along a roadway, seek to capture the attention of an audience casually passing by. The ads are therefore placed in appropriate sections of the periodical and composed of relatively large panels with copy, photographs and graphic symbols to capture attention and convey message.

Display ads take up a relatively large amount of space and tend to cost a lot of money. Consequently, metrofarmers use them most often for tactical purposes, such as selling mail-order products, and promoting U-Pick harvest specials, farmers markets and special sales.

Classified advertisements and hybrid classified-displays address an audience actively seeking information. They are usually located in the back pages of the periodical and are composed of several lines of specific information.

Charles Heyn uses classifieds to advertise U-Pick specials at his Heyn's Country Gardens. Jeff Larkey sells many of his products through local farmers markets. Following is an example of a classified ad placed in the "Produce For Sale" section by one of Jeff's farmers markets:

Farmers Market
Cabrillo College
SATS. from 8 till noon
Fresh fruits & vegetables,
dried fruits, nuts, melons
cut flowers, mushrooms, honey & eggs.

The following criteria will help determine the effectiveness of an advertisement placed in a periodical.

Flexibility: Computers are increasing the flexibility of producing advertisements to an extent unimaginable only a few years ago. Nearly anyone with a small computer and a little imagination can now create a captivating print advertisement. The technological revolution has also increased the flexibility of publications. Where a daily newspaper might have required a lead time of days, it may now require a lead time of hours or even minutes.

Increased flexibility notwithstanding, certain guidelines remain intact: Daily newspapers printed in black and white on pulp paper tend to be very flexible; annual magazines printed in color on coated stock tend to be very inflexible. The more flexible a medium is, the more likely it will be used for tactical campaigns, like "On Sale, Buy Now." The less flexible a medium is, the more likely it will be used for strategic campaigns, like "Locally Owned, Locally Operated."

Selectivity: The larger the market, the greater are your chances to find a periodical which selects concentrations of prospective buyers. There are dailies, weeklies, monthlies, quarterlies and annuals which target general and specific-interest audiences in neighborhoods, metropolitan areas, regions, states and nations. Distribution and editorial content are the means by which periodicals select audiences.

Periodicals distributed to a specific audience, such as residents of a neighborhood, are more selective than periodicals distributed to general audiences, like residents of metropolitan area. Where budget is a concern and the market well-confined, it makes sense to use periodicals which provide the greatest selectivity.

Editorial content further selects audience. Even neighborhood weeklies provide topical sections to concentrate audiences. Traditionally, the greatest concentration of women between the ages of 25 and 49 is found in "Food and Living," whereas the greatest concentration of men 25 to 49 is found in "Sports and Business." Periodicals are valued by how well they select target audiences. Select those which select best.

Cost: Two factors determine the cost of investing in a print advertisement: advertisement size and audience size.

Periodicals sell ad space by the column inch, so determine how many column inches it will take to get your message across and add the cost of the inches. The objective, as always, is to generate the maximum amount of interest with the minimum number of inches. Next, determine how much the audience will cost. Though publications may sell column inches, smart advertisers buy audience. Some publications charge more for audience than do others and what may appear to be the least expensive audience may in fact be the most expensive.

One way to evaluate the cost of a periodical's audience is to make a Cost Per Thousand (CPM) estimate. The formula for finding a newspaper's CPM can be seen in Figure 10.12.

Cost Per Thousand Estimate for Newspapers

THE FORMULA

$$\frac{\text{Cost of Ad}}{\text{Circulation}} \times 1{,}000 = \text{Cost Per 1,000 Impressions}$$

FIG 10.12

Try this estimate on a "Pumpkins Now Ready For Harvest" display ad. A neighborhood weekly, the *Gazette*, will print this ad for $50, while its competition, the *Times,* will charge $70. Since budget is limited, you must invest where it will generate the most efficient return. The *Gazette* claims to reach 6,500 neighborhood readers each week and the *Times* claims to reach 9,800. Compare CPMs to find which publication will provide the most efficient buy:

Cost Per Thousand Comparison for Display Ad

THE GAZETTE	THE TIMES

$$\frac{\$50}{6,500} \times 1,000 = \frac{\$7.69}{CPM}$$

$$\frac{\$70}{9,800} \times 1,000 = \frac{\$7.14}{CPM}$$

FIG 10.14

The CPM comparison tells us, if all other factors are equal, the *Times* will be more efficient than the *Gazette*. The estimate also provides a means to compare the weeklies with alternative mediums, like local dailies or radio stations.

DIRECT MAIL: An increasingly popular way to reach a target audience is to send a message through the mails. Metrofarmers use postcards, letters, flyers, discount coupons and catalogs to advertise U-Pick harvests, sales, special events and direct order gift items like fruit baskets.

Flexibility: The flexibility of direct mail is limited by the need to present a good image and by the logistics of assembling and mailing the pieces. Needless to say, it takes a considerable amount of lead time to produce a well-designed piece and to prepare the piece for the mail.

Direct mail campaigns are seldom used for strategic messages. Only rarely does one find a "Locally Owned, Locally Operated" message in the mail box. Most direct mail contains tactical messages which seek to move buyers toward specific objectives, like "On Sale, Buy Now." Direct mail also allows for inclusion of coupons, which provide a measurable way to evaluate the effectiveness of an offering. Send out coupons and see how many come back.

Selectivity: A direct mail campaign uses the mailing list to select audience. A good list contains carefully collected names and addresses of legitimate prospective customers. Many advertisers compile a list by having satisfied customers sign a guest register.

Other advertisers purchase lists of names collected and collated by professionals. There are two kinds of mailing lists for sale, compiled and house. Compiled

FIG 10.13 Examples of display and classified display ads.

lists are those compiled from outside sources like public records and auto registrations. House lists consist of names collected from actual sales or inquiries.

The mailing list selects the audience. A good list will enable you to hit your target audience. A bad list will put you into the wrong mailbox.

Cost: Direct mail is sold by the piece. You can pay for 100 pieces of direct mail or 100,000 pieces. However, smart business people do not buy pieces; they buy audience. One way to measure the cost of a direct mail audience is to make a Cost Per Thousand (CPM) estimate. The formula for finding a CPM for direct mail can be seen in Figure 10.15.

Cost Per Thousand Estimate for Direct Mail

THE FORMULA	PUMPKIN PATCH
$\dfrac{\text{Cost of Mailing}}{\text{Circulation}} \times 1{,}000 = \begin{array}{l}\text{Cost} \\ \text{Per} \\ 1{,}000\end{array}$	$\dfrac{\$575}{12{,}000} \times 1{,}000 = \begin{array}{l}\$47.92 \\ \text{CPM}\end{array}$

FIG 10.15

Use this CPM estimate to evaluate a direct mail campaign for a pumpkin patch. Plan a direct response coupon to give holders a fifteen percent discount on any large-sized pumpkin in the patch. A local direct mail company will produce this coupon and include it in a packet with others being mailed. This coupon will go to 12,000 homes and cost a total of $575. Is it a cost-efficient investment?

The CPM estimate reveals it will cost approximately $47.92 to insert a coupon into the mailboxes of 1,000 well-targeted neighborhood homes. This estimate also provides a way to compare costs charged by other direct mail companies and with alternative media, like newspapers and radio.

RADIO: Radio is a broadcast medium. It casts its programs out over the air to whoever will tune-in and listen. Though its status as an advertising vehicle has been diminished somewhat by television, local radio has held up remarkably well as an advertising medium. There are several reasons for this continued strength.

First, radio is pervasive. It wakes people up in the morning, entertains them as they prepare for the day, informs them as they commute to work, soothes them as they work and informs them as they commute home. Second, radio is easy to use. A commercial can be readily produced which can induce a target audience to imagine the taste a fresh strawberry and then encourage the audience to act while the taste is still fresh in the imagination. Finally, radio allows for the building of frequency; frequency creates urgency; and urgency moves buyers.

Metrofarmers often use local radio to advertise sales, special events and the availability of U-Pick crops. Charles Heyn appeared regularly on a local radio talk show called "Country Gardens."

Flexibility: Radio is a flexible medium. In fact, a radio commercial can often be changed in the middle of a schedule simply by asking the station to produce another commercial. This flexibility allows radio to be used for both strategic and tactical advertisements.

Strategic commercials, like "Locally Owned, Locally Operated," are often booked as sponsorships for local news and weather. Sponsorships, which typically give the advertiser a commercial and a "brought to you by" announcement, keep the advertiser's message in front of the public and build a favorable position for the advertiser by providing a service to the community.

Tactical commercials, like "On Sale, Buy Now," are flighted. This means commercials are bunched together as an individual flight during a critical segment of time. A special closeout sale, for example, might be flighted with commercials bunched closely together immediately preceding and during the sale to build frequency and create urgency.

Selectivity: Radio is a selective medium. It is so selective, in fact, people actually develop a personal attachment to an individual radio station. When a person tunes in their personal favorite in the early morning and keeps it tuned in until evening, it will likely have a lot of influence on the person's purchasing decisions.

A radio station selects audience by the power and position of its transmitter and by the content of its programming. The transmitter selects geographical audience because listeners tune in strong, clear signals and tune out weak, static-filled ones. Programming selects demographic audience: In general terms, young people listen to fast-paced contemporary music while older people listen to slower-paced fare like easy-listening, classical, country and newstalk.

When the radio audience is filtered through power, position and programming, there remains a concentrated group of people who live in the same general area and who share similar tastes and interests.

Cost: Radio is sold by the second. There are 10, 15, 30 and 60-second commercials for sale at most local radio stations. Smart business people, however, do not buy seconds; they buy audience.

Since commercial production is free for most local advertisers, the cost of radio advertising is determined by audience size and by competition among radio stations for the advertiser dollars. One way to determine whether a radio station will be a cost effective investment is to conduct a Cost Per Thousand (CPM) estimate.

Radio Script Techniques

☑ **ACTOR**	Script focuses on an event or "slice of life" experience in which an actor's need is well satisfied by the product. Script action begins with a crisis (boy meets girl and girl says "no"), continues with the introduction of the product (boy applies deodorant) and ends with the successful resolution of the crisis (girl says "yes"). Requires a believable experience and evocative actors.
☑ **ACTOR-ANNOUNCER**	Script consists of an announcer supplementing or reacting to an actor's message. Requires an evocative actor and a friendly, believable announcer.
☑ **ANNOUNCER-ONE**	Entire script delivered by a single announcer. Requires a friendly, believable voice.
☑ **ANNOUNCER-TWO**	Script divided into sequences for delivery by two announcers. Requires an exciting, fast paced delivery, a newsworthy presentation or a special affiliation between announcers.
☑ **COMBINATION**	Script combines two or more script techniques into a single commercial. One often used combination is a customer testimonial followed by an announcer tag. Combination scripts require more technical ability than single technique scripts.
☑ **CUSTOMER**	Actual customer testifies about a favorable experience with the product. Requires a believable customer and a fascinating real-life experience.
☑ **HUMOR**	An actor, announcer or customer dialogue— often a slice of life experience— delivered in a humorous vein. Humor is used for products which are low-priced, which are purchased for enjoyment, whose primary appeal is taste or whose presentation requires special treatment for one reason or another (*e.g.* every business on the street is advertising the same product). Humor is not used for expensive or serious products. A humor campaign will require a number of scripts because a single humorous treatment of a subject will lose its viability with repeated exposure.
☑ **JINGLE WITH DONUT**	Musical jingle with a timed segment for voice presentation. Requires a pleasant, easily-remembered musical presentation and a friendly, believable voice.

FIG 10.16 Commonly used techniques for creating effective radio commercials.

The formula for finding a CPM for radio can be seen in Figure 10.17. (Average Quarter Hour Persons— or AQH— is the average number of persons listening to a particular station for at least five minutes during a 15-minute period. It is an estimate of how many people will actually hear a commercial when it is broadcast.)

Cost Per Thousand Estimate for Radio

THE FORMULA

$$\frac{\text{Cost of a Commercial}}{\text{Average Quarter Hour Persons}} \times 1{,}000 = \text{Cost Per } 1{,}000$$

FIG 10.17

Use the CPM estimate to evaluate a radio buy for the pumpkin patch. Plan for a flight of 18 "Hurry, Final Closeout Sale" commercials for the days immediately preceding Halloween. Aim for a target audience of women between the ages of 25 and 49. Station KXXX will sell an AQH of 1500 Women 25 to 49 for $40, while station KZZZ will sell an AQH of 1650 for $55. Which station represents the most efficient way to reach women 25 to 49?

Cost Per Thousand Comparison for Radio

KXXX	KYYY
$\frac{\$40}{1{,}500} \times 1{,}000 = \$26.67 \text{ CPM}$	$\frac{\$55}{1{,}750} \times 1{,}000 = \$31.43 \text{ CPM}$

FIG 10.18

The CPM estimate reveals radio station KXXX is the most efficient way to reach women 25 to 49. It also provides a way to compare the costs of KXXX with other radio stations and with alternative mediums like newspaper or television.

Many smaller radio stations do not subscribe to an audience ratings service. To help determine if they are efficient mediums, make a list of local advertisers. Call the listed business owners and ask, "Would you buy advertising on this radio station again?"

TELEVISION: Television is the only medium to combine sight, sound and motion into one message. Because costs have been high, the use of television has been limited to businesses large enough to benefit from economies of scale. However, television is also the medium of technological change, and two changes have made television available to small-scale advertisers as well.

The first change was the introduction of alternatives to broadcast television. A few years ago, three broadcast networks and their local affiliates owned the nation's television sets. Today, television sets are wired into hundreds of alternatives, including cable, satellites and video tape machines. The second change was the introduction of new production technology. A few years ago, television shows and commercials were produced with large cameras costing well over $100,000. Today, programs and commercials are taped by amateurs using pocket-sized equipment costing only a few thousand dollars.

The good news is that more channels and less production costs have made television advertising accessible for small businesses. The bad news is that target audiences now have hundreds of places to hide, instead of only three! Thus the advertiser's cost to achieve reach and frequency objectives is still a significant consideration. Nevertheless, many small businesses, like metrofarmers, now use television to advertise. Consider the following criteria before you invest in television advertising.

Flexibility: Television is the least flexible advertising medium. It will always require a substantial lead time and considerable expense to produce a commercial which delivers a good impression of your business and its products.

Small-scale advertisers often compensate for costs by making one commercial to deliver both strategic and tactical messages. The technique is simple. A "Locally Owned, Locally Operated" strategic commercial is taped and produced. This strategic commercial is then run occasionally to establish a favorable position in the mind of the buyer. Then, when the need arrives to move buyers with an "On Sale, Buy Now" message, the strategic commercial is made into a tactical one by changing its sound and graphics.

Selectivity: Television selects audience by delivery and programming. Satellite systems not withstanding, there are two ways in which television programming is now delivered. One is broadcast and the other is cable.

Broadcast television, like radio, selects audience by the power of the transmitter and position of the antennae. Power and position determine exactly which geographical area will be served by the station. This selectivity allows advertisers to target broad areas like metropolitan regions. Cable television selects audience through a system of cables which deliver programming directly to communities of license. Indeed, with a system of cables connecting station and individual sets it will soon be possible to select and reach individual homes with messages, like personal mail.

Television also selects audience through programming. In very general terms, adult males are selected with sports, traditional homemakers with "soap operas," young adults with action dramas and children with cartoons. Ratings services provide a fairly accurate estimate of where target audiences can be found.

Cost: Television commercial time is sold in 10, 15, 30 and 60-second increments. Smart business people, however, do not buy seconds; they buy target audience.

The cost of reaching a target audience with a television commercial can be determined by adding costs of producing the commercial and the rate charged by the television station for its commercial time. The efficiency of this commercial can then be measured through a Cost Per Thousand (CPM) evaluation. The formula for finding a CPM for a television commercial can be seen in Figure 10.20.

Cost Per Thousand Estimate for Television

THE FORMULA	PUMPKIN PATCH
$\dfrac{\text{Cost of Commercial}}{\text{AQH}} \times 1{,}000 = \text{Cost Per } 1{,}000$	$\dfrac{\$117}{6{,}000} \times 1{,}000 = \$19.50 \text{ CPM}$

FIG 10.20

Use a CPM estimate to compare a local television buy with the radio buy discussed above. Plan to run 18 "Hurry, Final Close-Out Sale" commercials to reach women 25 to 49. A local broadcast television station, KYYY, will charge $300 to produce the commercial and $100 for each commercial to run on programs which deliver an AQH of 6000 women 25 to 49 in the target marketing area. To determine if this is a cost effective investment, add total production costs and total commercial time costs ($300 + $1800 = $2100). Next, divide this sum by the number of commercials to find the average commercial cost ($2100 by 18 = $117).

Television Script Techniques

☑ **ACTOR**	Script presents a "slice of life" experience in which an actor's need is satisfied by a product. Action begins with the establishment of the actor's need (boy meets girl and girl says "no"), continues with the introduction of the product (boy applies deodorant) and ends with the successful resolution of the actor's need (girl says "yes"). Requires a believable experience and evocative actors.
☑ **ANIMATION**	Animation is a series of inanimate drawings brought to life on sequential frames of motion picture film. Simple animation is often used to help illustrate complex procedures, such as the growth and development of a plant. Although animation is a costly procedure requiring extensive lead time, simple movement employing a few characters on an elemental background can be used to effectively supplement conventional photographic techniques. Rotoscope is a combination technique in which animated objects appear "in person" with real objects.
☑ **ANNOUNCER**	Announcer stands in front the audience to present, display and demonstrate a product. Announcer may be situated in a neutral setting or in a setting appropriate to the product. The successful announcer is friendly, believable and subordinate to the product.
☑ **CELEBRITY**	Well known individual lends credibility to a product by delivering a friendly and believable personal testimonial. The successful celebrity spokesperson is one who is— in some way— related to the product and who will not dominate the presentation to the exclusion of the product.
☑ **CLOSEUP**	Takes advantage of television's visual capabilities to show an attractive and appealing closeup of how a product will work. Dialogue is typically delivered by an off-screen announcer. Requires the very best visual quality for a successful extreme closeup presentation.
☑ **COMPARISON**	Takes advantage of television's visual and auditory capabilities to compare features and benefits with the competition. Indirect comparisons use a "Brand X" competitor; direct comparisons use a named competitor. Direct comparisons must be credible for two reasons: First, potential customers may see through a faked presentation and sympathize with the competitor. Second, the competitor may initiate legal proceedings against the

	attack on its name, in which case the claims made must be substantiated in a court of law.
☑ CUSTOMER	An actual customer testifies about a favorable experience with a product. A variation is the technique in which the customer compares two products on screen to see which is best (see below). Customer testimonials require believable customers and fascinating real-life experiences.
☑ DEMONSTRATION	Takes advantage of television's visual and audio capabilities to demonstrate how the product or service will work. Requires a relevant and real demonstration because a faked one may be perceived as such by potential customers and may instead become subject to legal scrutiny. (Advertising agencies often require commercial actors to sign an affidavit signifying events in a commercial actually occurred as they appear on the screen.)
☑ GRAPHICS	Takes advantage of television's visual capabilities to make use of product related still photographs, charts and illustrations. The skillful use of cameras, including zooming in and out and panning back and forth, can give movement to the graphics. Requires top quality graphics to impart a favorable impression of the product.
☑ HUMOR	A "slice of life" experience delivered in a humorous vein by an announcer, actor, customer or story. Humor is used for products which are low-priced, which are purchased for enjoyment, whose primary appeal is taste or whose presentation requires special treatment for one reason or another (e.g. every business on the block is advertising the same product). Humor is not used for expensive or serious products. A campaign based on humor requires a number of commercials or a short run because the humorous treatment of any subject quickly loses its viability with repeated exposure.
☑ STORY	Presents the commercial as a 10, 15, 30 or 60-second story complete with a well-defined beginning, middle and end. Narration with product features and benefits is typically delivered off-screen. Requires a well-developed story line to hold the attention of a restive audience.
☑ VIGNETTES	A series of fast-paced "slice of life" scenes showing customers enjoying the product (e.g. families happily picking berries). Scripted message is delivered with voice over or with a jingle. Requires friendly, evocative actors and believable situations.

FIG 10.19 Commonly used techniques for creating effective television commercials.

Now estimate the CPM.

The CPM estimate says television station KYYY will deliver 1,000 women viewers between the ages of 25 and 49 for an average cost of $19.50. This estimate also provides a way to compare KYYY with other television stations and with alternative media, like radio and newspaper.

CPM estimates are not the only means for measuring the efficiency of an advertising purchase. Larger television and radio stations offer Gross Rating Point projections for reach and frequency levels. However, CPM is a good place to start and it offers a basic look at the real cost of investing in advertising.

ALTERNATIVE MEDIA: Other advertising opportunities for metrofarmers include printed flyers, packages, point-of-purchase displays, recipes, labels, sides of pickup trucks, business cards, stationery, hot air balloons and so on. Some mediums, like sides of pickup trucks, are suitable for delivering strategic messages such as "Locally Owned, Locally Operated." Other mediums, like point of purchase displays, are suitable for delivering tactical messages such as "On Sale, Buy Now."

The smart advertiser always looks for alternative ways to "climb a tree and holler" for the consumer dollars. The successful advertiser finds the ways.

OBSERVE THE RESULTS: The objective of investing in an advertising campaign is to use the right weapon, to hit the right target, with the right amount of force. The final step in an advertising campaign is to observe the results.

Observation consists of two related steps. The first step requires expectations to be defined in a precise and measurable way. Do you expect to increase the awareness of your metrofarm by 25 percent? Do you expect to increase the customer traffic at your end-of-season closeout sale by 50 percent?

The second step is to observe results of advertising to see if your objectives have been achieved. Large-scale advertisers use objective measuring devices like test markets and consumer panels. Small-scale advertisers continually survey customers and prospective customers for results.

Advertising is one way to campaign for consumer dollars. Keep track of your advertising campaign: If it achieves favorable results, continue advertising; if not, go back and discover where you did not use the right weapon, hit the right target, or use the right amount of force.

Publicize

Advertising is direct: You pay money and deliver a message to an audience. Publicity is indirect: You send a message— as a news item— and the message is then used as the media see fit.

Publicity is news about change in a business and the significance of the

FIG 10.21 Other advertising mediums used by metrofarmers include point of purchases displays like the one pictured here.

change for consumers. Since publicity is not paid, publicity submitted will be used only if an editor thinks it will be of value to an audience. Furthermore, the submission will likely be altered to suit an editor's needs and requirements.

Publicity's strength is that it is more readily accepted than advertising because there is an editorial filter, however thin, of objectivity applied to its use. Publicity's weakness, conversely, is that it cannot be controlled because it has not been purchased. Consequently, most publicity campaigns are directed toward achieving strategic objectives of establishing and maintaining a favorable position. Few publicity campaigns are directed toward moving buyers toward specific tactical objectives like an "On Sale, Buy Now!" event.

Metrofarmers featured in Part IV, Conversations, demonstrate the effective use of publicity. The Flower Ladies have been featured in television, radio, newspapers and magazines. Charles Heyn and his Country Gardens have been featured in local newspapers and a weekly radio show called, appropriately enough, "Country Gardens." Jeff Larkey and his partners at Pogonip Farms have been featured many times in local newspapers. If the inches and minutes of this publicity had been priced at commercial advertising rates, it is likely none would have been purchased. By cleverly leveraging time and know-how, however, the metrofarmers succeeded in using publicity to establish and maintain favorable positions for their respective businesses.

This section will help you campaign with publicity. It provides steps to developing a story, targeting the right mediums, releasing the story and tracking results.

DEVELOP A STORY: The fact that many believe publicity is free can be seen at the offices of newspapers and radio and television stations, where there will likely be garbage cans filled with press releases from local businesses seeking "free advertising." Very few of the attempts achieve their objective; the ones which do begin with a good "angle."

Angle is the distinctive element of a story which makes it worthy of consideration by a target audience. Other names for this element of interest are "peg" and "slant." Finding a good angle is not as difficult as the number of rejected attempts would lead one to believe. All successful angles have one element in common: Each contains news which will, in some way, be of significance to a target audience. Following is a list of angles which will help you avoid the media garbage can.

New and Different: New is news. Make the editor's job easy and find something new to interest the audience. Grow a purple pumpkin and win the headline, "Always-the-Best Grows World's First Purple Pumpkin!"

Timing: Something happening this week is always more important than something which happened last month or which will happen next year. Plan to release pumpkin stories near Halloween and watermelon stories around Independence Day.

Prominent: Big names make big news. Host a charity event and invite a celebrity to be the honorable guest, chairperson or judge. Connect the celebrity's name with the event and your business as many times and in as many ways as possible.

Local: Few headlines are more certain to capture attention than "Local Business Makes Good." The Flower Ladies (Part IV, Conversations) understood this certainty and marched into editors' offices with giant floral arrangements and said, "Do a story on us." The Flower Ladies have consequently been featured in all the local newspapers, as well as regionals like the *San Francisco Chronicle*, *San Jose Mercury News* and *Sunset* magazine.

Emotional: Find a way to call up joy, fear, humor, sorrow, pride or love. Children are an efficient way to summon an emotional response. The Gizdich Ranch, the apple and berry U-Pick mentioned earlier in this chapter, provides free field trips of the working farm to area school children. In addition to creating good future customers, this free service provides editors with great image and copy.

Conflict: One way for a business to generate news is to become identified with a political issue or movement. March up and down the street with a banner and you will get attention. However, this attention will likely alienate as many prospective customers as it pleases, so most people avoid attaching their business names to political causes. However, one relatively safe cause for metrofarmers is the cause of small versus big. Find a way to demonstrate that your David-sized farm is beating Goliath Farms Inc. and the media will beat a path to your door.

Social Issues: Another sensitive arena for a business is a social issue like the conflict between rich and poor. Half the audience always seems to support one side of the issue while the other half supports the other. One relatively safe place to tread is employment. "Always-the-Best Provides Inner-City Youth with Summer Jobs" is a headline which always seems to work.

Future: Find a scientific or ecological aspect to your business which has implications for the future of the community. New crop varieties, culture practices and pest control techniques are good places to look. For example, find a way to eliminate the toxic clouds of pesticides wafting over the neighborhood, and you will earn a good headline.

TARGET THE MEDIA: Local media include newspapers, tabloids, magazines, business publications, newsletters and radio and television stations which service your target market with news. There are two kinds of media to consider: public and private. Public media consist of government and industry publications established to provide farmers with the means to publicize their businesses.

Public mediums generally do not require a large capital outlay because they are subsidized with tax money to provide farmers with "free" publicity. One example of public media is the *Farmer-To-Consumer Directory*, which is published by the California Department of Food and Agriculture and the California Direct Marketing Association. This catalog lists names of farmers who sell direct to consumers, as well as their addresses, business hours, products and special notes. Another example is *Country Crossroads*, published by Santa Clara and Santa Cruz County Farm Bureaus. This publication features a map which notes the location of farmers who sell direct, as well as their products, business hours and special notes.

Private media, like local newspapers, require a larger capital outlay because they need to earn a profit to survive. Conversely, they will likely be more effective to use because each must provide their audience with quality news and entertainment.

One way to guarantee your publicity effort will land in the garbage is to give it to the wrong person at the wrong time, like a sales manager one hour before deadline. Garbage cans are filled with this silly mistake. Learn working characteristics of the medium you want to use. Learn its various positions and how they function, its working routines and its deadlines. Following are general observations which will help you get the right information to the right person at the right time.

The Editor: Positions at commercial media are divided into three areas of responsibility: editorial, business and production. The editorial department is staffed with individuals whose job it is to manage news-related items. Your daily newspaper, for example, will likely have individual sections like "Food," "Business" and "Gardening." The editor of each is responsible for finding interesting stories to fill his or her section. The smart publicist identifies the editors who can benefit from receiving metrofarm-related news. Editors of all three sections mentioned above are likely targets for items regarding local metrofarms. Who are the individuals and when is their deadline?

Establishing Contact: If you mail a publicity item and then wait for it to be published, you will likely still be waiting this time next year. Editors receive many proposals. Fight yours through the competition and sell it just like you would any other piece of business. Following are three ways to win the struggle:

Person-to-Person: Whenever possible, make personal contact with the editor either through a visit or a telephone call. This will enable you to "pitch" your story and to establish a good working relationship for the future. The best time to visit an editor is right after a deadline has been met; the worst time is when a deadline is being met. Follow The Flower Ladies and march into editors' offices with samples of your product.

Mail: The mail will enable you to contact many editors for little cost. This explains why most local news media receive scores of letters every week from individuals seeking publicity for their ventures. To increase your chances of being seen by an editor, make certain all of your materials are in the proper form, contain the right information and are visually impressive.

Meetings: In certain cases it may prove fruitful to set up a formal meeting in which you present a story idea to an editor, a writer, photographers and related staff. This "Dog and Pony" show, as it is called in the trades, will require you to make a well-organized presentation complete with background materials and proposal.

RELEASE THE STORY: Publicity is proposed to editors with a press release. This release is a highly formatted letter which provides the who, what, where, why and when of your news in a way which can be understood quickly and efficiently. There are two kinds of press releases. One release is for hard news; the other is for soft news.

The hard news release calls attention to a specific item— like an event— which has a limited time of relevance— like a holiday— and is therefore flagged with a "To be Released" date. "Always-the-Best to Harvest World's Only Purple Pumpkin!" is a headline for a hard news release.

The soft news release calls attention to a human interest item and is less likely to post an urgent release date. Soft features are often called "evergreens" because they can be picked up and used just about any time they are needed. Soft news releases are more difficult to write because they lack the edge of hard news and must therefore be presented to sell the human interest side. "Always-the-Best Regularly Hosts Inner-City School Children," is a headline for a soft news release.

```
                              FOR IMMEDIATE RELEASE

Always the Best Pumpkins
4444 Fairweather Boulevard
Anywhere, U.S.A.
39393


Contact:  April or Roger Jones
          999-8888


    ALWAYS THE BEST TO HARVEST PURPLE PUMPKIN


   Always the Best will harvest the state's first
purple pumpkin at 1:30 PM on Friday, October 22nd
at their 4444 Fairweather Boulevard pumpkin patch.

   The giant 150-pound purple pumpkin is thought to
be the result of a genetic mutation in one of Always
the Best's standard orange varietals.

   Anywhere's Mayor Alice Walker will help harvest
the giant purple pumpkin Friday afternoon.  The
pumpkin will then be carved into a giant Jack-o-
lantern by students from Anywhere High School and
raffled off Halloween night.  Proceeds from raffle
ticket sales will go to Anywhere High School's
marching band.

                         ###
```

FIG 10.22 The press release is used to reach local media with your news item.

Press Release: Professional publicists submit highly stylized press releases consisting of "must-includes" which communicate the news clearly and quickly. The must-includes are listed below:

Name and address of sender: Place your business name, address, phone and contact name in upper left hand side of paper.

Release date: Include a "Release Date" headline in the upper right hand corner which states when the information is to be published.

Headline: Center and underline a headline in capital letters approximately one-third down the page. This headline should summarize the story.

Copy: Start the copy one-half way down the page and double space the lines so it can be corrected and annotated. Place subheadings on the left and underline them. Skip an extra line between paragraphs and keep them short and to the point.

Continued: When text runs past one page indicate so with a "continued" at the bottom center.

End: Indicate the end of the release with the notation "###."

Other Releases: Though the press release is the heart of a publicity proposal, other materials can be used to make the pitch more salable, including query letters, fact sheets, profiles, articles, photos, charts and cover letters. Following is a discussion of these materials:

Query Letter: A typed page which asks if an editor would be interested in a news item. Use a standard business letter and, in simple sentences, sell the editor on your idea. If you succeed in establishing an interest, follow up with additional materials; if you fail, query another editor.

Fact Sheet: A typed page which summarizes the who, what, where, why and when in a simple, easy to understand way. List the facts and omit the fluff. A fact sheet should provide enough information to help the medium write or produce a story about your event, product or service.

Profile: A typed page which provides a history of a business or its principals. You may use human interest angles to add color to business and personal histories but keep the profile simple and easy to understand.

Articles: A regular article on a subject covered by your area of expertise submitted to the appropriate local media. Become

FIG 10.23 Nita Gizdich of the Gizdich Ranch near Watsonville, California, makes news people happy by providing images of a kitchen filled with deep dish pies; of orchards and fields filled with customers picking apples and berries; and of children hungrily dipping fresh strawberries into whipped cream.

a local "noted authority." Charles Heyn used this technique to great effect by becoming the produce expert on a local radio talk show called "Country Gardens."

Graphic: A printed photograph, chart or graph which tells the story for you. Suggest a caption or a brief description to help the editor follow the right angle.

Cover Letter: A one page introduction to the material which follows. The busier an editor is, the more important the cover letter becomes. Each time you submit material to an editor, use a cover letter to sell the editor on looking at the material which follows. If an editor has to invest time to figure out what the material is, about who sent it and why it was sent, chances are the material will end up in the garbage can with all of the other inconsiderate submissions.

FOLLOW UP: After a potential news story has been submitted, follow-up with a call to the editor (after his or her deadline). Ask if the submission has been received. Ask for an impression of the submission and a critique. Ask if you can help by providing additional information.

Nita Gizdich, of the apple and berry U-Pick mentioned earlier, provided editors with a kitchen filled with deep dish apple pies; of orchards and fields filled with customers picking apples and berries; and of children happily dipping fresh strawberries into whipped cream. It would cost advertisers a fortune to place the images during prime time news; it cost Nita time, know-how and some deep-dish apple pie a la mode.

Publicity is not a commodity which is bought and sold like advertising. However, publicity is not free. Nita Gizdich, who is often invited to speak at marketing seminars for small-scale farmers around the country, said, "Hard work is the answer."

Promote

A promotion is an event of limited duration designed to focus attention on a business or its products. If advertising is direct and publicity indirect, promotions are a combination of both approaches.

Like advertising, a good promotion will enable you to reach a target audience. Like publicity, a good promotion will not have the appearance of a commercial message and consequently your message will more likely be received. A bad promotion, on the other hand, can reach the wrong audience with the wrong message and cost too much money and time.

Metrofarmers participate in many kinds of promotional activities, including community events, service organizations and contests. The following list will help you determine which activity will be best suited for delivering your marketing campaign.

COMMUNITY EVENTS: The promotional calendar of your community is likely filled with events designed to bring people together and encourage commerce. The larger your community, the more events you will have from which to select. Community events most frequently used by metrofarmers include county fairs, harvest festivals and how-to fairs.

County Fairs: Though established and supported by state governments to promote local agriculture, county fairs have become incorporated into the life of major metropolitan areas. Fairs provide residents of the metro with an opportunity to mingle and metrofarmers with two ways to promote their farms and farm products.

One way to use a county fair is to compete in the fair's contests and displays. The officially-sanctioned events provide an excellent opportunity to display a product. A mouth-watering display of blue-ribbon apples, for example, can generate a more favorable impression than an entire flight of television commercials.

The other way to use a county fair is to rent commer-

FIG 10.24 County fairs provide a crowded marketplace in which metrofarm products can be displayed and sold.

FIG 10.25 Viable communities, like Gilroy, California, celebrate their local harvest with festivals, like the Gilroy Garlic festival.

cial space and display and/or sell your products. Booths or stands provide a good opportunity to meet the community "face to face." Friendly conversation can be very effective and you can use the opportunity to collect names and addresses for a mailing list.

Harvest Festivals: Many communities sponsor harvest festivals to celebrate and publicize local crops. Unlike state and county fairs, which are government sponsored institutions, harvest festivals tend to be loosely organized events sponsored by groups like the Chamber of Commerce.

One famous event is the Garlic Festival in Gilroy, California. This festival, though a relatively new event, attracts about 150,000 people annually to the little town of Gilroy (population 22,000). Numerous garlic confections are sold, music is played, people dance and all concerned have a great time. This event has promoted Gilroy garlic into the thoughts and onto the tables of many new customers.

Inspired by Gilroy's success, neighboring California towns and villages have organized their own harvest festivals. Consequently, there is a zucchini festival in Hayward, an artichoke festival in Castroville, a brussels sprout festival in Santa Cruz and so on.

Take advantage of your community's harvest festivals. Set up an attractive display and sell your metrofarm products at a good price. Collect names and addresses for future direct mailings. Use local harvest festivals to build a favorable position for your metrofarm business.

How-To Shows: Another way to promote a business is through participation in How-To shows. How-To shows are usually sponsored by commercial promoters to advance interests of a particular industry. There are home improvement shows, lawn and garden shows, brid-al shows and many others.

When a How-To show has been well organized and advertised, it can concentrate a large, highly targeted audience in one location. Take advantage of a well-organized show. Rent a booth and set up an attractive display. Offer prospects a free sample and provide them with a tangible reminder of your business.

Another way to take advantage of a How-To show is to give seminars. Be the expert you have become. Offer to give a free seminar on "Organic Pest Controls" or "Edible Landscaping" or whatever subject might be of interest to the show's audience. The Flower Ladies promote their wedding floral business by participating in area bridal shows. Their booths, displays and seminars give them the opportunity to meet individual brides, provide demonstrations and give flower samples to each prospective new customer.

SERVICE ORGANIZATIONS: Each community is served by nonprofit organizations. The organizations and the events they sponsor provide many ways to promote a business to the community.

Although few small-scale business people have the time needed to actively participate in all the events sponsored by organizations, most can participate in one or two. Volunteer some of your precious time or product. The quality of what you volunteer will be directly associated with your business.

One group common to most communities is the Chamber of Commerce, which is an organization dedicated to promoting local business. Each Chamber sponsors informal after-business-hours mixers at which local business people can meet and talk. Business leaders are a significant target audience.

June Smith, of California's Roudon-Smith winery, is a liberal donator of her company's product to local Chamber of Commerce mixers. This promotional activity has helped establish a favorable position for the Roudon-Smith label in nearby communities. As a consequence, local business people are inclined to pick up a Roudon-Smith label for important business or social occasions.

CONTESTS: One always popular way to attract attention is to offer a target audience the chance to win something for nothing. There are many variations on the contest theme. The best focus on the objective of establishing and maintaining a favorable position for the business and its products.

One contest used by small-scale marketers is the Guess-The-Date, in which a prize is offered for guessing a correct date of something special. For example, a Guess-The-Date could be used to promote the coming sweet corn season at a roadside stand. In this contest the stand's existing customers would be encouraged to guess the date the first sweet corn is harvested and then to write this guess down on an entry form (thus providing names for next year's mailing list). The winners could be paid with a few dozen ears of sweet corn, a small price to pay for establishing a favorable position as the area's best purveyor of fresh sweet corn.

Designing, producing and executing a good promotion is an art which requires imagination, uniqueness, initiative and experience. Again, as Nita Gizdich said, "Hard work is the answer."

CLOSE THE SALE

A business cycle has a beginning, middle and end. The beginning consists of developing a competitive strategy, the middle of producing a marketable product and the end of selling the product to a buyer. Our discussion of metrofarming has now reached the end of its cycle— the closing.

Closing is the term used by sales people to describe the completion of a sales transaction. This completion is when legal rights are exchanged between buyer and seller. Closing requires dedication because if legal rights are not exchanged, all of the investments you have made in planning and production will go for naught.

There are two ways to successfully close a sale. One is with a contract; the other is with a smile.

Close with a Contract

Selling to brokers, wholesalers and retailers often consists of exchanging a promise of a product for a promise of a payment. A contract is a legally-binding agreement in which a seller and buyer agree to keep promises.

To understand the need for a legally-binding agreement, consider the experience, which is not uncommon, of a small-scale shallot farmer in Southern Oregon. This farmer received a verbal commitment from a large retail chain on the East Coast to purchase 1,000 pounds of shallots. The purchase price, agreed upon during a telephone conversation, was $2.00 per pound. Since she did not have the full 1,000 pounds, the farmer went out and purchased a large quantity of shallots from her neighboring farmers for $1.50 per pound. Then several shiploads of shallots arrived from overseas and were put on the market for $.90 per pound. The large retailer quickly broke his agreement with the farmer, leaving her holding many bags of very expensive shallots. Had she signed a contract with the retailer before buying the extra shallots, she could have forced him to keep his promise and avoided her loss.

Intermediaries require a dependable supply, a uniform quality and low prices. If the needs can be satisfied, a long-term relationship can be established in which transactions can be closed with a smile, a handshake and a "Thank you." Without a contract, however, intermediaries may quit a transaction at will to capture a gain or avoid a loss.

A contract may be written or verbal and must satisfy the following requirements to be legally-binding:

COMPETENCY: All parties to a contract must be legally competent. Neither a minor nor a person who is mentally incompetent has the legal capacity to enter into a contract. If either party is incompetent, the agreement may be voided by the incompetent party but not by the legally competent one.

LEGITIMACY: A contract is not valid if based on illegal activity. For example, a contract between an apple farmer and distributor to sell quarantined apples to markets of a distant state would be invalid because of the quarantine placed on the apples.

CONSENT: The terms of a contract must be willingly consented to by all parties to the transaction. This is evidenced by the acceptance of an offer made by the buyer to the seller. This acceptance suggests agreement on any terms which may affect the transaction.

The terms commonly covered in contracts between farmers and intermediaries include quantity, quality standards, storage, delivery dates and points, transportation costs, basis for dockage (deductions for poor quality) and when and how payment is made. A mutual understanding and agreement must be present on all pertinent points for a contract to be legally binding.

CONSIDERATION: Something of value must be exchanged by the buyer and seller for a contract to be valid. It is important to note, in the absence of fraud or duress, courts will not question the value of the consideration. Thus, an agreement to sell two apples for two pennies or two hundred dollars would constitute a valid contract.

Close with a Handshake and Smile

Selling direct to buyers in farmers markets, roadside stands, special events and other direct markets consists of the exchange of product for payment. Direct transactions are best closed with a smile, a shake of hands and a friendly "Thank you."

The friendly close of a direct transaction, however, is difficult for some sellers because they lack the patience and fortitude to serve the public. Following are discussions of the three personality traits of a good closer.

ENTHUSIASM: Nothing will close a buyer more surely than an enthusiastic seller. Intelligent buyers, however, recognize two kinds of enthusiasm. One kind of enthusiasm comes from the possibility of making a large profit on a small investment. This get-it-while-you-can kind of enthusiasm, however, creates a sense of uneasy mistrust in an intelligent buyer.

The other kind of enthusiasm comes from the pride one feels at having produced a high-quality product at a competitive price. When a product has been grown from seed and is proudly held up for a customer to see, it generates the kind of enthusiasm which will close a buyer without need for an additional push. If you have earned this kind of enthusiasm, let it shine when dealing with a buyer. This is not to say you should stand on a soap box and "hawk" the virtues of your wares. (Many farmers markets actually prohibit hawking.) It means to simply let the belief in your product be communicated to the buyer.

Visit a farmers market and allow successful sellers to demonstrate this enthusiasm. Though the sellers may do little more than stand quietly behind their products, their enthusiasm will shine through with a slight smile, a twinkle in the eye and an eagerness to serve. Buyers know genuine enthusiasm and are closed by it.

CONFIDENCE: It has been said animal predators can actually smell fear. Buyers certainly can! In fact, a buyer who smells fear will become a price predator and a very tough customer to close.

Be confident. The best way to develop confidence is to have the right product and the right information. Avoid uncertainty by satisfying all of the requirements needed to close a sale. Know prices and post them on signs. Dust off scales and arrange counters to facilitate fast, friendly transactions. Have plenty of change on hand. It is easy to be confident when you are prepared.

FIG 10.26 Close with
enthusiasm, confidence,
courtesy and a smile.

FIG 10.26 Continued.

Use this confidence to take charge. This aggressiveness should not be confused with pushy, greedy behavior. It means find out what the buyer needs and then satisfy the needs. Keep going! Find ways to satisfy additional needs by making additional sales. ("Hi, Fred. How's the family? You might try some carrots with this lettuce. They were harvested this morning.")

Consider a youthful example of this take-charge confidence. My five-year old recently went on her first neighborhood commercial venture, which consisted of selling candy bars for a kindergarten fund raiser. Instead of asking prospective customers if they would like to buy a candy bar, she held up one in each hand and asked, "Would you like to buy one or two?"

If sincere, a confident attitude will readily increase your volume of sales; if false and abrasive, it will turn customers away.

COURTESY: A good closer practices elemental courtesies, such as keeping appointments, extending a helping hand to people in need, conducting smooth, businesslike transactions and using expressions like "please" and "thank you."

One local supermarket chain has become successful by going against the trend of low prices for cold impersonal service. The chain extends many different kinds of courtesies, such as a clerk to push each customer's grocery cart to their car. Though the chain charges more, its aisles are filled with happy shoppers. Be courteous. Serve each customer in order and make certain each is treated fairly. If you send customers home feeling as if they have received a fair deal, they will return to be closed again and again.

If you have developed a well-conceived and coordinated metrofarm strategy, produced high-quality products at fair prices and successfully marketed your products, then closing will be the most gratifying experience in the entire business cycle.

REVIEW

When demand is high and supply low, selling is fun and profitable. When demand is low and supply is high, selling is tedious and often without reward. When demand and supply have achieved an equilibrium, as they will tend to do, selling is a competitive business.

There are three ways to succeed when selling to a competitive marketplace. They are pricing, marketing and closing. The effective use of each requires a diligent, focused effort. If you become proficient with the skills, you will gain a significant advantage in the competition for consumer dollars and, as Lincoln envisioned, enjoy "the making of a comfortable living from a small piece of land."

CHAPTER TEN EXERCISES

i   Start a new subsection entitled "Pricing," and place it in the "Sales" section of your metrofarm workbook. Survey your market and identify three ways in which price changes are used to control the volume of sales.

ii   Survey the market and determine which pricing strategy would be best for selling your crop products: low-spectrum pricing, mid-spectrum pricing or high-spectrum pricing.

iii Start a new subsection entitled "Marketing." Give a one sentence definition for the *right* message, the *right* target, and the *right* amount of force. Next, survey the advertising mediums in your market and determine which can deliver the largest number of prospective customers for the least cost. Sketch an advertisement or write a commercial for this medium and include this work in your "Marketing" section.

iv   Survey publicity mediums in your area. List ways other metrofarmers have used the mediums to generate "free" publicity. Write a press release for a grand opening of your metrofarm.

v   Visit a farmers market or other direct market in your area and take a job selling products for several days or weeks. Use this opportunity to practice closing customers in ways which will have them returning to be closed again and again.

PART **IV**

CONVERSATIONS

INTRODUCTION

Y ou must prove-up. At some point in time, you must prove your profits justify your investments of time, money and know-how. "Prove-up" is an expression born with the Homestead Act of 1862. Under this legislation an individual could become the owner of 160 acres of land by living on the land for five years, improving it through cultivation or construction and registering the claim with the government. When the requirements had been satisfied— or proved-up— the individual received fee-simple title to the land. Though homesteading has gone the way of the buffalo, proving-up is still the law of the land for all business persons.

Part IV of *MetroFarm* contains conversations with five individuals who have proved-up. The individuals prove you can succeed at metrofarming whether you are male or female, young or old, married or single. You can lease land or own it outright. You can farm full-time or part-time. You can succeed in the benevolent climates of California or in the more harsh environs of Montana.

The individuals featured in Conversations prove it is possible to compete with some of the largest corporations in the world and win the consumers' dollar. Most importantly, the individuals featured in the following pages tell about the personal investments you must make to prove-up.

Gerd Schneider Nursery

Gerd Schneider
Aptos, California

I discovered Gerd while flying a small airplane over California's Central Coast. From 2,500 feet, Gerd's metrofarm looked like one of the microprocessors manufactured in nearby Silicon Valley— a small chip of intense activity plugged into the south-facing slope of the Pacific Coast Range.

Gerd is a propagation specialist. He "starts" ornamental plants for commercial nurseries throughout California and "nurses" some 200,000 of them on an acre of leased land. There is nothing easygoing about Gerd. He is likely to generate a thousand cuttings while discussing the weather and then, when you pause to enjoy the weather for a moment, he might load up his truck with rooted starts and head out on a delivery to Southern California.

Gerd does much of the nursery work himself and is therefore something of an enigma to larger growers in the area. One professional, the manager of a large local nursery, remarked, "I don't understand Gerd. He could have been very big by now if he wanted. All he would have to do is hire other people to do the menial work. Instead he does everything himself. Maybe he just likes the independence."

Gerd is also a well-respected professional— the growers' grower. His knowledge of plants makes him a valuable asset to professional growers in the

Pajaro Valley and throughout California. "That Gerd, he's a gem!" was a common observation from area professionals.

Like many independents, Gerd was reluctant to discuss his operation on tape. If I called on Monday, Gerd would say, in his crisp German accent, "We are absolutely hectic. It will be like this for weeks. Call later." But after persistent letters and calls, Gerd consented to sit down for a conversation. "Be reasonable," he said, "no glamorous stuff."

It is difficult, however, not to express a little admiration. Gerd earns a substantial income on his small parcel of leased land. He has developed some unique ornamentals, like the pink-flowering strawberry tree, which have the possibility of becoming genuine best sellers. And, shortly after this interview had been completed, Gerd took a 50 percent interest in a huge 16-acre nursery operation and paid for it with his technical know-how and industry-wide goodwill.

You are part of an industry which produces and markets ornamental plants. What part do you play in this industry?

My lines are clearly defined. I produce the basic product of young ornamental starts. These starts then go to commercial growers who transplant them into larger containers and sell them to the ultimate consumer, which is either a landscape contractor, a retail nursery or the state. Occasionally I will cut across this chain of command and do custom work for the state.

Would you explain the nature of this custom work?

For example, I grow some plants here for the Candlestick State Park up in San Francisco. One time the state came to me and asked for some Abrams Cypress. The Abrams is a native California tree which is not grown commercially; consequently nobody knew about it except some clever landscape architect. I had to go through all the trouble of getting permits from one agency to collect seeds in a restricted area for another agency. That is the kind of custom work I do for the state.

How did you get your start in horticulture?

It was before I was even in school. I was four years old and living with my grandparents in the Saar region of Germany. They loved to garden. They had a garden of about one-fourth acre; it was their great thing. They subscribed to a gardening magazine. I think it was the only publication they ever received. They had stacks of them lying around the house. As I was a little boy, I could not read but I felt an atmosphere coming out of those magazines. I used to go outside into

FIG IV-1.2 "People always think that they need to get bigger, but that is not necessarily so. My work area is 140' by 40'; it will hold a half-million cuttings at one time."

the woods, dig up little seedlings and bundle them together. I would play at marketing; if you had a bundle, you were an operator.

Later, when I was junior-high age, my uncle asked me to serve as an apprentice in his nursery. Back then we didn't have fancy aptitude tests or anything like that. So when I took stock in myself, I found that I understood what soils do and nutrient elements in water, so I accepted the apprenticeship.

I started out growing vegetables and cut flowers. Then I went into nursery tree production. By then I had grown extremely fond of this subject matter and I read. Then I went to Switzerland, Holland and Denmark to see what all those people were doing. Then came the "Big Deal."

I became acquainted with a student from the University of California at Davis. That person got me so excited that I just had to come to work in the United States. Davis was 5,000 miles away from home and things were entirely different.

I wanted to stay at Davis for a year as an observer out in the field. I met many people I would not have met working in any other place. I shared my knowledge with them and learned more than I could by working in any nursery. I worked there five years and then moved here to work in propagation, which is my specialty.

How do you propagate plants?

We do some propagation by germinating seeds but that is only a small percentage of our work. In seed propagation, we do not create a genetically uniform population of plants. This may be insignificant if you are propagating pines and eucalyptus because these species don't have much of a variation in individuals. Most ornamentals, on the other hand, have distinct features that we want to duplicate. To accomplish this, we have to clone the plant through vegetative propagation. We have 40,000 cuttings on this table. These cuttings can be traced back sixteen years to a single parent plant. From that parent plant we took five cuttings. A couple of years later, we had fifty plants to take cuttings from. This is the principle of vegetative propagation.

Before we go further into vegetative propagation, I would like to ask a seed question. What is the best kind of seed to use when starting a crop?

The quality of seeds that you buy off the shelf is not really the best. Originator seeds are the best. If you were to buy a package of originator carrot seeds, it would cost you $3 a pack, compared with the 25¢ a package that you would pay for KMART seeds. Cheap seeds are harvested from mass-produced varietals and not from selected strains. There is a lot of variance in a crop of plants growing in the field. If you were real fussy, you would go out and mark one plant out of 200 and use those seeds for your crop. You can't mass produce seeds like that and there's the difference in price.

How many methods of vegetative propagation are there?

There are about ten grafting methods and about eight non-grafting ones. I would say that at least 80 percent of vegetative propagation is done by making cuttings. Cutting and grafting are the important techniques; the rest are fringe methods.

When is the best time to take cuttings from a plant?

You want to cut the soft new growth. You want the leaves to be soft, the stem pliable and the internodes relatively long. You want to cut the plant when it is in its vegetative growth— when it's concerned only about growing. Then it will have a high rate of cell division and it will be producing a lot of auxins. The trick is to get the plant before it goes into its reproductive phase. When the plant has hard stems and leaves and is already producing flower bud sets it is too late for good rooting. This plant is in its golden age and is not thinking about growing anymore.

Would you explain grafting?

Grafting is simply the uniting of a scion, which is the part of the plant above the ground, with a root stock. One has to understand these two plant parts in order to understand grafting because they each perform independent functions. After the union, the root stock will constitute the root part of the plant and the scion the above ground part. Grafting is a clonal procedure. We do not just graft random populations of plants; there is no reason to do that. By grafting a scion to a root stock, we are reproducing outstanding individuals.

Why make grafts when you can take cuttings?

There are two reasons. The primary reason is that some plants have a resistance to rooting, so the only way to reproduce them vegetatively is by grafting. The second reason is to duplicate a function of the root stock. For example, if we wish to maintain the dwarf growth characteristic of a certain plant, the only safe way to accomplish this is to have a clonal root stock with the proven dwarf characteristic. Whatever scion you graft on top of this root stock will exhibit dwarf characteristics. There are dwarf lemons, oranges and grapefruits. They could all be cloned from cuttings but they would grow up to be rather tall. But if you graft a scion from one of these trees onto a clonal dwarf root stock, you will have a dwarf tree. Another example of duplicating a root function is to reproduce a resistance to disease. This is the common practice for grapes. All the commercial vineyards are grafted onto disease-resistant root stock.

Your business card mentions you sell rooted cuttings, liners and grafts. What are liners?

In our industry, a liner is a small plant in a container. Normally it refers to starts that

FIG IV-1.3 "They are good men and I treat them well. It's very smart for me to treat them well instead of trying to be the bigshot."

have been rooted and transplanted into small pots that range in size from 1 1/2 to 4 inches.

What kind of soil medium do you use for your starts?

Peat moss and perlite. The mixture has no nutritional value; it's totally sterile. It does have a nice balance of air and moisture and that is important for rooting cuttings.

I noticed you have heated beds for your flats. Does heating the soil help the rooting process?

Yes. The combination of elevated soil temperature, air and moisture is the stimulus for maximum rooting activity. The night temperature should be 15 to 20 percent higher than day temperature. However, the temperature should never exceed 80 degrees. If we have 90 degree days, we do not heat the soil beds at night. In the winter we heat the beds almost every night but only rarely do we heat in the summer.

The effect of temperature can also be seen in the time it takes for cuttings to root. If we do not heat the beds this fall, when the greenhouse temperature reaches 65 degrees during the day, it will take the starts five months to root. But if we bring the soil temperature up to 80 degrees at night, then we can produce roots in sixty days.

Let's go to the business side for awhile. Did you have enough money to buy your land and build your greenhouses when you first started?

Impossible! What would I have left to eat? I worked for other people. Then at night and on weekends, I began scraping this thing together.

I knew that I was going to sell propagation products. By propagation products, I mean starts. So I built a little greenhouse; then at night and on weekends, I made cuttings. I didn't have any money so I needed to do something with a real fast turnover. With one sack of perlite, one bale of peat, a hundred flats and the small greenhouse, I could raise 20,000 cuttings. At that time it brought me $1,200 in ninety days. It was no fortune but you could buy the bale of peat for $3.50 and the perlite for two dollars. That is how I made my first cash flow.

The world is full of people in this kind of fix. When somebody asks me if they should invest their savings in a piece of property, I tell them that it is foolish to spend their savings this way. If you put all of your money in a down payment you will have to struggle like mad just to make your payments. It ties you down and paralyzes you. Lease the land and start producing and creating customers. Production and marketing make you strong and are, therefore, very important.

What about other ways of starting up an operation?

FIG IV-1.4 "Small operations are successful when they create their own little niche. To protect that niche, you have to be on top of the situation all the time— totally aware."

I have some friends who are a young married couple. They both went to school at Cal Poly, which is a pretty good school for horticulture. He works as a landscape contractor for forty hours a week and gets paid pretty well. She stays home and does the nursery work. When she gets the nursery to the point where it will break even, he will cut back on his hours. This is one way for a young married couple to get a start.

What about the people who have money to invest in starting-up an operation? What approach might they consider?

I have some other friends, a husband and wife over in the valley, who knew nothing about the nursery business. But they did own thirty acres of land and, of course, thirty acres is capital. The bankers just love to see them come in! He quit his high-paying job in the electronics industry and they hired a good man. They were sharp and after that man was there for three or four years, they knew the scoop. They paid for everything. No question about it. They are my best customers now. They are doing really good— $1.5 million.

Have you ever considered borrowing money from the Federal Land Bank program to expand your business?

No. The Federal Land Bank only lends money to farmers and nursery people. They do not lend to anybody else. I am on the pay-as-you-go program.

What are the most important factors to consider when establishing a horticultural business like yours?

It takes time, money and know-how. If these three ingredients are put together properly, they will at least create the potential for a successful operation.

From my point of view, money is the most difficult ingredient to come by but it can be compensated for by time. Know-how is the most important ingredient because it cannot be compensated for. Well, if you have enough money, you can even do that! But my situation is typical in that time and know-how compensated for and finally overcame the lack of money.

Would you elaborate on this for us? How can time and know-how compensate for lack of money?

It's quite simple. If you want to produce $100,000 worth of plants year after year, you will need, for example, $35,000 worth of greenhouses. Now if you have $35,000, you can put up those greenhouses in three weeks and immediately start to produce. But if you don't have that kind of money, you have to build a small greenhouse, realize a $5,000 profit, then use all of that net profit to build an additional greenhouse. That is what I mean by time and know-how compensating for the lack of money.

This is where know-how becomes so important. My background and knowledge give me the strength not to panic when things get rough. You have to know that you still have your resources, even though it doesn't look good. If you know that you still have divisions lined up you will not capitulate. Do you know what I am saying, in military terms? Even if it looks gloomy and grim you will not give up because you can evaluate your own potential. This is why you can go through a long period of time and compensate for the lack of money.

What was the land like when you first started operating here?

It was terrible, un-

FIG IV-1.5 "The combination of elevated soil temperature, air and moisture is the stimulus for maximum rooting activity."

believable! First of all, it was much steeper than it is today. I did a lot of grading work. Then there were the weeds! Thirty years ago, somebody figured that the only thing this land was good for was raising pigs. Pigs like Johnson grass roots, so they planted the whole area in Johnson grass. Its the damndest weed you can ever imagine!

What criteria did you use in selecting this land?

If you are trained in pomology (the culture of fruit trees) you learn to look at the terrain. It just automatically clicks. In fact, if you are studying at a school, they will give you an exam with different kinds of terrain on it and ask you where to put the orchards. The biggest mistake is to put the orchards down on the flat land because that is where the cold air stays. You have to locate on a hill or on the half-hill. That is exactly where I am, on the half-hill. In November the temperature difference between here and a mile down into the valley will be ten degrees. Cold air behaves like water; it flows down hill.

I noticed most of your land is not being used. How much space would you need for your operation?

People always think that they need to get bigger but that is not necessarily so. My work area is 140' by 40'; it will hold a half-million cuttings at one time. Now if I had absolutely nothing else to worry about, I could, theoretically, turn these half-million cuttings four times a year. That is two million cuttings. You could have a nice gross off this work area if everything went right. I have never reached this level of efficiency.

Which plant did you select for your first commercial crop?

I started with junipers because they were hot and I could sell large quantities. The way they are propagated is simple: you take cuttings of little branches, apply some rooting compound, plant them in flats with the right media, then put them into the greenhouse to root. Commercial growers liked this service because they did not want to fool with it themselves. I sold them for a nickel each when I first started; today I get a dime.

What are the most popular varieties of ornamental plants?

As with everything else, popularity changes all the time. A small operator like myself has to be flexible and listen closely to what's happening in the market. We do have a core of about fifteen or twenty reliable varieties that always seem to be popular. Among this group are the Star Jasmine, Indian Hawthorne, Photinia fraseria, Diosma pulchra and Nandina domestica.

FIG IV-1.6 "Most ornamentals have distinct features that we want to duplicate. To accomplish this we have to clone the plant through vegetative propagation."

Did you have any trouble making connections with the nursery industry when you first began marketing your starts?

I had been in this industry for a number of years and everybody knew me. They knew me because I was a foreigner and well trained— and not just trained to hold an office job, either. I had academic credentials and real practical know-how. I could understand what the boys from the university were talking about and I could plant plants faster than any farmworker. This combination made me a hot number. I could go into somebody's nursery and they would talk to me. I didn't have to talk to any manager; the boss would come out and take me home for coffee. When you go right to the guy who signs the check, you've got a big advantage. I have let people down many more times by not producing all of what they wanted rather than the other way around. So my marketing is not a problem most of the time. I can always sell, even if it is bad. This is because there are very few people doing what I am doing.

Why is your service so valuable to commercial growers?

There is always a good demand from growers because their money can't buy them grafting know-how. Grafted plants are awfully scarce because grafting is a diminishing art. In the past thirty years there have been significant advancements in propagation technique. Where you used to graft certain plants to reproduce them vegetatively, you can now propagate them by making cuttings. But there are still some plants that have to be propagated by grafting and that is where I come in.

If grafting is a diminishing art, what about craftsmanship in general? Are skilled hands becoming a thing of the past on the American farm?

Well, yes. To understand this loss of craftsmanship, you have to consider the separation in the working force of American agriculture. The American wants to be the manager; he wants to drive around in a pickup and herd cheap labor— like Mexican nationals— into doing the work. He doesn't want to get down and do the work himself. Isn't that correct? Those farmworkers work hard and do a good job but they don't have any goal-directed ambitions. When I became an apprentice nurseryman in Germany, I knew that I would have to work real hard for a whole year before I would be shown how to graft. I anticipated that event and it became a great moment in my life. Here, you have managers who want to delegate but don't want to get down and do the work and you have cheap labor that works hard but doesn't really care. This may be very well but somewhere in this system you lose the craftsmanship.

I noticed you employ several farmworkers. Do you have any trouble finding good help?

I have two men that have been with me for seven years. They are good men and I treat them well. It's very smart for me to treat them well instead of trying to be the bigshot. We are small and we are all interested in what we are doing here. I don't have much of an employee turnover at all. I guess I am lucky.

Why do you make deliveries yourself? Couldn't you improve your productivity by hiring somebody else to do this work?

I have a personal relationship with every person I sell to. When I show up, we can talk. He can ask me things— pump me for information. I can tell them what the other guys are buying. When you have ten good customers, you have to stay in touch all the time. Small operations are successful when they create their own little niche. To protect that niche, you have to be on top of the situation all the time— totally aware.

How do you determine which products will be hot?

What we grow from year to year is determined by the commercial grower. The grower is the speculator. There are two markets for the grower: He can sell direct to retail nurseries or he can sell to commercial landscapers. Each of these markets has a different character. Landscape contractors buy large quantities at one time and are not as quality conscious as retail nurseries. It's a highly competitive business. Retail nurseries, on the other hand, buy smaller quantities and, since they have to sell to a lot of customers, are more quality conscious. Where one of these markets might have a good year, the other may not. For example, there are not many new houses being sold right now, so the landscaping business is not very good. But if retail nurseries have the right plants at the right price, they can do pretty good.

Do you try to stimulate demand for your new products in the retail market?

Yes. We've got to encourage the retail nursery to be persuasive and direct the consumer's attention toward some of the new or improved plants that we are working with. We always have new things coming on line; it is a mild form of having a new model car each year.

In your job as propagator, you have the opportunity to "hit it big" by developing and marketing new varieties of ornamental plants. Would you explain how this works?

There is a technical aspect to your question and an economical one. Let's look at the economical one. I would say that out of twenty-five new plants being made available to the public, two will make it big. It depends, of course, on the plant's market appeal, which is very hard to determine in advance. The guy that brought

FIG IV-1.7 "We do have a core of about fifteen or twenty reliable varieties that always seem to be popular."

out the Edsel thought he had a hot thing on his hands but he didn't. We don't know in advance either. The biggest thing to hit the market in the last five years is the clonal redwood. The nursery industry here in California had raised redwood trees from seed for a hundred years. The trees always turned out to be scraggly looking specimens. Then we started to pick certain trees and clone them. All of a sudden, we could produce beautiful looking redwood trees. These clonal redwoods are now in the best-seller class. They made it because they look good to the final consumer. The market will never become saturated because clonal propagation of redwoods is not easy. Now roses are another story. They are like Toyotas— another year, another five models. There is a saturation point where the consumer becomes numb to new varieties.

How do you go for a best seller? Do you anticipate future demand, then build your stock to meet the demand?

Yes and it is a very costly process. Take my pink flowering strawberry tree for an example. Exciting! I have no idea at all if it will take off but the indications are that it will. It is the first real clonal strawberry tree that has pink flowers. I propagated it from a sport I found on a regular, white-flowering plant. It's a gamble! I have been working with this plant for six or seven years already and it's still going to take a lot of pushing to get it to take off. My neighbor gambled with the clonal redwoods and they took off. He did not patent it, either. His philosophy is that if he can provide a better plant for the public, it is reward enough for him. He does not get any royalties and 80 percent of the people that grow his tree do not even know that he was the originator and selector!

Is it possible to protect a new variety from competition? Is it possible to patent the variety or register the tradename?

You can have your plant patented and have your name in the Patent Registry; then you can hang everything in the living room. But everybody is going to grow the damn thing anyway because there is simply no gumption for policing that kind of thing.

You mean everybody makes their own cuttings, regardless of the documents the originator might possess?

Yes, but not everybody. Look, if my red strawberry tree takes off it will take the next guy at least five years to propagate a market-sized supply because it is not an easy plant to propagate. If a plant is really good and hard to propagate it will push itself. I can grow all I want regardless of the patents because the demand will be there. Do you see what I mean? Mums are a different story because they root in four days. So people register the new mums and get their name on the tag and sell the tag for a little money and prestige. Roses, on the other hand, are sold by the millions, so you can make some money on the tags. Dr. Lambert, who created the Chrysler Imperial, Charlotte Armstrong and Queen Elizabeth, lives right across the hill. He still gets royalties of a couple of cents or so but that is on sales that run into the millions!

Would you care to comment on the substance-value of ornamentals?

If you looked around and found that all of the vegetation had disappeared, it would be hell. In some of the subdivisions in East San Jose there was not enough money for vegetation. The only yard some of those people have is a garbage can and five cars. It's ugly and the people will eventually turn ugly. They will respond to television and beer and driving in their car. But if you are surrounded by the beauty of plants it has an affect on you. It really does! Plants have a way of making our lives a little more human. Just think of this place as a barren desert. It would be terrible.

If you could splice some genes together and develop a super-plant, what would you make?

I guess I would try to develop a plant that would grow in the Sahara and be rich in protein. Seriously, drought, insect and smog-resistant plants would be good directions. But when it comes to ornamental plants, I am very much against creating a couple of super-doopers that would limit the development of a broad variety of plants. That quiet, beautiful plant can be just as important as any super dooper when it comes to ornamentals.

Do you have any regrets about choosing this occupation to make your living?

I could have become a farm adviser or something along that line. Had I worked for somebody else, it would have been much easier and I might have made more money but it wouldn't have satisfied me. I like this hair-raising life where it is hot most of the time. The fact that I am in total control gives me a great deal of satisfaction. This is the way I am. I cannot overemphasize it; I would never want to change or trade. You couldn't give me a $100,000 a year job anywhere else and I mean this. Where else could I get this feeling of fulfillment?

How much income do you generate here at Gerd Schneider Nursery?

That's a hot potato! Nobody wants to disclose that kind of information. So let's talk about gross, rather than net; and let's speak in terms of what my goals are, rather than specific dollar amounts. My goals for the near future are for a gross of $125,000 a year and for the distant future for a gross of $250,000 a year. Now, you must remember that it has been a very bad year for the industry, so I fell short of my goal. However, next year looks much better.

Do you have any comments for people who would like to develop metrofarms of their own?

I'll tell you one thing in all honesty. It can be done if you are willing to pay the price. Let me emphasize that you have to take your own person- ality into consider- ation. If you can handle all the pain— if you can live mad all the

FIG IV-1.8 "I like this hair-raising life where it is hot most of the time. The fact that I am in total control gives me a great deal of satisfaction."

time— it can be done.

There must be a lot of magic involved in propagating 200,000 plants in this small greenhouse. Do you have a green thumb?

That is just a saying. I don't think I have a green thumb. I think it is just baloney! When I was fifteen, I had to make a decision. I just took stock and found that I knew what soils do and nutrient elements in water and I knew how to put them together. I had a feeling for that kind of thing. If you want to call that feeling a "green thumb," I guess it might be okay.

The Plant Carrousel

Jan LaJoie and Family
Medford, Oregon

I traveled to the Plant Carrousel on the recommendation of an associate, Nancy Tappan. "It's a small nursery headed by an energetic lady named Jan LaJoie," said Nancy. "This lady is really something. You should definitely pay her a visit."

My initial reaction, upon pulling into the driveway, was to slip the van into reverse and back away. The small greenhouse beside the driveway was very small, indeed! But a first impression is not always the best impression, so I slipped the van into park and climbed out for a conversation.

The Plant Carrousel is a part-time nursery business operating from the home of the Jan LaJoie family in suburban Medford, Oregon. I toured the Carrousel with husband Jack, who is a full-time employee of the Forest Service, daughters Jamie and Janelle and toddler son Jay while waiting for Jan to return from her full-time job at a large local nursery.

The little greenhouse was lush and tropical on the cold February afternoon. Healthy green foliage filled the small space to capacity and provided a pleasant contrast to the stark leafless trees outside.

When Jan returned from work we adjourned to the house for tea and conversation. Though the questions were directed to Jan, the entire family sat around in support. Without the slightest reluctance Jan put everything on the table, including her aspirations, dreams and gross income. I soon forget about my misgivings because this woman and her family were definitely doing something special!

Soon after our conversation, Jan quit her full-time job at the local nursery to go all-out with the Plant Carrousel. She increased greenhouse square footage from 800 to 4,500 square feet, expanded her line of houseplants and doubled the number of product sizes available to her customers. She also developed a line of geraniums with which she hoped to increase her off-season business.

Income also grew. The Plant Carrousel earned $25,000 in gross income during its first full year and close to $100,000 during its second year. Then Jan expanded production from 4,500 square feet to 150,000 square feet!

The real bottom line for the Plant Carrousel, however, is evident in the entrepreneurial spirit of daughters Jamie and Janelle. During summer vacation the sisters rented a neglected two-acre strawberry patch from an absentee landlord, hired labor to clean things up and then advertised U-Pick strawberries. The sisters brought in $3,000 on their investment of $1,500, and promptly invested the earnings into building bedrooms of their own.

What kind of business is the Plant Carrousel?

A wholesale nursery. We deal in foliage plants, which is just another way of saying houseplants.

Who works at the Plant Carrousel?

The LaJoie family: Jack, Jan, Jamie, Janelle and Jay. We also have one other employee who comes in two days a week to make our deliveries.

How much land do you have for the Plant Carrousel?

We are a little shy of three-quarters of an acre, which is not a lot but a good commercial greenhouse operation should be able to gross around $100,000 a year per acre. So we have enough land for a while.

When did you start this business?

We broke ground for the first greenhouse in May, which was about eighteen months ago.

How did you learn the wholesale nursery business?

I worked about a year and a half for a man who raises miniature roses and miniature geraniums. He taught me how to propagate those two kinds of plants. Roses are more of a woody, hard stemmed plant than most of the foliage we deal in. I worked eighteen months for him and pretty much operated his greenhouse. When he first started it was a wholesale operation; he went wholesale and retail the last year I was there.

Did you have any previous experience propagating plants? Any formal schooling or class work?

I don't think I had ever been in a greenhouse until I went to work for Small World Miniature Roses.

The owner taught me a lot about propagation, which consists of taking cuttings and planting them. When our speed was up, someone would bring the cuttings to me, empty the trays and take the full ones away. I could propagate about 1,000 cuttings an hour. I still have the calluses!

Did you start the Plant Carrousel when you left the Small World nursery?

When I left there, my husband and I decided to do something on our own. We talked about several different things before trying to buy a retail nursery here in town. We thought we made a reasonable offer but the owners turned us down.

We had been thinking about going into plant rental, which is big in California. We did a lot of research. We figured how many potential accounts there were in town; how many restaurants, doctors' offices and other businesses that would possibly rent plants. At that time there were four other people in the business, so we divided the potential accounts by five. If we were good at it, we would have more but you can't count on that when you are first starting. We would have had $7,000 in inventory before we had any income. It would have taken seven to ten months to pay for each plant. We would have to account for our people and our time.

To be profitable, we would need a greenhouse for servicing plants. And if we were going to have one, it would have to pay for itself, not just be a holding area for plants going out on rental. That's how we decided to grow something.

We did a little talking around the Valley and friends suggested that we raise foliage plants. We did a market survey to see how many two and one-quarter inch pots we could sell in a week. We came up with about 2,000. So to be on the safe side, economically speaking, we built the greenhouse so we could grow 1,000.

What led you to do a market survey?

I had gone to a SBA (Small Business Administration) course and picked up

FIG IV-2.2 "We are a little shy of three-quarters of an acre, which is not a lot, but a good commercial greenhouse operation should be able to gross around $100,000 a year per acre."

information on marketing. Then we talked to some people who owned a wholesale florist business here. They said the plants they had the hardest time getting shipped in were two and one-quarter inch stock— it arrived upside down or rolled over. It just didn't ship well.

Where did this two and one-quarter inch stock come from?

Most of it came from the Half Moon Bay area in California. Today, we are the only wholesale nursery that grows foliage south of Cottage Grove (Oregon) and north of San Francisco. We found we could have a corner on the local market. We saw a need that we could fill. This is part of the reason we have been successful.

How did you do a market survey?

By phone and in person. We called every grocery store and chain store in the area. We even sampled a few "Mom and Pop" stores to see if they were interested. They were.

What kind of questions did you ask in your survey?

We asked people whether or not they sold foliage in their store and, if they did, approximately how many plants in each size they sold and what their cost was. I was surprised. We didn't find anyone who wouldn't tell us. This gave us an idea on how to figure our costs and what to charge.

How did you finance your start-up costs?

We had some cash. We had $5,000 from an insurance policy and that's what we used for our initial investment. We have been operating for 18 months and don't owe anyone anything.

Did you have credit established?

Initially, we tried going strictly cash. But when we reached the point where we were buying more than usual, such as for Mother's Day, we had to use credit. We pay all our accounts in 30 days. Our cash flow is quicker than that of most businesses. Our terms are net 10 days on all our accounts. We're usually paid within 15 to 20 days after delivery.

Where did you get the "Mother" plants for starting your first crop?

We purchased them from a couple of wholesale nurseries. We got them in six and eight inch sizes. We started with about 25 and we still have some in the greenhouse.

How did you determine which plants to propagate?

We knew we should be dealing in what's called "soft goods"—

FIG IV-2.3 "We deal in foliage plants, which is just another way of saying houseplants."

FIG IV-2.4 "We wanted plants that would root and grow quickly. We needed them to be ready in six weeks..."

soft plants like Wandering Jews, Charlies, Coleus and Piggybacks. "Soft" describes the stem. We wanted plants that would root and grow quickly. We needed them to be ready in six weeks from the time we made the cuts until we delivered them to the stores and they had to look good in two and one-quarter inch pots. We do have some harder stemmed varieties now— Golden Pothos and Nephythytis.

How did you learn which plants would be the most profitable to propagate?

Mostly by trial and error. For example, we did Wandering Jews in two and one-quarter inch pots and after they had been in the store for two weeks, they wandered all over the place. They were long and leggy; no one would buy them. Now we try to stick with things that are compact like Peperonias. Soft plants are the ones that take a beating in shipment so ours look better than the ones that come in on trucks.

How do you propagate your plants?

We make stem cuttings, dip them in a commercial rooting compound and then root them under a misting bench. Through trial and error we learned to dilute our rooting hormone differently for the hard and soft varieties. For the soft ones we dilute one to twenty and, for the hard, one to ten.

Does propagation require any special greenhouse equipment?

Yes. Part of our success is due to the type of misting system we have. It's automatic and works on evaporation. It has a screen and the water evaporates off the screen at the same rate it evaporates off the leaves. When the screen is dry, the water comes on until it is wet again. In the winter that may be three times a day; in summer 30 times a day. The system keeps the leaves wet so they don't dry out, thereby allowing the plant to root. It worked well, even when it got up to 114 degrees last summer.

What are some of the problems you face in getting plants to grow in a greenhouse environment?

Not having more than 24 hours in a day! Seriously, we've had some water problems. Our water is very hard and was making the leaves leathery and leaving a white deposit on them, although the plants weren't hurt. When they came out from underneath the misting system, the new growth would be bright green and healthy but the older growth would be yucky. We've also had a problem with transplants wilting and are trying a new product, Cloud Cover, that is designed to prevent this.

We don't know yet how it will work because we just started experimenting.

What are some of the expenses you have in operating greenhouses?

We mix our own soil, which is slave labor. Our soil has no dirt in it. It's a soilless mixture composed of peat moss, redwood compost and perlite. We have to buy pots, fertilizer, electricity and labor. We do pay our children for their work in the greenhouses. There is transportation and heat, which comes from propane.

What are the heating requirements for an Oregon greenhouse?

We have a six month long heating season. The fog comes into the Valley and doesn't leave for six

FIG IV-2.5 "When our speed was up, someone would bring the cuttings to me, empty the trays and take the full ones away. I could propagate about 1,000 cuttings an hour!"

months, so the outside temperature stays around 30 degrees for the entire season. We use an infrared heating system that is powered by propane. Infrared heat is radiant heat that warms plants much like sunshine does. It's a relatively new way of heating greenhouses and we've found it to be very successful for our foliage but you can't use it on all of the plants.

Are you experimenting with any ways of reducing your need for additional heat?

We are trying some passive solar techniques such as the barrels of water you saw in the first greenhouse. Even though we double-walled the second greenhouse for additional insulation, it still costs more to heat than the first one.

Are you able to compete with the large commercial greenhouses in California?

We are very competitive in the two and one-quarter, three and four inch market but we do have trouble meeting the six inch prices because we don't have the room to grow the quantity we need to give a price break. Our light levels are lower than those in California, which means higher energy costs. Because of the light levels there, six inch plants can be grown and shipped cheaper than we can grow them here.

Do the additional costs make it hard for you to compete?

Although our prices are a little higher than our closest competitor's, it doesn't seem to bother our customers. I proved to them that they could sell the plants. One of the men was very skeptical. So I said, "Okay, I don't do this very often but you take these six plants and hang them up. If they are not sold by the end of the week, I'll take them back at no charge." He called me midweek and needed more plants. The price difference between what he paid our competitor and what he paid us was one dollar— a big difference. If he pays three dollars per plant, he can sell it for $4.99; but if he pays four dollars, he has to sell it for $5.99. But the plants sold and that's because of their quality.

Are there any other differences between you and the competition?

I think another reason we have been successful is our service to the stores. We offer the personal touch. Most of the customers know who I am when I call. I don't have to say, "This is Jan from Plant Carrousel." Sometimes I don't even have to say, "This is Jan." They know who we are and that makes a difference. Most of the big places don't care; they are so big they have trouble caring— which is understandable. We have had two or three of the produce men we work with out here to look around. Now they know where the plants come from and this makes them feel comfortable and important.

How did you start marketing your plants?

The first six weeks were probably the worst for the Plant Carrousel. We knew we had a product that would sell; we knew we had people that would buy from us. Medford is a very hometown community and the people here prefer to buy from locals rather than from somebody somewhere else. We knew we had a market but we didn't have a product yet. Until you have something to show people, they're not going to tell you they'll buy it. We propagated 6,000 plants— 1,000 a week for six weeks— and then had two more weeks until the first crop was ready.

Did you have any disasters in the greenhouse during the first six weeks?

No, we were fortunate. We've had very few crops that failed; out of our first crop of 1,000, we lost only 10 plants.

What was your first sale like?

Our first account was with the local Safeway. I went in with a shoe box filled with twelve cute little plants. It was like taking candy from a baby! It was so easy to sell this stuff, I couldn't believe it. It was just great! I remember coming out of the store because I was about three feet off the ground. Jack was waiting for me in the parking lot and I came floating out and said, "We did it!"

We service three Safeway stores and they wanted us to handle all eleven stores in the area. But they are spread out from Klamath Falls (OR) to Yreka (CA)— a long distance. It would mean expansion. We've picked up so many other accounts that we no longer have the room to grow the quantity that they would need. It would take everything we grow right now to fill those eleven stores on a weekly basis.

Where did you go after the first sale?

From Safeway we went to Shop-Rite, which has only two stores in the Valley. They weren't carrying plants at all, so Jack designed and built plant racks for them.

Next we picked up the Richard's accounts— nine stores. That was a big step for us because it meant buying new stuff. We were no longer working with just two and one-quarter inch pots but also with three and six inch pots.

The Richard's chain did a Hawaiian luau promotion. They wanted large plants and I was able to find them at a good price, so we brought them in. The nine stores in this area bought all their plants from us. This is the only time we've had to borrow money.

We called an uncle and said, "Help!" We had to because we didn't have an account with that particular grower and we had to pay for the plants as they came in. But we had a fast turn around. The stores paid in ten to fifteen days and we were able to send the money back to uncle quickly. That was our big week. We did $2,500

FIG IV-2.6 "We knew we had a product that would sell; we knew that people would buy from us. Medford
is a very hometown community..."

in one week— it was mind boggling!

The Richard's promotion involved wholesaling other grower's plants? Do you still work with other growers' plants?

Yes. We buy from a company outside of Brookings and seven cases is their minimum order. When you have a plant rack in a small store and bring in seven cases of plants, you have a rack that is jammed full for two weeks. The next two weeks you'll have a bunch of dead stuff because people forget to take care of it. We've been able to service these small stores every two weeks. If they only need three hangers and two uprights, that's OK. We're out there pushing hard so we can afford to deliver a small quantity. I have no minimum order for my stores. We just haven't reached that point yet. I figure that if they know they can buy $12 worth from me this week, they'll buy the rest from me later.

How do you service your accounts?

When we first started, we delivered the plants to the back door— just opened the grocery door and slid them in. Then we found that some of our accounts were having trouble. I went into the Shop-Rite one day and saw that our plant rack looked horrible. The owner's wife happened by and said, "Something should be done about this. I was going to buy a plant for a friend but went out and bought a book instead."

Jack and I talked it over and decided to approach them on a plant rack service program. We go in once a week and groom, clean and supply the plants on the rack. We charge them a flat fee for this service. Their sales increased dramatically and their losses went down to almost nothing. In some of the stores that we work with, our plant rack is a stepchild for the produce men; consequently, the plants do not get the attention they need and go downhill rapidly. Our rack service program was a spin-off to help the stores reduce their losses. But with most of the stores, I just call them every Tuesday morning and take their orders over the phone. Then we drop the delivery by the back door.

Do you have any trouble collecting accounts receivable?

No. So far, we've been very, very lucky. Most of our accounts pay regularly. I used to bill every week but that became too much work, so now I bill every two weeks. Since we have been in business, we have had only two invoices that didn't get paid.

What are some of the important things to keep track of in marketing your plants?

You have to be aware of the next market and what

FIG IV-2.7 "You have to be out there well in advance lining up crops and customers. If you don't large commercial growers will come in and sell the chains for the entire season."

happens at different times. For example, holidays are a big flower time. You have to be out there well in advance lining up crops and customers. If you don't, large commercial growers come in and sell the chains for the entire season. If the customers want bloomers, they will buy them from someone else if you can't supply them. We buy those plants that will keep the customers satisfied.

When is your season for selling houseplants?

We're backward from everybody else in the nursery business. Houseplants become very popular in the fall, when people are moving back inside after being out on the patio all summer. Our best season is from September until May. Our last big whingding is Mother's Day. June, July and August are slow. The hotter it gets, the slower it gets; people just are not interested in houseplants then. When the kids go back to school and the moms and pops move back into the house, then business starts to pick up again.

Are you developing any summer specialties?

Not at this time. We are looking into expanding into geraniums. They are big around here but are more of a spring product than one for summer.

How many plants do you sell from your 800 square feet of greenhouse space?

I really don't have any idea. We programmed ourselves to do about $1,000 a month, figuring that amount would keep our head above water and pay for the business. We built our greenhouse in May and sold our first crop in July. At the end of December, when we did our taxes, we had done $5,000 worth of business in five months; in the next seven months, from December through July, we did $20,000. In a year's time we did $25,000, which was twice what we programmed ourselves to do. That's off 800 square feet of greenhouse.

Economically speaking, are you better off now than when you started?

I think our financial situation has improved considerably. If you were to look over our books, you would see that the Plant Carrousel has never paid wages other than to the woman we hired to deliver plants. The business pays her and our children but has never paid us any wages. It does pay for our telephone, utilities and vehicles—there's a lot of spin-offs that way. And we've seen the business grow. We keep pumping whatever we get back into the business and we are continually improving and changing.

Where do you get your business-related information?

We get one magazine that is pretty good. Its a new publication called *Greenhouse*

Manager. I like it because you can sit down, read an article and absorb it in twenty minutes— which is all we ever have— without getting lost in technicalities. The articles are written about people like us and they talk about how they work things out through trial and error. It's easy reading and informative.

Do you talk with other growers?

Not very much because there are not many around here. There is one who helped us when we first started. He owns a wholesale nursery up at Butte Falls. He grows outdoor stock now but did grow foliage at one time. We had not met him before so we called him up and told him what we wanted to do. He asked us to come up, so we did. He told us his whole story, which was extremely helpful and informative; he even gave us the grand tour of his operation. He told us what to grow, what not to grow and why.

Do you have aspirations to expand into something big?

We have been talking about growing the leather leaf fern. The leather leaf is the heavy-green, almost plastic-like fern used by florists in all of their arrangements. Almost all the leather leaf fern used in the United States is grown in Florida. The rest is grown by a place down in California. The people who own that business are retiring soon and we'd like to be in a position to pick up the business. It's a very easy crop to grow. Our basic requirements would be 200,000 square feet of greenhouse and $750,000 to get it off the ground. Given the nature of today's economy, we've decided to hold back on that one for the time being!

What kind of advice would you give a best friend who wants to start a metrofarm?

Start small and let it grow. The Plant Carrousel grew much faster than we thought it would. When we put up our first greenhouse that May, we thought it would be a year before we would build another but by Thanksgiving we were putting up another one because we were out of space. The demand for our product was so great!

I think you also need to be aware of your market. The main thing about metrofarming is that you have to fill a need. That is part of the reason for our success. It's difficult to compete against somebody who is already established. People were already buying and selling foliage before we started but they were bringing it in from California. If you can fill a need locally, which was what we did, you can be successful.

Another important thing is to stick with it. Listen. I had several produce guys say, "I'll buy them this time but this will never last. You'll never be back." But for 18 months now, we've been there every Tuesday morning on the telephone. When the store's phone rings at 8:30 a.m., they know its Jan from The Plant Carrousel.

FIG IV-2.8 "I went in with a shoe box filled with twelve cute little plants. It was like taking candy from a baby! It was so easy to sell this stuff, I couldn't believe it!"

Do you have a green thumb?

I don't know what a green thumb is and I don't think I have one, unless you see me just after fertilizing, in which case all of my fingers are turquoise. But I was never interested in houseplants. In all the years that we've been married, I've never had houseplants in any quantity in my home. I don't think it takes a green thumb to grow plants. I think it takes luck, knowledge and caring.

Heyn's Country Gardens

Charles and Lois Heyn
Billings, Montana

Word of "Country Gardens" came from my mother, Irene Olson, who lived in Billings, Montana. "They grow the best cantaloupe and strawberries I have ever tasted," she said. Taking her observation to heart, I decided to pay Country Gardens a visit.

Charles and Lois Heyn are truck farmers. This "truck" word comes from the French word "troquer," which means "to trade or barter." The Heyns trade in farm-fresh vegetables and small fruits. Their business, Heyn's Country Gardens, is located on a small parcel of prairie bench land in the Yellowstone River Valley of South Central Montana.

The Heyns are of the old school. A busy city-dweller making a casual purchase at their roadside stand might look around and think, "Hmmm, 'American Gothic.'" But I am familiar with the old school and so, on a cold January afternoon, I found myself sitting down to lunch and conversation with the Heyns.

It was the kind of meal I remember from my days as a Montana ranch hand. We had roast beef, mashed potatoes, vegetables, salads, homemade bread with preserves and strawberry shortcake with berries from Heyn's Country Gardens. Though the berries were frozen, then thawed for our meal, they were the best I have ever tasted and I live in Central California's strawberry-growing region. "They're a cross between one of those fancy California hybrids and a native berry," Charles said.

The Heyns' telephone rang repeatedly through our meal and ensuing conversation. Lois answered each call and then made notes in a ledger. Finally, I asked about all the phone calls. "Oh, we're just taking reservations for our cantaloupe and strawberries," she said.

"In January?!" I exclaimed.

The old-school conservatism was also evident in the appearance of the Heyns' farm. Indeed, the busy city dweller speeding down the highway might glance over the fence and see only a small triangle of land with some plain buildings tucked into a corner. But a very close look revealed seven acres of rich, well-textured earth, several tractors with implements, trucks— one old, one new— a pickup, an automobile and facilities for processing and marketing truck produce. It was definitely an income generating metrofarm. "How much income?" I asked.

"Well, you'll have to ask the boss," Charles quipped.

After several moments of quiet, Lois answered, "Enough!! "

"You have been farming for many years and are both healthy. Why don't you relax and enjoy the fruits of your labor? Why don't you sell your land, move into the city and collect Social Security?" I asked.

"Social Security is a tax we pay to help those that need it. We pay into it but we don't draw from it. Heyn's Country Gardens is our Social Security," Lois answered.

When did you start farming?

Charles: 1921. I was born in 1919 and my dad moved the family to the farm two years later. When you grow up on a farm, you start working as soon as you can.

How much land do you have under cultivation?

Charles: We've got about 32 acres here with about seven in produce. It's not that much land if you have to sell it on the open market.

Why did you select this particular piece of property to farm?

Charles: When we came out to look at it, Pa said it was a pretty good piece. We also asked all the neighbors around about it too. That's the best way to find out about land. Take somebody who really knows land and ask the neighbors. But catch

FIV IV-3.2 "First we have to find the best varieties. If we are always trying to find the best variety, our quality will improve."

several neighbors, not just one, because one or two may also have an interest in seeing it sold.

How did you finance this property when you first started farming?

Charles: I worked nights. Winters too. We worked and put money in the bank and then drew it out in the summer. I worked a lot when I should have been farming because my employers were short on help or something. I didn't seem to mind it at the time. Now I'd scream like a dying bird. I like my eight hours if I can't get ten.

Lois: We traded our house in town and got a farm loan for $2,000. The house in town paid for the rest of it.

Charles: We paid it off as fast as we could, which was kind of stupid because we borrowed money at four and one-half percent from the FHA. Federal Land Bank loans are guaranteed by the government but you borrow the money from a bank. We paid our loan back as fast as we could because I never liked the idea of a mortgage.

The first few years it was tough because it hailed every year. We would go out into the field to find everything lying flat and pounded into the ground. That hurt, especially when we had just started harvesting; in August it is too late to reseed. I never bought hail insurance because it would never pay off. They'll take the premium payment right now, no problem. But they won't pay off in the right percentage to what you've lost. I carry my own hail insurance.

Lois: One year we lost the entire cantaloupe crop. It froze just before we started the harvest.

Would you list the crops you grow here?

Lois: We have sweet corn, tomatoes, potatoes, cucumbers, onions, cabbage, cauliflower, broccoli, cantaloupe, watermelon, squash, pumpkin, strawberries and apples.

Charles: There are two types of melons—Crenshaw and Honeydew, two types of cabbage and three kinds of squash.

Lois: We also raise barley on the acres out back.

How did you determine which crops to grow?

Charles: By what sold. I don't know of anything that didn't work for us. Well, pickling cucumbers didn't quite work out. They would have if we'd have wanted to go in for them but they consume a lot of time. When you pick them you have to turn right around and sort them. When I was down home with Dad, we picked three rows about 30 feet long. Then we had to bring them in and sort them into sweets— which were the little tiny ones— medium dills, large dills and cucumbers. We didn't finish until around nine o'clock at night. It was long hours and a lot of work. At that time the money was there but down the line they brought less money than they were worth.

Are there any formulas for deciding which crops to grow?

Charles: No. There may be items that sell hotter and faster than the ones we've got. These Walla Walla onions, for example, and this elephant garlic. But we haven't stumbled onto formulas for figuring that out yet. It is just like anything else. You ease into it slowly. If it works, fine; if it doesn't, why, it's easy to slough it off.

Lois: His father had most of this stuff. We tried what his father had and what worked, which was most of it, we kept.

Would you give us other examples of crops which didn't work and why?

Charles: Beans and beets require too much labor because of the picking involved. We eliminated them. Also carrots and beets are not too good because of our soil. They grow too far down to pull them out easily. Feed corn is also a problem. I have a chopper and everything else and I could sell the corn all right; but if I sell it to these fellas that have feed yards they have a problem. They can only pay you so much and sometimes they don't even have that. You can sell them corn in July and be lucky if you got your money by the following June.

You run a reservation list for your cantaloupe crop. Is cantaloupe your most popular product?

Charles: Not everyone likes cantaloupe. It gives some people heartburn. Sweet corn is the crop that draws people in. People love fresh sweet corn! When you do go into a supermarket, you're not going to run for the sweet corn and run back out the door again, unless the price of corn is right. You're going to pick up a cucumber and maybe some lettuce and tomatoes to go along with it. People usually buy for a week when they come out here.

When you listed your crops, you mentioned apples. Does anyone else grow apples around here?

Charles: They used to have some huge orchards out west a little ways but not anymore. Change in times, I reckon. Folks want to walk into a store and buy a Delicious or Rome Beauty. There have been big changes in the apple business.

Where do you buy your seeds?

Charles: Most of it comes from Joseph Harris or Burpee.

Do you use the open-pollinated varieties or hybrids?

Charles: I use the hybrid seeds. When I first started out, I tried both the open-pollinated and hybrid varieties. You can tell the difference; the hybrid plant was more vigorous. I tried it for two or three years before I decided to pay the extra money.

Even though I used to buy open-pollinated varieties, I never saved my own. I have seen other vegetable men go with the open-pollinated varieties and have

FIG IV-3.3 "Sweet corn is the crop that draws people in. People love fresh sweet corn!"

FIG IV-3.4 "In the winter I haul about three loads every other day... I am getting as much nutrition out of these well-fed dairy cattle as I would from anything else."

problems. I think they saved the seed too long.

Is there a taste difference between the open-pollinated and the hybrid varieties?

Charles: I don't think so. With this cantaloupe there might be a difference because it is a variety that you cannot get in the local pollinated seed. You can't get it because it is crossbred. It is a little larger than the old Milwaukee Market variety, which was a well-known local open-pollinated variety. The texture and skin of this hybrid are quite a bit like the Sugar Rough. It has a very rough skin, vein centered. It's a different kind of cantaloupe because it has been crossed from two sides.

How do you start your seeds?

Charles: I start early by using the hothouse for things like eggplants, peppers and tomatoes. For example, we will put three tomato seeds in one pot. When those three seeds come up and produce true leaves, we transplant two of them into a cold bed and leave one in the pot. Then we begin heating the bed immediately. We heat electrically with thermostatically controlled cables. When it gets warm enough the electricity kicks off and, when the weather gets cold, it kicks back on. After the plants hit the true leaf stage, they are ready to be set out.

Describe the condition of the soil when you first began working it.

Charles: Sterile and foul.

How did you go about changing the condition?

Charles: With a hoe and a fork and that's just about it! Now we plow under all of the crop residue, even from the sweet corn. Everything goes back under and supplies a certain amount of humus. If you can keep a good supply of manure on hand and use it, you'll have no problems.

How can you test the fertility of your soil?

Charles: You go through the field and take soil samples from various locations and mix them together. Then take some from the lower end of the field and some from the upper; that way you can tell if there is too much of a difference. The lab analyzes the samples and tells you how much protein, nitrogen, phosphorus and potash you have.

Where do you get your manure?

Charles: Over at the local dairy, which has about 160 dairy cows. They produce a lot of manure. In the winter I haul about three loads every other day. We worked out an exchange; the owner gets corn silage or barley straw and we get manure.

Lois: In the past we used chicken manure from the Hager Brothers' farm but we had to load that by hand!

Do you use anything besides manure to increase the fertility of your soil?

Charles: No. I am getting as much nutrition out of these well-fed dairy cattle as I would from anything else. And the source is just across the road.

Have you run across any good commercial fertilizers?

Charles: I had some salesmen come through with this stuff called "Plant Two." Oh my! They found a seam of this stuff down in Utah so they crushed it and sent it up here in sacks. They would fly couples from Montana to a well managed and manicured farm in Illinois to see how Plant Two worked. Really wonderful! Very elegant. So anyway, they took the proceeds from this "Plant Two" and put it into the bank. Then they took it out of the bank and put the money on this farm. So I guess they were right; "Plant Two" really worked well— for them.

How do you work the soil through a growing season?

Charles: After the harvest we start right in with the discing and chopping. This puts the crop residue back into the ground. With corn, for example, we chop all the stalks and blow it back on the ground. Then we go in and spread heavy with manure. We spread as heavy as the old New Holland spreader will put it out. Then we plow the whole works under. That's it for the fall. I wish I could do more.

I wish I could work it flat so that in the spring I could go right in and start planting. But I can't because the wind gets to blowing. If I worked the soil tight enough in the fall, that ditch would be full of my topsoil. When I plow in the fall, I leave it rough. I could show you land over there that has blown. You can see it along the edge of the field where the weeds and stuff are. It has shed anywhere from two to seven inches of topsoil.

FIG IV-3.5 "By working the soil like this, I don't have to step up ten to twenty horsepower worked last year...my seventeen horsepower worked last year and this..."

What happens in the spring?

Charles: In the spring when the plants are in the hothouse, we work the ground with the harrow. This tool pulverizes the soil and gets it down to a firm seed bed so that your seeds won't fall down a hole to China. The tool also aerates the soil, which helps plants grow. By working the soil like this, I don't have to step up ten to twenty horsepower every year to pull the plow through the same ground. My seventeen horsepower worked last year and this and it will work next year because I put my five dollars into the soil.

You are located near the 46th meridian of north latitude. Do you have trouble growing fruits and vegetables this far north?

Charles: If you were to look at a map you would find that this area compares favorably with the Bitterroot Valley. This surprised me. There is an area from here through Bozeman to the Bitterroot that has pretty much the same growing conditions. Except that here it is much colder. A little farther north and it's even colder yet. There are some good microclimates in these areas. In fact, we have better growing conditions here than in some parts of Colorado. I can show you where there are peaches and pears growing in Billings! They grow against the south facing wall of a garage or a house. Those little microclimates are real fine for growing peaches and pears or almost anything.

Besides the hothouse, do you use any other techniques which give you a head start on the short Montana growing season?

Charles: Hot caps give me a considerable head start on the season. I like to peak just as the other vegetable men are starting to harvest. There is the cold weather! We had 26 degrees here the other morning. Some of the cantaloupe leaves were touching the frost caps and froze but the foliage that was down a ways made it through just fine. Even the foliage in hot caps already opened made it okay. Foliage outside of the frost caps froze but foliage inside made it fine.

Are there any other benefits to using hotcaps?

Charles: Hot caps protect plants from insects and mice. Mice love cantaloupe, watermelon and squash seeds. Just love to eat them! Cantaloupe seeds are expensive— a handful will run us about thirty cents. If mice eat those seeds, it costs money, so we cap them and the mice won't bother them. It works.

Are there any problems with using hotcaps?

Charles: Some. We had a little pepper plant growing in a hot cap. It grew about three times as big as the rest of the pepper plants. Lois decided it would be a good idea to take the cap off that pepper plant. It died because the change was just too great. Another problem is cut worms. Occasionally, you will cap a cutworm egg in with a cantaloupe seed. When you uncap it, out comes a miller and there won't be any seed left.

Please describe how you work the hotcaps.

Charles: It takes us a week to put about four acres under hot caps. First, we plant seeds into the ground; then we water them and cover them with the frost caps. That way you have seeds in there with moisture and heat from the sun— hopefully. You have all the ingredients necessary to bring those seeds out of the ground. They'll last in those frost caps for quite a while; but if you took hardened-off hothouse

FIG IV-3.6: "The little yellow tractor does help...when it comes to weeds though, the best chemical in the world is a hoe with a good stock end."

plants and put them under a hotcap they would die. It's just too hot for them. But when the plants grow up under the environment of a hot cap they are used to it and will live for a long time— three or four weeks if necessary.

Do you need hotcaps or are they an expense you could do without?

Charles: That would be a tough one. You have to use new ones every year. If you buy plastic caps that would last a few years, you can't break them open for the plants. You're not after a plant in a hothouse; you want these hot caps to get that seed up, to get it established. After that you start opening them as fast as you can. If you were to leave the plant completely covered, you would have all kinds of problems. You would pull the cover off and find the plant has gone to pieces. It's cheaper and more effective for us to buy paper hot caps.

What other techniques do you use to cope with adverse weather conditions?

Charles: Prayer helps a lot. Sometimes.

How much water do you give your plants?

Charles: Enough. If you have a real hot day here, above 100 degrees, plants are going to wilt. But along about six o'clock they will come right back up. If they need water, however, they're not going to do that. The plants will hang down until the next morning before they come back to true shape.

How do you water your crops?

Charles: I use a gravity flow irrigation system— ditch type. We have an irrigation ditch above the field and it's gravity flow from there. I would like to try some drip irrigation though. A person could put a pontoon float in the ditch and have a water wheel to run a pump, which would force water over the bank. It doesn't take much pressure to run those things. We could run the whole strawberry patch under drips. It would be fun to do it. You would have some headaches, for sure.

This water comes out of the Yellowstone River and is pretty dirty. You could put a screen in the pump filter and use the turning water wheel to clean the screen. But still, you have to figure a way of filtering and then cleaning the filter. That's the hard part. Then you could just turn on a faucet or set a timer and water the strawberries twenty-four hours a day. If you had the lines running down the middle of the patch, people wouldn't even know they were there. And it wouldn't make the rows muddy for the pickers. When you cultivate with a tractor you could disconnect the drip system. Your best deal would be to build beds. Two rows of strawberries to a bed and then run the hose right down the middle.

How do you control problem insects?

Charles: The only insects I have problems with are the cabbage worm and, to a degree, the cabbage looper. I raise about as much cabbage as anything else, not in acres but in sales. Those two insects are the only ones I have had to knock out with chemicals. I use malathion. According to the literature, it has a three-day residue period but I usually go with seven.

Why don't you have more problems with insects?

Charles: I think it is because I have healthy plants. If you have healthy plants, you don't seem to have the bug problems that you do with unhealthy ones. If I am wrong on this, I'll let you know. But so far we haven't had any problems except with the cabbage bugs.

All bugs have natural enemies. Take this wasp, for example. It's a tiny black wasp that lays its eggs on strawberry roller worms. This worm rolls a leaf together and spins it shut with a kind of spider web. Then it stays inside and eats. We used to do a lot of spraying and dusting with things like Parascreen. But they are not available anymore. So it's impossible to get this roll worm out of its leaf with anything except this wasp, which is smart enough to send in a maggot which eats the worm. Very convenient.

Have you ever had to fight off grasshoppers?

Charles: We have used what we call the "alfalfa special." It goes something like this. In the evening grasshoppers crawl to the top of the alfalfa plant. I don't know why they do it, maybe for the cool breeze or humidity or something. We have this tray which is a galvanized pan with a screen on top and one in the back. We put oil in the pan and pick it up with a tractor and carry it through the alfalfa. When the contraption comes through the field, the grasshoppers jump. But they hit the screen and fall into the pan and you have oiled grasshoppers.

What percentage of your crop do you lose to insects?

Charles: None. If we see anything start to happen we go work on it right away.

You run a very clean operation here. Do you think cleanliness helps control insect populations?

Charles: Yes, it does. Take host weeds, for example. If you come in and spray a crop for the flea beetle, you will still have a whole population breeding on the weeds. So you're defeating yourself when you have a lot of weeds.

Are there any other reasons for controlling weeds?

Charles: Yes. Take potatoes for example. Spuds are like anything else. If you have fifty plants on a square foot of ground competing for the same amount of available moisture and nutrients, the potato is going to come out second best because it is only one plant among forty-nine others.

Do you use any machinery to control weeds?

Charles: I use my old "hand run cultivator," at least on the young plants in the spring. In this type of farming you really don't need big equipment but the little yellow tractor does help. It's an International and I've got six cultivator tools for it. When it comes to weeds though, the best chemical in the world is a hoe with a good stock end.

What kind of equipment did you need when you first started farming?

Charles: Well, at first it was horses. Actually, we started with both horses and a tractor. But it was a lot different back then. If I were going to do this from scratch, I'd have my ground custom plowed. I would like to have a small tractor for hoeing, cultivating, discing and things like that. All you really need is a small tractor and a strong back.

Yours is obviously a very self-sufficient operation. Is there anything you need to

FIG IV-3.7 "We had 26 degrees here the other morning. Some of the cantaloupe leaves were touching the frost caps and froze, but the foliage that was down a ways made it through just fine."

buy from the outside world?

Charles: Seeds, hot caps, potting soil and a gallon of malathion.

Other than the manure you get from the dairy down the road, do you barter for any supplies?

Charles: Not too much. See, bartering takes time because there is a lot of give and take. We just don't have the time for it.

How many people does it take to work Heyn's Country Gardens?

Lois: Charles and I. Our daughter Joy also is helping this year.

Charles: It's been a long summer though and it's gonna be a lot longer before we are done!

Where could you get additional help if you needed it?

Charles: We could go to the Montana Rescue Mission or to Eastern Montana College. Quite a few of those students would be glad to work. Oh, there are places that we could pick up labor all right, short labor. It was real nice at one time because a Greyhound bus came through. Help could catch the Greyhound up and back. Now the Greyhound goes on the super and we are on a secondary.

Is your land producing at its maximum capacity?

Charles: Definitely not. But we are doing all that we can with the time allotted us. We are making a living and paying the bills.

Would you like to expand your operations?

Charles: No. This is just great. We have more than enough resources on our thirty acres of land. You've got problems enough without asking for more. If you make more money, Uncle Sam comes around and taps you on the shoulder and says, "Boy, that will be another $4,000 here." There's your incentive killer. It's almost as bad as socialism.

A hypothetical question. What would happen if you did decide to expand?

Charles: The land is available, right across the road in either direction. And it would actually be cheaper to rent than it would be to buy— by far. I would have to go with the headaches of hired hands and more equipment, although I think we have enough machinery right now to do twice as much as we are doing. I would need more help but not more machinery. But I would have problems and a lot of them would be labor problems. I can't get tough enough. If they don't do it, then I have to. So we have more than enough land. We tailor our resources to fit our needs.

You have been farming for many years now. Do you ever try any new techniques?

Charles: Every day. When we run across something that looks good, we try it. If it works, fine; if it doesn't we go back to what we were doing before. That is how we stumbled onto this popular

FIG IV-3.8 "More people came out all the time. Finally so many were willing to come that we didn't have to go into town anymore."

Photo by Sharon Richardson

cantaloupe. Lois bought me some seed for Christmas. I wouldn't buy it because it was so darned expensive. All you need is one package to make a good comparison— more than one or two plants. You see if it did any better than what you had before. This year we are trying a new tomato. We have three rows of our old variety to one of the new. We'll see how the customers like them.

Where do you get information?

Charles: A lot of my information comes from the county assessor but the kind of information I get from him I just leave in my hat. But seriously, there is the state college and the county agent. I've known our agent, Don, for a long time and it works out well. If I ever want to know anything, I call him up. There are also periodicals, pamphlets and books. We used to get the *Farm Journal* but it filled up with ads. There are a lot of magazines like that. I used to take the *American Fruitgrower*, too. But it started specializing in big things that would be nice if I had another 1,000 acres. The machinery they talked about was huge. I would need this place and four others like it and then the will to survive.

Do you belong to any trade associations?

Charles: No, but I do trade information with other farmers all the time.

How did your direct marketing program evolve?

Charles: We started out selling to small food stores like the Food Liner, Johnny's Super and Tenth Avenue Market. I would shy away from the big ones like Buttrey's, Albertson's and Safeway and I'll tell you why. They want 134 dozen today and forget us for the next two weeks. My stuff isn't raised that way. I have it coming along every day.

The smaller stores were taking three or four sacks, not much, but steady. Every day that I picked, I had a pretty good idea of what they would want. Each morning we decided what to pick. If they needed more when we delivered, we would have to pick more and then deliver it. But the waiting in line! I had to stand in line at the cash register to get my money and sometimes the line was four or five deep.

Lois: We had water running at home and work to do and he would be standing in there waiting to get paid!

Charles: So I figured it would be cheaper to sell it out here wholesale than it would be to take it into town.

How did you go about setting up a roadside stand?

Charles: It took us a period of years to set up the roadside stand. We started out by cutting out the stores farthest away and servicing the closer ones.

Lois: More people came out all the time. Finally so many were willing to come that we didn't have to go into town anymore.

How far from town are you?

Charles: About ten miles.

How did people find out about the availability of your produce?

Charles: Through word of mouth. When people bought our produce from the grocery store, the produce man would tell them it was from Country Gardens. They knew where the good stuff was coming from.

How can the quality of produce be judged?

Charles: I used to work in a produce house and did a lot of sorting so I know how to judge from that angle. There are little things to watch out for with everything. Take watermelons, for example. I guess there is a buzz on them. The only reason for plunking a watermelon is to tell whether it's overripe. You can't tell the degree of ripeness by plunking it, period. But you see these guys in the stores: plunk, plunk, plunk.... "Hmmm, what do you think?" You have to know your farmer. We guarantee everything that we sell. It's amazing how lucky we are in that respect.

Then how do you protect the quality of your produce?

Charles: First we have to find the best varieties. If we are always trying to find the best variety, our quality will improve. Our soil also affects quality so we are always building it. And we have to take care of the plants and harvest them at the right time. Then we have to sell the produce when it is ready to be sold.

Have you ever done any advertising?

Charles: When we first started, a lady had a program on the radio called "Country Gardens." It was a little program where she advertised local stuff. It helped us get things moving. We also put an ad in the classifieds that says, "Strawberries now available at Heyn's Country Gardens."

What else do you do to let your customers know what is available?

Charles: Our Country Gardens sign has a place to hang eight smaller signs. We can advertise what is in season at that time.

FIG IV-3.9 "Social security is a tax we pay to help those that need it. We pay into it, but we don't draw from it. Heyn's Country Gardens is our social security." Photo by Sharon Richardson

How do you work the roadside stand?

Charles: We harvest everything but the strawberries and tomatoes ourselves and haul it into the garage. Our garage is actually our sales room. During the summer our equipment is stored somewhere else and there is nothing in there but the counters, scales and so forth. The produce goes in there and is sorted and displayed. Then the customers come during the day.

Are you satisfied with your life on this farm?

Charles: If I didn't like it, I wouldn't be here. I'd be somewhere else. You have to like what you're doing. There are some people who are just doing it for the money but money is not enough reason to be a farmer.

If you had some good friends who wanted to start metrofarming along the same lines as Heyn's Country Gardens, what advice would you give?

Charles: The first thing you have to do is establish a market. I'll tell you a little story about what happens when you don't have a market. There was a salesman from town who decided he had to have a place out here in the country. He came out one day and bought himself 40 acres and decided to put it into spuds.

He asked me for advice. I told him to first establish a market and then get a contract— either a written one or, if nothing else, a word-of-mouth deal. Well, he had a connection with a large wholesale house and just knew he could sell everything there. So he went ahead and put in 40 acres of spuds. Since he had a good job in town, he bought himself a tractor with a double-row spud planter and a cultivator. But as soon as he put his spuds in, up came the weeds!

So he hired himself a foreman to run the place and the foreman hired a bunch of hands. They ate up all of the profits from the spuds still in the ground! I didn't see much of him that summer because I was busy here. But that fall he put the spuds into his cellar because he couldn't find a buyer. One day a hired hand went down to light the oil furnace so the spuds wouldn't freeze. He couldn't get it lit right away, so he threw another match at it and before anybody knew what happened, the salesman from the city had a lot of baked spuds for sale.

One last question. I notice you are sporting a green thumb. How did you get it?

Charles: It comes from pulling thistles. Seriously, I would have to say that it comes from doing the right thing at the right time, which means having the ability to tell what will work and what won't.

There used to be people around here that planted by the moon and they would do much better than those who weren't planting by the moon. People started talking about how important it was to plant by the moon. But these people were also doing everything else right, too, so the moon might not have made any difference. Those who were having trouble were just come-alongs. The moon does make a difference, though. You can see it in the way it freezes. If you get a freeze in the light of the moon, it won't freeze as hard as it will in the dark.

Pogonip Farms

Jeff Larkey
Santa Cruz, California

N ews of Pogonip Farms came via the local grapevine. Word had it a farm had been established on a small parcel of greenbelt a few blocks from downtown.

Pogonip is an apartment farm. Four young men lease an eight-acre vacant lot on a year-to-year basis. The lot is then subdivided into four parts, with each co-lessor controlling two acres. As the farm develops, the co-lessors take on partners of their own. When I first saw Pogonip, there were many people working many different crops on the four small plots of land. It resembled, in character, a crowded apartment complex in the city.

The Pogonip farm complex promised an interesting conversation: Here were four hard-working young people farming four very small parcels of land just a few miles down the road from the Pajaro and Salinas Valleys— California's agribusiness heartland. How could small businesses like the ones contained within Pogonip Farms compete?

Individual Pogonip farmers, of course, were reluctant to discuss their own business affairs. Their reluctance, however, did not extend to speculating about the

*Pogonip is thought to be a Coastanoan Indian term for fog.

neighbors' profits and losses. Said one of another, "I think he makes about $10,000 to $15,000 an acre."

Said another of two others, "Those boys are buying a lot of toys with their sales."

I watched their enterprise for awhile, drew a name from a hat and asked Jeff Larkey, one of the four, to sit down and talk.

Though Jeff is young, he is every bit the professional, proudly earning 100 percent of his living by farming. Jeff is also a student of agriculture and is constantly experimenting with new crops, techniques and markets. Through the experiments he builds an understanding of what will work and why. This knowledge is Jeff's principal asset because, unlike his leased land, it grows in value and cannot be lost to the whims of a landlord or politician.

Jeff was in the process of expanding his business by leasing an additional eight acres of land a block or two down the road, expanding his product line to include tree crops and increasing the size of his market by helping establish a new farmers market in nearby Monterey.

Later I met Jeff at a red light in town and he leaned out the window of his pickup truck and exclaimed, "Guess what? I just sold 250 bundles of sweet basil at the farmers market!"

Where did you get your start?

I guess my first experience was in growing you know what. But I really didn't know what I was doing then. My first real learning experience was at Cabrillo College here in Santa Cruz.

Cabrillo is a junior college. What kind of agriculture program did it offer?

None when I started. The school was trying to make a parking lot out of one section of the college. All the topsoil had been bulldozed off when the students decided that they didn't want another parking lot. They had a referendum and talked the administration into starting an organic garden program there. It began as an organic community garden and it had an instructor named Richard Merrill, who was new to teaching. He had extensive experience in Santa Barbara at the New Alchemy Institute. Some other students decided they wanted an alternative energy program. The abandoned parking lot became the site of the school's environmental studies program, which combined solar energy and agriculture.

What did you do in this program?

At the time, I was bummed out with the way my life was going and didn't know what to do. I knew that the program would be helpful. It was also a new field, which made it exciting. So I studied organic gardening and solar technology for two years. Then I became the teacher's assistant and became involved in the organization end of things.

What did you do after you completed your studies at Cabrillo?

I moved into a small commune-like place in a suburban setting. We had about an acre of land with some cows, chickens and goats. We also had about an acre of garden space that was not in production.

Did you put this acre into production?

Yes. We started to dig and build intensive soil beds. We grew a few things that we could sell direct, like lettuce and broccoli. At that time a lot of things started to happen at once. One of them was a farmers market that started up about two blocks away. I was involved in organizing that. We took a lot of our produce there every Saturday from June to October. Places in town like Community Foods— natural food stores— would buy from us direct. Our acre was next to a mobile home park and a lot of those elderly people didn't have the space to grow anything. We sold a lot of produce right over the backyard fence.

Was it your intention to begin metrofarming right away?

That was the idea. First, we wanted to feed ourselves. We had a small community of six or seven people. You know what happens when you are living in a situation like that. Some people are really motivated to do the gardening. Two or three of us were excited about doing the work and the rest had their own things going.

How long did this enterprise last?

About four years. Then we were evicted. The person who owned the land had a vision for it. He thought that it could become the start of a new way for feeding Santa Cruz. It would be filled with French Intensive beds. We moved in on the tail-end of this dream. Then he moved to Sacramento and took a job with the state. His brother didn't share this particular dream and doubled the rent.

Where did you go from there?

Lucky for me, something better came along. Three other people and I got a lease on eight acres of land along the San Lorenzo River. My friend Bruce Dau had been farming about an acre and a half over in La Selva Beach and was looking for a better piece. He came upon this land, which was a couple of hundred yards outside the city limits. It had a "For Lease" sign on it, so he called up the owners and found that

FIG IV-4.2 "When we started it was covered with willow trees and pampas grass. No disc could get those things out; we had to go in with axes and picks."

they were willing to lease for the price of taxes, which was about $1,800.

Bruce didn't feel he had the equipment or the energy to work all eight acres, so he asked Scott and me to go in with him. This happened about two months before we were evicted. We had some money saved and decided to go for it. The land already had a pump and well, so we purchased an irrigation system. The land hadn't been farmed for two years but had a history of agriculture that went back to turn of the century.

It is probably the best farm land around. It is an alluvial flood plain with very good drainage. The topsoil is forty inches deep! It has a high water table but not high enough to drown any kind crop or trees. When we started, it was covered with willow trees and pampas grass. No disc could get those things out; we had to go in with axes and picks. But it took only a week of full time work to clear the land.

What kind of lease did you negotiate?

Our lease is for one year at a time, which is not too great, especially for an organic farmer. You put all this energy into building your soil and may not be around to see the end product. This land has a high speculation value for builders. The owners have been offered $300,000 for the eight acres but at this time nobody can build on it because it is on flood plain. We've been on this land for three years now but the politics could change at any moment.

How did you organize your partnership?

There were four partners at first. Our agreement stated that we would share the cost of irrigation pipe, the electricity bills and the cost of discing the land. Also, if one of us were to leave before the end of a year's lease, the other partners would get first claim on that section of land. There wasn't much else. We didn't feel any great need to have a legal statement written up because we were friends.

How much start-up money did you have?

I started out with $4,000, which mostly went into buying pipe for everyone because they didn't have any money. They paid me back over the course of the first year. It took a total of $5,000 for us all to start.

How did you divide the land?

Even on that eight acres there are different kinds of soil. It's all good, though, so we just flipped coins and drew lines. We didn't run into any real problems.

How did you and your partners decide which crops to plant?

We decided not to compete with each other. There are so few organic growers around and they can only cover a few crops. There are an unlimited number of things that we could have planted, so competition was no problem.

What did you decide to grow your first year?

My first crops were garlic, cucumbers, bell peppers, leeks and statice. Bruce grew spinach, cilantro (coriander) and lettuce. Scotty grew golden zucchini, beets and some other things I can't remember.

How did you do

FIG IV-4.3 "We decided not to compete with each other. There are so few organic growers around and they can only cover a few crops... so competition was no problem."

your first year?

We had another partner then and that was the first year of the medfly (Mediterranean Fruit Fly) debacle. He was growing tomatoes, cherry tomatoes and one of the many other medfly hosts. He couldn't market his product legally and the illegal market for those products wasn't very big. It meant direct marketing. He tried to do an acre and a half and hold down a part time job but it didn't bring in enough. He just couldn't make it pay. He didn't know the medfly thing was going to happen and he picked the wrong crops so he bit the dust and isn't with us anymore. Three out of four made it, though. As far as agriculture goes— on the larger scale—that's a pretty good percentage.

To what do you attribute this percentage?

We didn't have to borrow money the first year. We shared resources and equipment and didn't put out a lot of money. That's what kills many beginners. We didn't need to buy any large equipment. Bruce already owned a large rototiller and that was all we needed for furrowing beds. Our biggest expense was for the irrigation system.

How do you get along with your co-lessors?

Some are friends. I get along well with others on a business level but not on a personal level.

How do your resolve your differences?

There are not many differences. We have an agreement saying we will not use pesticides because that would affect the others. We do our own thing but if there is something that needs to get done, we get together and do it. The common needs are dealt with easily and from there we go our own way as far as what and how we grow. We do try not to compete with each other and we've never tried to power someone out.

Where do you buy your seed?

I buy from several different seed companies and some I save myself. I buy from Stokes and Johnny's Select Seeds. I get flower seeds from Berkeley but I try not to buy vegetable seeds from them because I can buy the same varieties from smaller companies. Redwood City Seed Company is one.

Cantaloupe is one crop that we grow here that no one else has grown commercially. I buy cantaloupe seeds from a company that has forty different varieties. One of those is sure to grow in this microclimate.

How do you know which seeds to use in your microclimate?

I experiment. You have to experiment because any land you farm is going to be different. You have to try different varieties and see what happens.

What are some of the criteria you use for selecting good varieties?

Susceptibility to diseases and pests is important. I have some beans that were given to me by this eighty-year old Italian who has been growing them in this valley for many years. This variety came from Italy with his family and it's better than any other pole bean I am growing. There is a reason for this quality. Seeds have to be protected and selected; you can't take everything off your land and sell it.

How do you decide which company to buy from?

We deal with some unique varieties and some companies would rather deal in mainstream hybrids because that is where they make the most money. Some will supply you with a pretty good variety but there are so many they can't carry every one. I have to select seeds for my own microclimate and these large nationwide companies don't always have what I need. That's why I prefer to deal locally. People like Craig Dremann at Redwood City Seed Company have been focusing on selling seed that has been produced locally by other organic farmers. I might even grow seed crops next year. Redwood City Seed Company would probably buy them or I might take some of their varieties and grow for them. You could probably make more money growing seed for a seed company than by growing food for the edible market but I just don't want to get hung-up on them.

What have you been growing since your first year?

Leaf crops like lettuce and spinach, also onions, broccoli, cauliflower and garlic. We even made garlic braids. We also do specialty crops like purple broccoli. Flowers are another one; they are perhaps the best crop you can grow in terms of square feet. Customers shop at the farmers market because they can get fresh food at a reasonable prices. But people also love seeing food and flowers next to each other. I grow a lot of flowers and have a big variety. We sell fresh cut flowers and dried ones. We do well with both.

What kinds of flowers?

Statice, helichrysum and the star flower, which are sold dry. I always do a few fresh cut flowers like zinnias or daisies. I sell these flowers wholesale to flower shops or retail at the farmers market.

You have a well-diversified crop list. Anything else on the list?

We also grow cantaloupes, strawberries, leeks, sweet corn and popcorn. We are the

FIG IV-4.4 "What you do is start the plants in a greenhouse before it's time to set them out, not necessarily because it's cold at the time, but because you want to grow two, three or even more crops in succession."

only people growing carrots in Santa Cruz County. The rest are brought in from Idaho or somewhere else. I grow some herbs like basil, which is a very marketable crop. You can't ship it any distance and have it get there fresh, so it does well at the farmers market. I also grow oregano, which is sold either fresh or dried.

What is your most profitable crop?

For an annual crop, cucumbers are one of the best legal things you can grow. They do take time because you have to pick them every day or every other day. Per square foot, flowers are probably more profitable than any vegetable.

What about the specialty crops? How are they received?

There are oddities like purple broccoli, white and purple radishes, golden zucchini, golden beets, golden bell peppers, yellow tomatoes. They are especially popular at the farmers market because the customers can come up to you, talk about them and know where the produce comes from. You can even sell them for a little more, although I usually don't because people like to see a golden pepper costing the same as a green one. Keeping the price the same also helps when I'm trying to develop a market for a new product. Specialty crops started me thinking about expanding because there is a large market for things you can't normally get from a store. For a small grower, specialty crops are where it's at.

You mentioned Pogonip had forty inches of topsoil. Did you do anything to the soil before you planted your first crop?

Initially, we thought the soil might need more organic matter so we planted a cover crop of bell beans and rye grass. In the spring we plowed it under and applied some lye and chicken manure. We used twice the recommended rate of bell bean seed and manure as is recommended. We could afford it. It's not good to have too much nitrogen but it's hard to get too much if you're doing it organically.

What exactly do bell beans and rye grass do for the soil?

The bell beans are a legume. They give you nitrogen and some organic matter. They get about seven feet then we plow them in just before they flower. The rye grass doesn't supply nitrogen like the legumes but will accumulate the nitrogen that is already in your soil so you can plow it back in. The main reason you use it is to add organic material. It has very fibrous root systems; every adult male rye plant has about four miles of root

FIG IV-4.5 "Crop rotation is another technique. You grow different crops on the same piece of land, which is a good way of keeping soil borne pests and diseases under control."

hairs. That's a lot of organic matter.

Fava beans are another popular legume. How does the bell bean compare with the fava?

Bell beans are like fava beans but they're cheaper to buy. You can't eat them but they're good for planting. Fava beans are expensive. I think the reason is because they're marketed as an edible crop so the seed is in demand and hard to find. They run about $1.50 a pound. Bell beans are just grown for cover cropping.

How much fertilizer do you use?

For my acre and a half I use about twenty to twenty-five yards per year of chicken manure at $10 a yard. I don't use it all at once.

What does the term "organic" mean to you as a farmer?

It means that you are dealing with your soil instead of with your plants. By that I mean you feed the soil, not the plant; in turn the soil feeds the plants.

How do you deal with pest problems?

We don't use pesticides. We try to take everything around us into account. For example, I stayed away from radishes because you get root maggots with them in the summer. That's one of the few pest problems we have around here. We like to keep a lot of different crops going so we don't lose it on one. There are some pest problems and some crops are more susceptible than others but we haven't used pesticides because we have a diversity of crops. I could foresee having to use chemicals if we had one big area in one crop. But even then there are ways to avoid pesticides.

How does diversity in cropping prevent bug problems?

The strategy is to develop an ecosystem that has a balance of natural forces in it. If the natural forces are out of balance, pests will get out of control. If you have fields and fields planted in one crop, the natural forces get out of whack and the bugs take over.

What are the basic elements of organic growing?

That's a pretty broad subject but mainly it's using unsynthesized, naturally occurring fertilizers along with an integrated pest management program (IPM). IPM can be elaborate or simple, depending on how you want to do it. We don't like to go through the whole IPM computer printout thing; we just try to maintain a strategy of diversity to control pests. Organics can get involved but we try to keep it simple as we can.

Does growing organically increase the productivity of your land?

I think so, although for the first year it might not help. It would improve the productivity of your soil considerably in the years after that.

Do you think organically grown produce has more food value than produce grown with chemical fertilizers?

I think that has been proven by Hardy Vaughtman in Switzerland. Water soluble chemical fertilizers have a high degree of salt concentration so plants take in more salts than they normally would. This forces plant cells to expand and the additional pressure of salt solids causes plants to freak out. They try to relieve that pressure by absorbing more water. With chemical fertilizers you get plants that have more water content and may, therefore, be bigger but in terms of vitamin and nutrient content— or quality— it's the other way around. To prove this, take the water out of five pounds of cauliflower grown with chemicals and five pounds of cauliflower grown organically and you'll find that you will have more dry matter in the organic cauliflower.

What have you found to be the most effective way of watering your crops?

It just depends on what we're growing. We use overhead sprinklers for the most part. We've also done some direct watering, which works all right, but is not the best for some plants because getting water to the roots can be a problem. When you water direct, the water soaks into the ground and forms an inverted pyramid. For plants that have a large root system, you have to soak a lot of ground to feed all those roots. There are problems with sprinklers, too. We live in a foggy area and some varieties don't enjoy being sprinkled on and will get mildewed.

Do you hire any additional help?

Yes, I hire help at certain peak seasons like the harvest. I have no problem finding help; I pay four to five dollars, depending upon the skill involved. But I try to do as much as possible by myself. Hiring labor will set you back. If you are starting out on a shoestring, I would recommend doing most of the labor yourself.

What about equipment costs? Do you need a lot of equipment?

At first we rented a tractor with a disc and ripper and a rototiller. Not counting the irrigation system, equipment rental was our major single cost. Now we have a tractor but it won't pull a disc, so we're still renting the big stuff. Ripping is done just once a year but we may have a piece of land tilled three, four or five times a year, depending on what we're growing. The less of this you do the better— especially rototilling— because if you do it too often, it destroys your soil structure.

Most of the cultivation I do is by hand. Our tractor will do some weeding but I don't even use it.

Your two-acre farm is, to say the least, rather small when compared to the more conventional farms of California. Do you have any ways of increasing the productivity of this small parcel?

I don't think I need to but there are all kinds of ways to get more out of your crop. Bruce makes as much per square foot as I do—or maybe even more—by cropping the same piece of land four times a season. By transplanting lettuce starts, for example, you can get four or even five crops in one year around here. What you do is start the plants in a greenhouse before it's time to set them out, not necessarily because it's cold at the time but because you want to grow two, three or even more crops in succession. As soon as you pull crop, you have another ready to go; you get a good jump that way. If I want to get a jump on something, I'll start in the greenhouse because I

FIG IV-4.6 "But I try to do as much as possible by myself. Hiring labor will set you back. If you are starting out on a shoestring, I would recommend doing most of the labor yourself."

FIG IV-4.7 "I hire help at certain peak seasons like the harvest. I have no problem finding help...."

can give the plants more individual attention and use less water. Using less water, by the way, is always a good reason. This cropping technique is called "succession planting."

Crop rotation is another technique. You grow different crops on the same piece of land, which is a good way of keeping soil borne pests and diseases under control. For example, I would never want to plant corn two years in the same ground because we would have all kinds of problems. And you don't want to plant any of the cabbage family in the same place twice because you will get root maggots.

You can also pump out more per acre by using trellises and things like that. But around here we use succession planting and crop rotation. Sometimes we just take the soil out of production; we put in a cover crop and let the soil rest.

Let's talk about your marketing strategy. How many different markets did you develop for your first acre?

We sold wholesale to the stores and produce distributors and we sold retail at the farmers market, where we took turns manning the booth.

What is a farmers market?

It's a direct market from the farmer to the consumer— no middlemen are allowed. There are about 200 in California now and more springing up all the time. The state is even involved. Somebody from the county agriculture commission comes out to make sure that you are growing the crops you say you grow; the state certifies the markets. Farmers markets also provide better prices, especially for the small growers.

What is involved in setting up a farmers market?

There are a lot of technical things to do but the State of California almost sets it up for you. We had to find a place for the market, provide insurance and then talk the community into supporting us. Our market was organized by farmers and is run by a board of directors who are farmers. I left the board of directors this year; guess I was kind of burned out.

How is the farmers market doing?

It's doing pretty good. We opened another in Monterey this year. Our organization is called the Monterey Bay Area Farmers Market. It is a nonprofit organization affiliated with a larger, umbrella nonprofit one called Rip-Tide. I hope to use these direct markets for a long time.

How did you establish your wholesale customer list?

One of my partners at the first garden had been a produce manager at a local store, so we had a lot of contacts through him. We also had contacts with trucking companies; they enable us to ship our surplus to San Francisco or any place we want.

How did you develop your local market?

We developed a local route consisting of three natural food stores. One of these had a trucking company that distributed to other natural food stores in the area, which was good for us because we didn't have enough volume to go there ourselves. We also sold to three conventional supermarkets nearby.

Did the supermarkets work out for you?

One place is noted for its willingness to buy direct from local growers, as long as you can assure them of a steady supply. That's always the condition that supermarkets want, a steady supply. If you can't keep a steady supply flowing to them, you have to sell to distributors and they are not going to give you much.

What is the distributor mark up on produce?

The distributor we worked with sold to the natural food market. They would mark it up fifty cents a crate to the retailers. If we sold to the San Francisco market, we would get more because things are higher there than they are here. San Francisco is not an agricultural community.

Do you get different prices from distributors of organic and commercial produce?

Lewis sold half of his produce to an organic distributor, the other half to a commercial distributor. He was growing mostly spinach and lettuce. At that time, you were getting the same price on the organic market as you were on the commercial one. The price on produce varies from year to year. One year you can get eight dollars a crate for organic spinach and three dollars for commercial. But this year the price was high for both markets.

Did you try any marketing strategies which didn't work out?

Most distributors won't give you much for your produce. We try to stay away from them and deal strictly with the local and the organic markets.

How do you survey the markets for information?

Just word of mouth. I talk to other local farmers and there's a local certified organic farmer's association. I don't belong to it but it is a good organization; it helps farmers figure out what to plant and members talk things over.

Do you have a basic strategy for marketing?

Yes. My whole goal is to sell everything local. I do not like shipping anything long distance so I look for opportunities in the local markets. There are a lot of things like carrots that can be grown here but are not. I try to find produce that is shipped in from far away. Right now it's economical to ship this produce in from far away but soon it's not going to be. I want to be in a position to work with the local markets when that day comes.

Did you make money during your first year as a metrofarmer?

I broke even the first year but I almost lost my shirt. My main crop was garlic. I bought non-certified garlic from this big company that grows about 80 percent of the garlic in the United States. They sold me garlic all right! It turned out that this garlic was infested with nematodes and I lost my crop. I still had time to plant other things though but I wasted a lot of time and I just broke even. One of my other partners went broke and the other two made money. Bruce farmed three acres that year because his partner quit. He ended up grossing about $24,000 on lettuce, cilantro and spinach. Those aren't even the big money making crops.

You just said your partner made $8000 an acre in his first year growing lettuce! You live right next to the Salinas Valley which produces most of the head lettuce grown in the United States. How could he do so well with this kind of competition next door?

He circumvented the three middle men that usually handle this produce and sold directly to the stores. He was getting about twice what a grower in Salinas would get for a normal head of lettuce and he was producing a better looking head.

FIG IV-4.8 "My whole goal is to sell everything local. I do not like shipping anything long distance so I look for opportunities in the local markets."

What makes your lettuce and other crops so attractive to local consumers?

It is fresher and it looks better. It hasn't been sprayed ten times with chemicals. It tastes better and has better quality.

What are some of your best money-making crops?

Flowers, established fruit and avocado orchards, cucumbers, basil, strawberries....

There are thousands of acres in strawberries right down the road in Watsonville. Do you think you can compete with all the money behind the acres?

If I had my way we'd have a whole acre planted in strawberries because we grow them organically and can sell them for a better price, especially if we sell direct.

Does an organic label in a supermarket have any attraction to the consumer?

Oh yes. At least in the stores that label it as such. The big stores we sell to don't so it doesn't make any difference in those places. California has a certification program for organically grown produce. This helps, especially if you don't know the distributor and he doesn't know you. Many want you to have certification before they will let you call your produce organic.

What does this certification mean?

It means that somebody has checked you out and certified you as an organic grower. In California there is a legal definition of organic. So that I can put organic on my label, I conform to those specifications but nobody has checked me out. I sell mostly to people who know me, know where I live and where I grow. They know my produce is organic. If I get bigger, it will be to my advantage to become certified because I will be selling to people I don't know.

How does one become certified "organic" in California?

You pay your dues and become a member of the California Certified Organic Farmers. The certification process takes a year. Someone comes out and checks to see if you're using any chemical fertilizers or pesticides. If you're not within the guidelines, you won't get certified.

Does organic produce sell for a premium?

The consumer has taste! A growing number of people don't want pesticides in their food and a growing number who haven't cared in the past are starting to care now. I can't grow enough of some of the crops I have here. But if I were growing them commercially, with pesticides and everything, I wouldn't be able to compete.

Financially speaking, how well are you doing on your two-acres?

I'm still doing it, still making a living. I can make more than I need, financially. I don't think I could grow all my own food here but I can easily make a pretty good living. I haven't even maxed out my land yet. I could make more than I am now.

Do you have any plans to expand the operation from the two acres you have now?

I just leased eight acres of new land! It's an old abandoned plum orchard owned by some people down in Southern California. I have an eight-year lease for the price of taxes and putting in a pump.

How did you negotiate this expansion?

People have contracted before to maintain the trees for the fruit. I've been pruning them for the past two years and getting fruit along with several other people. I called the owner one day and asked the right question. She was feeling hesitant about keeping the property since she hadn't planned to move there in the near future. I told her the orchard was in really bad shape and needed work. I offered to do the work in exchange for an eight year lease plus an option for an additional eight years.

What do you hope to accomplish with this old plum orchard?

I am not sure at this point. Initially we planned to mix annual crops and trees. So far we've taken out about half the trees. Now we need to take into account the lay of the land and go from there. Some areas are suited only for annual crops, some for trees, some for both. I think we'll grow a variety of specialty crops geared for the Santa Cruz market.

What do you mean by "specialty crops geared for the Santa Cruz market?"

We want to plant citrus that will produce in microclimates like this, like lemons, limes and satsuma mandarins. There are other different kinds of citrus that can be grown here but aren't. Avocados are another example. There are some Haas avocados down the road that are growing as well as any commercial tree I've seen. We'll make a go of those, too. We'll also try some figs.

Do you think you can make enough profit on your trees in eight years to justify the expense of planting them?

FIG IV-4.9 "I can easily make a pretty good living. I haven't even maxed out my land yet. I could make even more than I am now."

That's why we're growing annual crops too. Some of these trees won't bear fruit for six or seven years and some, like the figs and nectarines, will bear in three. Until then we'll need to make money to cover our expenses. We'll do about two acres in row crops.

Its stupid for people around here to buy their fruit from Southern California. Times are going to change. They are not making farmland anymore. It's going to be more and more profitable to grow food on land like this than to have an overpriced horse corral on it.

What do you see in the future? Would you like to expand to even larger parcels than your new eight acre piece?

No. I'm farming as much land as I would ever want. If I get any bigger, all I would be doing is hiring more people to work for me. Even on my two acres, my income is increasing every year. The more years you farm, the better you get. You learn so many things that increase your productivity. There are almost no limits to the creative potential in agriculture.

Where do you get your information? Do you read a lot?

Not much. I do have some reference books in case I get into trouble. I like to share information with other farmers but often that ends up in a lie swapping contest. You know, "My carrots were this lonnnnnnng!"

What advice would you give to a friend starting up a metrofarm?

I would encourage them to look at metrofarming as a way of life. I would also encourage them to have some money before starting, although there are always ways to do it on a shoestring.

What is your definition of a green thumb?

A green thumb is just a way of describing somebody who knows how to think like a plant. That's what the phrase means to me.

The Flower Ladies

Ruth Ann Birkman, Nancy Lingemann,
Marcia Lipsenthal
Bonny Doon, California

Discovering them was as easy as reading. The Flower Ladies have been featured in the *San Francisco Chronicle*, *San Jose Mercury News* and *Sunset*, as well as many other local and regional publications. Getting to them, however, was a long drive up a bumpy road.

The Flower Ladies live and work in an idyllic mountain valley tucked high in the Pacific Coast Range south of San Francisco. Their metrofarm is situated amidst lush ferns and giant redwoods on a gentle south-facing slope. You can see the sparkling waters of Monterey Bay when you stand at the top of the slope.

The Flower Ladies are a Texan, a Southern Californian and a New Yorker glued together with giggles. Their business is a vertically integrated one in which flowers are started in heated greenhouse beds, grown into mature blossoms in a two-acre terraced garden, crafted into formal arrangements in a rustic workshop and then delivered to weddings in the Monterey, San Jose and San Francisco metropolitan areas.

I watched as the Ladies selected and harvested flowers from their garden in the cool of an early morning. The site was a riot of color with wild fairy bells, forget-me-nots, corncockles, lilacs, Iceland poppies, larkspur, delphiniums, anchusa, tamarix, daffodils, Phoenician verbascum and hundreds of others competing for attention.

As the sun burned through the morning mists and toasted the slope, I watched the Ladies arrange the cut flowers in the cool of their workshop. The fragrance was nearly overwhelming. The Flower Ladies grow over 200 varieties of cut flowers for their wedding market. They can produce arrangements with 25 shades of blue flowers or ones which smell like icing on a wedding cake. When their first large arrangement of the day was completed, I was sold.

I tagged along as The Flower Ladies serviced a wedding in a nearby town. The Ladies arrived early and carefully placed their arrangements to highlight the wedding altar, pinned corsages on important guests as they arrived and then took the very happy bride on a tour of her flowers. Everybody was sold!

The Flower Ladies would not discuss their income. "I don't even pay any attention to the figures," said Nancy. They would, however, provide some ballpark figures: The Flower Ladies average four or five weddings per weekend during a nine-month season. Their prices start at $350 and may reach $2,000 and more for the extra special affairs.

What is The Flower Ladies?

Nancy: A florist business. We start flowers from seed and carry the process all the way to the end where we come out with finished products like bridal bouquets or huge arrangements. Then we take these arrangements to weddings.

Ruth Ann: Rather than going to the market for flowers as most florists do, we actually grow the flowers. Then we do what florists usually do, arrange and deliver flowers.

How does this business work?

Nancy: People call us when they are ready to get married. They order whatever they need. Some people want a big wedding and will order all the trimmings like flowers for flower girls and garlands for the hair and huge arrangements and reception flowers and throwing bouquets and regular bouquets and corsages. And some people want the simplest kind of ceremony. They may want a handful of wildflowers to carry or just a boutonniere for the groom and a couple of bouquets for the table.

What is unique about your service?

Marcia: The unique part is that we offer something no one else does.

FIG IV-5.2 "Nancy's garden gives you the feeling of profusion and abundance... Ruth Ann's way is to plant everything in neat rows... Marcia's garden varies with the time of year and whether she has a boyfriend or not."

Nancy: We grow our own flowers and they are different from the ones you find at a regular florist's.

Marcia: And the way we arrange them is different, too. We don't do the really stylized arrangements with a point in the middle and flowers going down from that. We don't count our flowers as frequently as florists do. They count the number of flowers they put into arrangements.

Ruth Ann: We make an arrangement until it's done.

Nancy: Until it looks right.

Ruth Ann: We can also provide more color. Take blue, for instance. If a customer orders blue, a florist might use an iris or maybe a delphinium. But generally he would use a dyed carnation.

Nancy: Whereas we can boast of having 25 different varieties of blue flowers because we grow them ourselves. They are not the sturdiest of flowers but they are not like dyed carnations either.

How did The Flower Ladies get started?

Nancy: I was studying German literature at UCSC (University of California, Santa Cruz) and happened to meet Alan Chadwick* one day. It seemed like a fated

*Alan Chadwick was a Shakespearean actor turned horticulturist. His garden at the University of California, Santa Cruz made popular the French Intensive Biodynamic Method. In building this garden and others

meeting. Alan Chadwick was an English gardener who that day had come to the university to develop a garden using the Biodynamic French Intensive Method. He was looking for a core group of students to help him get this garden started. We talked over lunch and spent hours together that day. After that I seemed to drop my studies in German literature because he was offering a chance to learn something new.

What did you find so attractive about this course of study?

Nancy: Getting in touch with nature. I had never understood the process of how a blossom develops into fruit. I had never seen a plant grow from a small seedling into a full, mature plant and then have seed. I had never noticed the change of seasons. I am from Southern California, where the seasons are basically summer and spring. It was a whole new awareness for me.

Was this Chadwick garden hard work?

Nancy: Yes, but I loved it. I was only 20 years old then; I am 35 now. It's still not difficult for us. I suppose when we become old ladies....

What happened after school?

Nancy: I started growing flowers at home. At that time I was getting married and having a baby, so it was time to quit school. My husband owned some land up here and I had a garden. I began to adore the flowers. I had never loved them before from an esthetic point of view. But I began to love having them around me all the time. Gradually, the vegetables took a smaller and smaller place in the garden. Now my daughter and husband have a garden of vegetables and I don't even bother with them. I had lots of flowers and started selling bouquets at a local gas station. That is where I met Ruth Ann.

Ruth Ann: I was managing the gas station. I was living with the man who owned the gas station and we were ending our relationship. I needed a place to hide out— someplace in the country. So I moved up to Nancy's land in 1973. It was natural for me to start gardening because I grew up with it. My parents have a nursery in Port Arthur and I was called the "washing pot." When I was a little kid, I would help out around the greenhouse by washing pots. So I started gardening and selling flowers down at the gas station with Nancy.

Nancy: Then we had a conversation with the man who owns Shopper's Corner (a Santa Cruz market). We started selling flowers there.

Marcia, how did you become involved?

Marcia: My ex-husband and I bought property next to John and Nancy. I had

FIG IV-5.3 "Someone came to us and asked if we would do her daughter's wedding. That's what started it... where the business started growing."

always lived in cities. I grew up in New York City. The only acquaintance I ever had with flowers was buying them on the street corner in front of Bloomingdale's. I traveled around quite a bit and finally settled in San Francisco. We came here in search of a more rural environment. We had a baby and built a house. I saw Nancy's garden and it was the most glorious place I'd ever seen. It was covered in leaves and it was in Nancy's own style, which meant you could never find the path. But it was wonderful. When I saw that garden, I knew that was what I wanted to do. So I came over and started following Nancy around her garden.

So your business started from your enjoyment of the land?

Nancy: The gardens came first and the business came from that. We developed a business from our home rather than going out and saying, "We'll buy this land and make a business out of it."

How many varieties of flowers do you grow?

Nancy: We advertise having over 100 kinds—

Marcia: —and it is easily over 100—

Nancy: —it must be close to 200. Our greenhouse is just packed with flats of different varieties.

What was your garden like when you first started?

Ruth Ann: We're totally incompatible.

Nancy: Ruth Ann likes to have her beds much more tidy than I and also to have a lot of one kind of flower in them. Somehow I decided to cross over the fence to the other side. We found that I could do my own style and enjoy it and Ruth Ann could maintain the other garden beautifully and she found it more satisfying. Then Marcia had an area that adjoined our property lines.

Do you each have your own gardens now?

Ruth Ann: Yes.

Please describe each other's garden.

Ruth Ann: Nancy likes to have weeds growing everywhere and not to have any defined paths. She likes to work her way through—

Marcia: —with hip boots. Nancy's garden really does give you the feeling of profusion and abundance. There are a lot of weeds but you get an opulent, rich feeling there.

Ruth Ann: She likes to put out a few treasures here and there and only Nancy knows where they are. She loves it and it drives me up a wall; psychologically, it is not my way.

Marcia: Ruth Ann's way is to plant everything in neat rows. It's what we've come to

FIG IV-5.4 "We're experimenting with new flowers all the time. We will find something... from some other place far away and then we'll send for the seeds and try to grow them."

call the "Birkman Way."

Nancy: Her mother is this way too.

Marcia: And her brothers too. The flowers are like little soldiers marching in neat rows. They are all standing at attention and saluting you. It's all neat. There is not a weed in her garden. It's perfect.

Nancy: Marcia's garden varies with the time of year and whether she has a boyfriend or not.

Ruth Ann: Yes, well-planned chaos.

What was your soil like when you first started tilling it?

Nancy: Most of it was wonderful soil. There were areas we had to build up by allowing the weeds to grow, then turning them over and manuring. There are still areas like that.

Ruth Ann: In my garden, too. How big is my garden? Maybe one-half acre, maybe even three-quarters. There are areas that have to be worked every year with manure and compost. But I can see those areas improve each year.

What kind of culture technique do you use?

Nancy: It's French-Intensive with raised beds. When we first started, we dug down two shovelfuls to loosen the subsoil. Then we turned it over and added manure, bonemeal and ashes. Before we plant anything we always put a lot into the soil. And even if we have fertilized previously, we do it again before we put in another batch of flowers. By adding more and more we keep improving the soil. We don't have to double-dig anymore because our soil is so friable. It's wonderful! We also plant the plants close together, which is unusual. People sometimes look at our beds and comment on how closely the plants are situated to each other. But we find that the productivity is really great with this technique.

How much manure do you use on your two acres?

Marcia: Dump truck loads.

Nancy: Yes, and we like to shovel it ourselves too. There is something nice about having muscles.

Marcia: It's true. We have fun. We go with Nancy's husband and load up the dump truck and have fun.

What else do you use for fertilizer?

Ruth Ann: Compost. We make compost piles and then we buy bone meal and blood meal.

Nancy: And we save ashes.

Ruth Ann: I buy fish emulsion. I like to use it when I put the young plants out.

Do you use any commercial fertilizers or pesticides?

Marcia: No sprays, no chemicals— nothing. An important part of our technique is that we raise everything organically.

Ruth Ann: That's important to us. I don't know that it's actually important in raising flowers, though. Someone raising flowers using chemicals could do the same thing. But these gardens are part of our life. We live and work here and we don't want to live with those chemicals. We don't believe in it.

How did Chadwick feel about the use of chemicals?

Nancy: He was very much against the use of pesticides or any kind of chemicals. He stressed the importance of not allowing those energies into one's garden. Chadwick maintained that gardens are very special places where fairies and all manner of visible and invisible creatures grow.

What effect does this "natural" approach have on your gardens?

Nancy: Psychologically, it feels better and we're not—

Marcia: —harming the soil or depleting it with chemicals. We are enriching it all the time.

Ruth Ann: And we are not harming ourselves.

Marcia: We don't have to breathe sprays or chemicals. We don't kill the birds and insects.

Ruth Ann: And I think another effect is our lack of an insect problem. We have very few aphids every year.

Nancy: And a few spotted cucumber beetles.

Marcia: And a few rose beetles. We pick them off by hand and feel a little bad about it.

When is your growing season?

Ruth Ann: We start our seeds in February and put the starts out in April. Then September, October and November are big months to put out spring flowers—

Nancy: —and perennial plants.

Ruth Ann: But it seems like there are always a few things going in the greenhouse and—

Nancy: —every month we sow something.

FIG IV-5.5 "We begin talking about color instead of types of flowers because we have so many different kinds (of flowers)."

Where does the process of cultivating a crop of flowers begin for you?

Marcia: We have a greenhouse that we use to start all of our seedlings. We do two major plantings, one in February and one in the fall. We start the seeds and once they get their second set of leaves, we transplant them into bigger flats to give them more space. While this is going on, we prepare the soil beds in the garden. Then we put the seedlings into the beds.

Where do you get your seeds?

    Marcia: Catalogues.

    Ruth Ann: Park Seed Company—

    Marcia: —the DeGiorgi Company and Thompson & Morgan.

How do you select your seeds?

    Ruth Ann: Every year we become more efficient in our seed buying. We used to be like children when the toys came along at Christmas. We would want to try everything. But we've narrowed it down over the years. We know the flowers now. We know which ones are good for cutting and which ones we like—

    Nancy: —for sentimental reasons, like Alan Chadwick grew it or something. Maybe its not really useful but we still grow it.

    Ruth Ann: One thing we've been trying to do is become more specific. For example, a mixed package of snapdragons will not have much yellow, so we try to order specific colors of yellow.

Do you grow any seeds for yourself?

Nancy: Sometimes we collect our own seeds. We are doing more of that now that some of our varieties are becoming harder to find. The larger companies are cancelling some of our favorites.

Marcia: I am going to begin taking seeds from the particularly beautiful colors that we like.

Ruth Ann: Like a selection of phlox. These flowers are predominantly rose colored but maybe you'll get two purples and one white plant out of 30. So we'll collect seeds from the white and purple.

Are you experimenting with any new varieties?

Nancy: We're experimenting with new flowers all the time. We will find something in a world service catalogue or a catalogue from England or from some other place far away and then we'll send for the seeds and try to grow them. Maybe we'll get one or two plants.

What does it take to cultivate a productive flower garden?

Marcia: You have to like it. And you need good soil and lots of water.

FIG IV-5.6: "Our flower arrangements are characterized by abundance. There is a profusion of flowers, colors, shapes, and textures."

FIG IV-5.7 "The flowers speak for themselves. There is a certain knack to putting them together and knowing what to do with them. But the flowers themselves are so gorgeous."

Ruth Ann: A good garden also takes lots of muscle and hard work. Here in Bonny Doon it doesn't get very hot in the summer and the nights are cool. I think that's our biggest advantage.

Nancy: Yes, we can count on cool mornings to cut flowers. The color fades quickly if you are in a hot climate.

Do you hire additional help?

Nancy: We never have in the past but this past week .... Yes, we've begun to hire people, at least to weed the garden. It was a little bit against our principles but once we tried it—

Marcia: —we loved it. We have just found the most perfect weeder, a strong woman. She's great.

Ruth Ann: I think in the last couple of months we've become more interested in having people work in our garden. We've met several people who have asked for work and our name is well known. They say, "Oh, The Flower Ladies, to be able to work in their garden!"

Nancy: That's the way our whole business has been. As soon as—

Marcia: —we are ready to do something, it happens, Magic!

Nancy: It is magic!

Ruth Ann: It really is!

Nancy: There was a man who came to us a few months back with a portfolio of his floral work. He had studied floral design at a local college and it all looked fine. We had this 10-wedding weekend coming up and we thought that we should hire him. But then we never called him back because we knew—

Ruth Ann: —he wasn't used to working with our kinds of flowers. He was used to working with the cut flowers that florists provide.

Marcia: And besides, we all work so well together. To take the time to explain to someone what to do would be more time consuming than to do it ourselves.

What improvements have you made to the gardens?

Ruth Ann: We put up a deer fence.

Marcia: It's six feet tall. We have two acres inside the fence and it's worth every penny.

Ruth Ann: It's a relief when you go out in the morning and find everything there. I used to walk into the garden every morning and say to myself, "My God, look what's been eaten." Now I open the gate and don't even look.

Do you have any plans to expand your gardens?

Nancy: I hope not!

Ruth Ann: No.

Marcia: No. We can fill it. We can grow more flowers in the space that we have.

Ruth Ann: We could make better use of the garden space.

How would you improve the efficiency of your two acres?

Ruth Ann: Well, my garden still has bare spots in it, places where there are no beds or flowers.

Marcia: We don't need to grow many more flowers.

Nancy: Oh yes we do.

Ruth Ann: I think we could make better use of our space by getting the flowers out on time.

Nancy: Yes. Everything happens at once. The seedlings are ready to go in April and the brides are getting married in April. If we could get the plants out without delay, without letting them waste in flats....

Marcia: That is where we are learning to use hired people. We enjoy the whole business process. We enjoy covering the greenhouse every year; we enjoy planting the seeds; we enjoy digging in the garden. We enjoy meeting the brides and arranging the flowers. And we enjoy working with each other. And then we enjoy delivering.

Nancy: Going to the weddings and getting kissed and hugged. That's the crowning touch.

Marcia: Yes, the crowning touch. Though we can hire people to do small segments of it, I don't think any of us would want to part with any piece of it.

FIV IV-5.8 "Truthfully, we don't have any competition. There is no one who grows flowers like we do."

Ruth Ann: I agree but there are some mundane tasks that could be hired out. Like weeding!

Where do you get the information you need to cultivate all these varieties of flowers?

Ruth Ann: We just add to what we know every year. When you garden, you never know it all. Every year the weather is different.

Marcia: It really influences what is going on in the garden.

Do you read any books or magazines?

Marcia: We don't read. We were much more interested in reading a few years ago. But after we read a couple of how-to's, we evolved our own style—

Marcia: —by doing it.

Ruth Ann: We are still interested in reading things though.

Do you learn anything from talking with other gardeners?

Marcia: Oh yes. It's always fun to talk to other gardeners. It's always fun to walk into a garden with someone who spends a lot of time and energy there and talk about how she plants things.

Ruth Ann: What I really liked was learning that there are no set rules for growing plants. One person may say that delphiniums do best with a half day of sun and a half of shade. But we've grown beautiful delphiniums in a full day of sun.

Nancy: People tell us not to plant certain things in the spring or summer. We purposely go out and sow the seed to see if it's true.

Where did you learn the wedding flower business?

Nancy: We take a great deal of pride in being untaught!

Ruth Ann: We were totally ignorant about it when we started.

Nancy: We read a book on corsages to see how florists tape and wire flowers—

Ruth Ann: —and how to make them last.

Nancy: We haven't read much. You see, the flowers that we grow are not the typical florist's flowers. Some of them are but most are not. So we've had to learn the qualities of each and how to make them work. We needed to learn which ones were good for taping and wiring. Many are just taboo. After a few years of doing this, we've come to know each flower very well and what we can expect from it.

Haven't you taken classes in flower arranging?

Nancy: We have never been to a floral design school. I went to a college once that had a floral design school but I studied architecture, which hasn't done me a bit of good in this business.

Marcia: We went to a trade show where there were designers showing how they put their arrangements together. We snubbed our noses at those but we did pick up a few tips.

Ruth Ann: We learned some mechanical things like working with wire and tape. But nothing about flowers. We feel that we have that area pretty well covered.

Nancy: We did get a compliment from a professional florist not long ago. He had seen some of our work at a wedding and commented about how professional it looked. We rolled our eyes at that!

When did you become The Flower Ladies?

Nancy: When we started selling flowers at Shopper's Corner.

Ruth Ann: We used to advertise in a local paper and to help people know who we were, we identified ourselves as "The Shopper's Corner Flower Ladies."

Nancy: And then one day we said, "Why don't we take off the 'Shopper's Corner?'"

Marcia: It got to be too much trouble selling at the market. It wasn't worth the money because our wedding business had grown so.

Ruth Ann: We would have weddings on a Saturday morning and use all of our flowers for them. Then we'd have to make up bouquets because we were obligated to the market. We'd go there with 15 bouquets or less.

Marcia: Compared to what we could make by doing weddings, the time it took to make and sell bouquets at the market wasn't worth it. And the satisfaction wasn't as great.

Nancy: At the market we were limited to plastic cups but when we're doing a wedding we can use vases and baskets and—

Marcia: —all kinds of things.

Ruth Ann: It's much more creative!

How did you market your flowers at the Shopper's Corner?

Nancy: We sat on a bench in front of the store. We had a big paper banner that said, "Flowers, 85 cents." It was fun to have conversations with people and talk about flowers.

Marcia: A lot of gardeners would come by and not buy anything.

Ruth Ann: This was two days a week— Friday and Saturday.

Nancy: We'd get up early in the morning and cut all these flowers and put them in little styrofoam cups. We just made bouquets, never arrangements or anything. Then people started asking us to do weddings. At our first wedding we saw this white wicker basket and suddenly realized that we were expected to do something that looked like a florist's arrangement.

Ruth Ann: Someone came to us at Shopper's Corner and asked us if we would do her daughter's wedding. That's what started it. I'll never forget the corsages. We didn't tape or wire them. We were so proud because they had natural stems. We had these tiny bottles that we put them in. It's embarrassing now but that's where the business started growing.

How did the first wedding work out for you?

Ruth Ann: Everybody was pleased.

Nancy: They loved the flowers! We still have pictures of those arrangements in our portfolio today. We've taken out many pictures that we thought were not representative of the kind of work we do. But we still have our original wedding pictures in there.

How did you come by other wedding work?

Ruth Ann: Our reputation grew as we started doing more weddings.

Marcia: People would see our flowers at a wedding and know that they wanted them for their wedding.

Ruth Ann: I would say that today about 70 percent of our business is still word of mouth.

FIG IV-5.9 "We've done so many weddings by now that we feel confident in meeting brides because we can suggest many possibilities that they've never even thought about."

How long do your flowers last in an arrangement?

Ruth Ann: If the flowers are cared for properly, they will last at least a week in a bouquet.

Nancy: We used to do restaurants and were always distressed when we would go into one and find our flowers all wilted. In those smoke-filled rooms no one would take the trouble to water the bouquets.

Marcia: We hate to go some place and find our flowers not looking good. We take it personally. I was glad that we stopped doing restaurants because it wasn't good for our reputation. We go to a wedding and we deliver everything and we know that it's perfect when we leave. We know the flowers will look beautiful for the ceremony.

Are you saying you do more than just deliver the flowers to a wedding?

Ruth Ann: Yes. I think that's part of the reason for our success. I've had brides express amazement when I say, "I am going to stay and pin on corsages and boutonnieres." I talked to Diane today, who is one of our brides for this weekend and she asked me to remind one of her bridesmaids to pin on the flowers. I said, "But that is what I am there for." She was really happy! Nobody expects that kind of service anymore.

Do you counsel the brides?

Nancy: On what? Getting married? We'd like to!

Ruth Ann: We've done so many weddings by now that we feel confident in meeting brides because we can suggest many possibilities that they've never even thought about. For example, if the bride has four or five maids, we can arrange a different color combination for each.

What else makes The Flower Ladies successful?

Ruth Ann: I think part of our business success is our personalities. I am not trying to be egotistical but I think a lot of people go to florists and don't get any kind of personal attention.

Nancy: It's not a happy affair for them. Our brides often tell us that they enjoyed meeting us. They call us by name. They call us on the phone. Marcia has had brides call her in the middle of the night because she is so reassuring.

How do your prices compare with the conventional florists in town?

Marcia: We don't quite know but we think we're comparable or maybe a little cheaper. But we do raise

FIG IV-5.10 "Our brides often tell us that they enjoyed meeting us. They call us by name. They call us on the phone."

prices all the time.

How does your product compare?

Marcia: It is totally different. That's what we feel. We really cannot compare the products at all.

Nancy: Our flower arrangements are characterized by abundance. There is a profusion of flowers, colors, shapes and textures.

Ruth Ann: And they're wild and—

Nancy: —they're really floral. If somebody says they want a Japanese arrangement, we say, "I'm sorry but that's just not our style." We have a way of diverting people's questions when they say, "What kind of flowers are you going to use?" We begin talking about color instead of types of flowers because we have so many different kinds. If we would list them, they would say, "I've never heard of those." They begin to trust us when they see pictures with all these unique color schemes. Our portfolios are what charms them, usually.

Ruth Ann: We have such a big variety of flowers and colors; when people see them, they're just taken aback!

Nancy: In that sense, it's hard to compare ourselves with other florists.

How do you judge the quality of your wedding arrangements?

Marcia: The flowers speak for themselves. It's hard to go wrong with those flowers. There is a certain knack to putting them together and knowing what to do with them. But the flowers themselves are so gorgeous. They're not the carnations and the roses that you see at a regular florist.

How many weddings do you do in a typical weekend?

Ruth Ann: Four is a good average.

Marcia: And ten is the most we've ever done and that almost killed us!

How do you go about planning your weddings?

Marcia: We have a meeting every Monday morning where we discuss the weddings for the upcoming weekend. We figure out what we need to buy. Even though we have such a profusion of flowers in our garden, we may need special things like gardenias or leather leaf ferns. We plan and make a list. Each of us gives her weddings for the weekend to the others. We take down notes. Then we make a shopping list. If we need to, we go to the San Francisco Flower Market on Wednesday morning. We have to be there by 3:30 a.m.!

How did the three of you do ten weddings in one weekend?

Marcia: We had our Monday meeting and talked. Wednesday we went to the San Francisco market, came home and took a nap, then went back to work. I think we set our minds to it by saying, "Okay, this week all we do is weddings. Forget you have kids; forget you have friends; forget you have families. You just have weddings."

Nancy: I memorized my corsages. I had thirty-four to do! Then I had bouquets....

Marcia: We'd say, "Do we have enough blue in the garden? Do we have enough peach?" Then we'd take a walk out to the gardens and see what we were talking about.

How many extra flowers do you need to buy to make your arrangements?

Marcia: I can't say in terms of the number. I can tell you in terms of the amount of money but I don't want to.

Five percent?

Nancy: Oh, not that much.

Ruth Ann: A very small amount. In spring it would be more. We don't have to worry about buying a week ahead of time and wondering what kind of business we'll get during that week. We only buy if we need it. The commercial florist buys what he needs for a week but he also has to buy for those orders that are going to walk into the shop in during the week.

What colors do brides go for?

Ruth Ann: That's a very important factor. The last couple of years it's been lavender—

Marcia: —with pink running a close second.

Ruth Ann: Now there was a year when peach—

Marcia: —right; it was peach and yellow that year. This year we've had only one yellow wedding and people just hate orange this year!

Nancy: This year we have a lot of lavender flowers. And blue is always valuable. Brides often ask for dusty rose, which is a color invented by the fashion industry. Roses are pink, red and yellow. To produce a dusty rose in our gardens, we would have to drive back and forth on the dirt road a lot!

How have you improved The Flower Ladies over the years?

Ruth Ann: We are more efficient now. We've learned over the years to have all the supplies where we need them. We're even more comfortable now. We are inside....

Nancy: We used to work outside at a picnic table where we were subject to the weather. We used to hang up sheets and bedspreads to protect us from the sun.

Marcia: And we didn't have any refrigeration then. Now we have a big walk-in, a big six-door and one little refrigerator to store everything and protect it from the heat. We used to take our arrangements and put them in the creek and that would keep them cool.

Nancy: The early Italians who settled this area did that with their dairy products!

Marcia: It was an incredible amount of work to carry our arrangements up to the creek all the time.

Ruth Ann: And we would always hope that an animal wouldn't knock over our arrangements, like a raccoon did one time!

Marcia: We could never enjoy a sunny day. We were never happy unless it was cold and foggy like a refrigerator. If it was cold and foggy, it was a perfect day for a wedding!

Ruth Ann: Now we can relax even on hot days because the flowers are safe in the refrigerator. We learned that modern appliances are much more efficient than creek beds.

Did you purchase a regular florist's refrigerator?

Ruth Ann: We thought about buying a regular florist's walk-in unit but after looking at the prices we decided that it was beyond our budget. So a friend knew someone who had an old ice cream truck that was not being used. At the same time the food co-op was moving and going to sell its compressor and motor. Between the two the price was right. It was pretty exciting.

Marcia: It was great! It was wonderful going to get it.

Ruth Ann: Yes. It made us feel so grown-up!

Nancy: You must edit this!!

Have the improvements increased your ability to earn money?

Marcia: I'm thinking we've made a lot more money this year than we ever have before.

How did you finance The Flower Ladies when you began working together?

FIG IV-5.11 "We like our product. It's right there in front of us... and it grabs you."

Ruth Ann: We didn't. Nancy cut a few flowers and I cut a few flowers and we made bouquets and took them down to the health food store and sold them.

Marcia: We've never worried about spending sixteen cents more than somebody else. We don't even think about it. The business is as fair as we can make it without putting a lot of stress on it. We get all we need anyway.

What kind of equipment did you need to start besides the normal gardening tools?

Ruth Ann: Snips, baskets—

Nancy: —and styrofoam cups.

Haven't you borrowed money to finance any of your purchases?

Marcia: No. We haven't financed anything. We don't buy something unless we can pay cash for it.

What kind of expenses do you have in making wedding arrangements?

Marcia: I was just figuring that out. If we do arrangements we put them in baskets because we like to give people something pretty to take home as well.

Nancy: We don't like paper-mache holders—

Marcia: —or plastic.

Nancy: We like to use glass vases or baskets.

Marcia: Right. So if we give someone an arrangement in a basket, there is the cost of the basket, the container inside the basket and the foam. Then there are the greens that go inside the basket. I was just figuring that out today. For the little 10 dollar arrangement, the basket would cost one dollar, the foam—

Ruth Ann: That little basket?

Marcia: Well, maybe 69 cents with the foam an additional 30 cents. Then there is 60 cents for a third of a bunch of greenery.

Ruth Ann: A third of a bunch! We better sit down and talk this over.

How do you manage your supplies?

Marcia: It used to be that every time we had a wedding we'd run out and buy supplies for that wedding. We'd try to track down foam, corsage pins, wires and things like that just for one wedding. Now we have a backroom that's stocked with all kinds of vases and wires and baskets.

What is your wedding season?

Marcia: We go from Valentine's Day through Thanksgiving.

Ruth Ann: But in this climate the garden work is never done.

Marcia: Yes, fortunately we have incredibly rainy winters so there are days we—

Ruth Ann: —have a good excuse not to—

Marcia: —work.

Ruth Ann: From the middle of November to the middle of February there are three months when we have the time to sit back and relax. But there is always that little guilt.

Who is your competition?

Marcia: We prefer not to think about our competition.

Ruth Ann: We feel sorry for them.

Nancy: But we don't ever say that.

Marcia: Truthfully, we don't have any competition. There is no one who grows flowers like we do. There are people who advertise "old fashioned" or "country" or "mountain," but we've never seen anyone who grows flowers like ours.

Nancy: That's not to say that someone won't. There are people who call us frequently, wanting to know about gardening techniques so they can do something similar to what we do. And we are fairly free with our information; we tell them what we know. I've even sent out seed catalogs. But somehow they haven't seemed to get it together.

Ruth Ann: That's not to say people aren't growing beautiful flowers and ones similar to those we grow. It's just that they don't have the florist aspect developed yet. And I don't know they want to.

Marcia: Also, one person couldn't do this. We do it together and that is important to us. We have a good time doing it and we cover for each other a lot. One person undertaking something on the scale that we do—or even one-third the scale— wouldn't be able to handle it.

Nancy: It would be very stressful.

Where is your market?

Marcia: From Carmel to San Francisco.

Who are your customers?

Nancy: From the rich to the poor.

Marcia: Right. We do funky barefoot beach weddings and elegant Carmel affairs.

Nancy: I remember one wedding we had this year. The bride came to us wanting a simple wedding on the beach. But she ended up having her wedding in the fanciest church and spending a lot of money on flowers. Then her photographer didn't work out, so she had us redo the flowers for a special shoot-

FIG IV-5.12 "We get to the wedding and suddenly the bride jumps up and says, 'These flowers are wonderful!' Then we know for sure that it has all been worth it."

ing session. It went from a very simple prospect to an elaborate affair.

Ruth Ann: Twice!

Marcia: There is a wedding designer in Aptos who does exclusive designs and sometimes she insists on our flowers.

Ruth Ann: She specializes in a lot of old fashioned weddings and our flowers go well with that look.

Do you have plans to expand your customer base?

Ruth Ann: Yes. I'd like to go further or get larger weddings with seven or eight bridesmaids. We like people who love flowers and are willing to spend the money for them. We don't care how far we have to go as long as we can get the flowers there fresh and be well paid for it.

Marcia: We charge for deliveries.

How do you advertise?

Nancy: We don't advertise much. There have been stories written about us and we do trade a local newspaper flowers for advertising but basically it's word-of-mouth.

Marcia: We've gotten lots of free advertising. We've been in most of the local publications around here by going into the offices and saying, "Wouldn't it be nice to do a story on us?" And they have.

Nancy: And we bring them a big bouquet. The articles started with *Sunset* magazine, which did a story on us. I think that gave us the idea to seek free advertising.

Marcia: And the *San Jose Mercury News* saw us in *Sunset*. Then we went to the local publications and traded flowers for advertising. We also advertise on radio stations but that's cold cash. But we do give them flowers and get a good rate. We've often fantasized about Flower Lady jumpsuits but we've never done it.

Have you arranged flowers for funerals?

Ruth Ann: Yes. One worked out OK because the deceased was the groom at a wedding we had done a couple of years before. His wife wanted our flowers because they had both enjoyed them so much at the wedding. But generally we don't get a very good response at funerals and we need that response to keep going.

Marcia: See, we thrive on response and no one is going to be effusive about flowers at a funeral.

Ruth Ann: At funerals people expect to see real formal, monochromatic kinds

FIG IV-5.13 "We enjoy the whole business process. We enjoy covering the greenhouse... planting the
seeds... digging in the garden... meeting the brides... arranging the flowers... working with each other...
and then delivering."

of arrangements. It's all so somber.

Nancy: Our flowers don't convey that feeling.

Marcia: They bring you up.

Have you had any brides reject your flowers or arrangements?

Nancy: Until recently we could brag that no one who had seen our portfolios
had ever not used us for a wedding. Now there are some who turn us down. I
remember one young bride who came in and said, "But I wanted red roses with
spider plants hanging down because I saw that in a magazine." We had no pictures
of that in our portfolio and I knew we were losing her so I turned my nose into the
air. We lost one other bride who wanted narcissus in June. I told her she couldn't
have narcissus in June. We found out later that she went to someone else who also
couldn't provide her with narcissus in June but who told her otherwise.

Marcia: Some people insist on that rose look. They want twelve red roses—

Ruth Ann: —and Baby's Breath—

Marcia: —in their bouquet. We are not interested in doing that.

Nancy: We had a 62 year-old bride who was getting married for the first time. She told us over the phone that she wanted yellow and white daisies. She'd never seen our work before but a friend had recommended us. I thought she would change her mind when she saw the portfolio. But she didn't. She wanted yellow and white daisies. We were so struck that we decided to do yellow and white daisies for her. We did a fine job, I suppose but it didn't give us a chance to be creative.

Ruth Ann: It wasn't our product. It was a typical florist product.

How long did it take for The Flower Ladies to begin earning a profit?

Nancy: We've always made enough.

Marcia: We've always made money! We've never operated at a loss or anywhere near a loss.

Nancy: The business has kept up with our expectations.

Ruth Ann: When we did the cut flower business we made a little extra cash. But I don't think it was until we got into the wedding business full-time that we started making money.

Marcia: And it took time for our reputation to grow and for people to see us as something besides hippies in front of a food store— as professionals they could count on.

Nancy: Now we go to the weddings in high heels and wear nice dresses. It gives us an excuse to wear nice dresses.

Marcia: And to clean our feet!

Ruth Ann: That is the first thing I think of every time I dress up for a wedding.

Nancy: I look at my hands because I usually pin corsages. Your hands must not be really rough and dirty.

Marcia: We always wash the dishes before we go!

How do you manage to get along with each other?

Ruth Ann: I think we compensate for each other on different days.

Marcia: Some days one of us will have something on her mind and be slow. We know that it's just the other persons' day to work harder. When one of us is feeling grouchy, the other ones will back off. We cover for each other that way.

Ruth Ann: We trust each other and don't ever have harsh words for each other.

Nancy: The personality thing is not a problem for us. We know there is a certain amount that needs to get done and we just stay there until it's done, even if it means working until midnight. And we all have the same intense sense of responsibility when it comes to doing a wedding. No one is going to slack off for any personal reason. Also, there is something so pleasant about working with beautiful flowers.

Ruth Ann: We like our product. It's right there in front of us. It's not like the printed word. It's right there and it grabs you.

Marcia: We can be up until two o'clock in the morning for two days in a row and be totally wiped out. Then we'll come in the morning to pack up and deliver and we'll look at all those beautiful flowers sitting there... it's exciting. We get to the wedding and suddenly the bride leaps at us and says, "These flowers are wonderful!" Then we know for sure that it has all been worth it.

What advice would you give to close friends who were starting their own metrofarm?

Nancy: First, I would tell them to make a nice long list of all of the flowers that would do well in their area.

Ruth Ann: I think I would recommend working the soil first. Take care of the soil, the beds. Condition the soil. Even if you're starting from clay, it can be built up over a period of time.

Nancy: Yes. Usually double-digging will give you good results. Especially if you take out that lower layer of clay or stone. Take all that out and dig underneath and put some good soil back in. Good seed is also very important. You want to buy seed from reliable companies. But even reliable companies can mess up.

Ruth Ann: I think patience is important. Don't expect miracles in one year. You learn a lot every year.

Nancy: Expect to work a lot with your body. It takes a lot of hard work out in the sun. Wear a hat and get some good gloves. You can't expect to take your vacation when the plants need you.

Marcia: That's right. When we put our new seedlings into the ground, we can't even be gone a full day because we have to be here to water them.

Nancy: I've been thinking about my seedlings all day long. They are new babies out there in the beds and you have to think of them and care for them.

One more question. What is a green thumb?

Ruth Ann: It's brown knees.

Marcia: It's an aching back.

Ruth Ann: Seriously, I'd say that it's a desire to grow.

Marcia: And a lot of hard work.

Nancy: When Ruth Ann and I first met Marcia, we didn't think that she would be talented at this thing. She just didn't look like ....

Ruth Ann: It seemed like a whim.

Nancy: Her first bouquets were not so great but now she is our champion big flower arranger. She makes the big, beautiful arrangements that we do. And it didn't take her long. So it isn't like you are born and raised with a green thumb.

Ruth Ann: But at the same time I don't think it requires talking to your plants either. I think it is just being sensitive to what they need. You don't have to be there all the time either but I think part of the magic of growing is being there.

Index

D

H

I

N

P

T

U

V

W

X

x-ray, 17
xylem, 11

Y

Yellowstone County, Montana, 63
Yellowstone River, 429, 439
Yellowstone River Valley, 429, 439
yield, 47, 229
yokel, 3
Yreka, California, 421

Z

zinc, 9-10, 227-28
zinnias, 455
zoning, 86, 87, 121
zucchini, 380

MetroFarm

BY MICHAEL OLSON

For Additional Copies:

Binding Options	Qty	Price
Trade Paper Edition @ $29.^{95}	_____	_____
Library Hard Cover Edition @ $39.^{95}	_____	_____
Subtotal		_____
CA residents add 8.25% sales tax		_____
Shipping and handling @ $5.^{00} each		_____
Total		_____

Order by Phone:

1-800-624-BOOK

Mastercard   Visa

Order by Mail:

TS Books
PO Box 1244
Santa Cruz, CA 95061-1244

Name _____

Street _____

City/State/Zip _____

____ Enclosed is my check made out to **TS Books**

____ Enclosed is my money order made out to **TS Books**

____ Please charge my ____Mastercard or ____Visa

expiration date_____   card # _____

Signature _____   Date _____

Personal checks allow 6 weeks for clearance.